ENVIRONMENTAL CONFLICT AND THE MEDIA

Simon Cottle
General Editor

Vol. 13

The Global Crises and the Media series is part
of the Peter Lang Media and Communication list.
Every volume is peer reviewed and meets
the highest quality standards for content and production.

PETER LANG
New York • Washington, D.C./Baltimore • Bern
Frankfurt • Berlin • Brussels • Vienna • Oxford

ENVIRONMENTAL CONFLICT AND THE MEDIA

Libby Lester & Brett Hutchins, Editors

PETER LANG
New York • Washington, D.C./Baltimore • Bern
Frankfurt • Berlin • Brussels • Vienna • Oxford

Library of Congress Cataloging-in-Publication Data

Environmental conflict and the media / edited by Libby Lester, Brett Hutchins.
pages cm — (Global crises and the media; v. 13)
Includes bibliographical references and index.
1. Mass media and the environment. 2. Environmentalism.
3. Environmental protection—Press coverage.
I. Lester, Libby, editor of compilation.
II. Hutchins, Brett, editor of compilation.
P96.E57E56 333.7—dc23 2013011821
ISBN 978-1-4331-1893-7 (hardcover)
ISBN 978-1-4331-1892-0 (paperback)
ISBN 978-1-4539-1146-4 (e-book)
ISSN 1947-2587

Bibliographic information published by **Die Deutsche Nationalbibliothek**.
Die Deutsche Nationalbibliothek lists this publication in the "Deutsche Nationalbibliografie"; detailed bibliographic data is available on the Internet at http://dnb.d-nb.de/.

Cover image by Espen Rasmussen/Panos

The paper in this book meets the guidelines for permanence and durability of the Committee on Production Guidelines for Book Longevity of the Council of Library Resources.

29 Broadway, 18th floor, New York, NY 10006
www.peterlang.com

Printed in the United States of America

Table of Contents

Series Editor's Preface

Global Crises and the Media

We live in a global age. We inhabit a world that has become radically interconnected, interdependent, and communicated in the formations and flows of the media. This same world also spawns proliferating, often interpenetrating, "global crises."

From climate change to the war on terror, financial meltdowns to forced migrations, pandemics to world poverty, and humanitarian disasters to the denial of human rights, these and other crises represent the dark side of our globalized planet. Their origins and outcomes are not confined behind national borders and they are not best conceived through national prisms of understanding. The impacts of global crises often register across "sovereign" national territories, surrounding regions and beyond, and they can also become subject to systems of governance and forms of civil society response that are no less encompassing or transnational in scope. In today's interdependent world, global crises cannot be regarded as exceptional or aberrant events only, erupting without rhyme or reason or dislocated from the contemporary world (dis)order. They are endemic to the contemporary global world, deeply enmeshed within it. And so too are they highly dependent on the world's media and communication networks.

The series *Global Crises and the Media* sets out to examine not only the media's role in the *communication* of global threats and crises but also how they can variously enter into their *constitution*, enacting them on the public stage and helping to shape their future trajectory around the world. More specifically, the volumes in this series seek to: (1) contextualize the study of global crisis reporting in relation to wider debates about the changing flows and formations of world media communication; (2) address how global crises become variously communicated and contested in both so-called "old" and "new" media around the world; (3) consider the possible impacts of global crisis reporting on public awareness, political action, and policy responses; (4) showcase the very latest research findings and discussion from leading authorities in their respective fields of inquiry; and (5) contribute to the development of positions of theory and debate that deliberately move beyond national parochialisms and/or geographically disaggregated research agendas. In these ways the specially commissioned books in the *Global Crises and the Media* series aim to provide a sophisticated and empirically engaged understanding of the media's changing roles in global crises and thereby contribute to academic and public debate about some of the most significant global threats, conflicts, and contentions in the world today.

Environmental Conflict and the Media, edited by Libby Lester and Brett Hutchins, expertly responds to the challenge of the series brief above as it dissects the deeply entwined roles and representations of media and communications in environmental conflicts around the world today. "In an age of global media and crises," state the editors in their opening chapter, "events occurring in local and national contexts are now embedded in transnational communications networks and shared by sizable communities of activists, supporters and citizens around the world." The chapters that follow proceed to examine diverse environmental conflicts in different parts of the world, including Australia, the UK, the US, Europe, Latin America, China, Japan, the Pacific Islands, and Africa. Together these "demonstrate how conflicts emanate from and flow across multiple sites, regions, and media platforms." This is a timely and considered intervention into not only the world of intensifying environmental conflicts but also the intensifying debates that surround the rise of new media and communication networks and their interrelationship with the former. The contributing authors to this volume have their preferred views and theoretical perspectives on these and other key debates, which represent some of the best scholarship on such issues. They all also demonstrate a shared commitment to improved understanding of the problems and possibilities of media and communications in respect of humanly pressing issues and threats confronting contemporary world society.

Readers may be interested to know that some of the chapters began as papers written for the Environmental Politics and Conflict in a Digital Media Age Symposium, hosted by Professor Libby Lester and Associate Professor Brett Hutchins in Hobart, Tasmania (17–18 November 2011). Alongside the assembled academics were a number of Pacific Islanders, journalists mostly, who sought to share the problems and difficulties they'd encountered when reporting the effects of climate change impacting on their small island communities, their livelihoods, and traditional ways of life. These islanders are paying a high price for climate change, a problem not of their making but now visited upon them by industrialized economies and societies, many of whom are still largely in denial about what's unfolding and what's at stake. The testimonies of the Pacific Islander journalists concentrated minds, all the more so when set against the reporting of much of the Australian press at the time of the symposium, which displayed a cynical stance of climate change denial—a stance woefully behind the times, out of step, out of touch.

Environmental conflicts today, whether played out at local or global levels, are increasingly entangled within and executed through multiple communication flows and media formations around the globe. Media systems and communication networks in Australia, as elsewhere, are not destined to always be aligned to the forces of conservatism, however; they also can be put to work in mobilizing for change. Christine Milne, Leader of the Australian Green Party, who also spoke at the Tasmanian symposium, powerfully reminds us in her Afterword to this volume how changing communications have performed an integral role in processes of political mobilization and policy change in the past and continue to do so in the present, often in creative and fast-changing ways. Media and communications, then, are not "outside" environmental conflict and politics but critically sutured *inside* their very dynamics and unfolding trajectory. This volume both grounds and theoretically reflects on the politics and possibilities of communicating conflicts in a global, media age.

To end I can do no better than rehearse the editors' own encapsulating statement about *Environmental Conflict and the Media* as my way of recommending this book to you: "This wide-ranging collection moves across continents, shifts between media platforms and technologies, considers activism and campaigns in multiple contexts, travels between the local and global, and traces the mediation of slow-moving crises and unexpected disasters. The ambition of this scope is informed by the knowledge that the future of the planet is unfolding before our eyes," and "witnessed in and through media." And so it is.

Simon Cottle, Series Editor

Acknowledgments

The editors would like to thank the Australian Research Council for funding the three-year research project 'Changing Landscapes: Online Media and Politics in an Age of Environmental Conflict' (DP1095173), which has included this book and related symposium held at the University of Tasmania in November 2011. Thank you to Janine Mikosza for her initiative, patience and painstaking work in preparing the manuscript, and to Stephenie Cahalan for her invaluable support with organizing the symposium. Thank you to the University of Tasmania and to Monash University for funding support for the symposium. The editors would also like to acknowledge the enthusiasm and collegiality of symposium participants, who travelled from thirteen countries and brought a range of media and journalism research traditions and industry perspectives to the gathering. Libby would also like to thank the University of Tasmania for a period of study leave that helped with the completion of this collection. Brett is appreciative of the support offered by his colleagues in the School of English, Communications and Performance Studies at Monash University, with particular thanks going to Shane Homan, Andy Ruddock, Kevin Foster, Sue Kossew, Robin Gerster, Jodie Wood and Kerry Bowmar. He also knows that nothing is really worthwhile without the love and support of Janine and Rowan. Finally, thank you to series editor Simon Cottle for his interest and participation in the research project and book, and to Mary Savigar of Peter Lang for her publishing support.

■ INTRODUCTION

Tree-Sitting in the Network Society

BRETT HUTCHINS AND LIBBY LESTER

Miranda Gibson is living on a small platform 60 metres above the ground in an ancient gum tree. The tree is located in the Styx Valley in the remote southwest of Tasmania in Australia, a state that is home to one of the three largest temperate wilderness areas in the southern hemisphere. A 31-year-old environmentalist and schoolteacher, Gibson has been there for 368 days without a break, surpassing both the Tasmanian and national records for the longest tree-sit.[1] A committed activist representing the grassroots group Still Wild, Still Threatened, Gibson is attempting to protect her area from logging and draw widespread attention to the destruction of old-growth forests in the island state. These immediate objectives are tied to a broader vision of environmental sustainability and ecological citizenship in social, economic and political systems worldwide. Gibson's teaching career remains on hold as she endures the trials of changeable Tasmanian weather that features sun, wind, rain and snow. Her last birthday was celebrated in the depths of winter, protected by only a small canopy. Sleep is difficult depending on the weather conditions, and there is no one immediately present to chat with for long periods of time, although her mother and sister have scaled the tree for short visits. The nearest township is Maydena (population 245), an old timber and mining town. Negotiating the track, forestry road and bush that leads to the tree is hard work for those visiting and delivering supplies to Gibson, underlining her isolation and distance from metropolitan space and the conveniences associated with urban living.

Gibson's tree-sit—in the Observer Tree (http://observertree.org/)—is a significant manifestation of environmental conflict and media in "the network society" (Castells, 1996, 2004). Her daily routine involves communicating via a mobile phone and laptop computer that enables access to Skype, online chat, email and social networking services. She converses with Australian and international journalists, fellow activists, political figures, students, researchers, interested onlookers, family members and friends. Her media use shifts from the local to the regional, national and international and back again, depending on who she is communicating with and where her heartfelt message about the Tasmanian forests is directed. Gibson accesses details about developments in news and politics through websites, Facebook and a range of online resources (Gibson, 2012a), motivated by an enduring commitment to the environment and the considerable time she has to fill each day. She also produces her own environmental media texts and images by writing a regular blog, maintaining a Facebook and Twitter profile, and distributing photographs and short videos shot from her platform.

The digital media practices described here function as both a "networking *agent* in and *window* on" Gibson's protest (Segerberg and Bennett, 2011: 200). This dual character was on display as the anniversary of her protest was marked by a "worldwide cyber event" involving live video streams conducted by Gibson from her platform. These streams saw her answer multiple questions sent by supporters in both the northern and southern hemispheres. These streams also linked up with community events staged in cities and towns across Australia, as well as in Tokyo, Seattle, and Bristol, UK. A Flickr account, "Observertree one year anniversary",[2] shows more than 200 photographs of groups and individuals sending visual messages of support. The anniversary of the tree-sit and these activities also saw Gibson attract national and international news attention. The Observer Tree is an ideal case study to consider at the start of this book. This protest demonstrates key features of environmental conflict in an age of global media and crises, highlighting that events occurring in local and national contexts are now embedded in transnational communications networks and shared by sizable communities of activists, supporters and citizens around the world. The next chapter, by Simon Cottle, offers insight into the interdependencies that produce this situation in the twenty-first century.

It is difficult to fathom the significance of Gibson's protest or the advanced media practices discussed here when standing next to the gum tree in which she is perched, a tree that is presently an assemblage of the natural and social worlds (Latour, 1993). The tree is both organic plant matter standing in a pristine natural environment and a temporary piece of communication infrastructure containing a laptop computer and mobile phone connected wirelessly to a nearby mobile service tower that is, in turn, linked to a global network of computing devices. Networking logic and power now reach deep into the wilderness and are used by those who seek to protect these settings from harm and others wanting to harvest and harness them for economic gain. Yet, for all the novelty of Gibson's protest and her extensive media activities, it is important to recognize that she is tapping into an established history of tree-sits in Tasmania and elsewhere, many of which have received extensive news coverage. This interplay between continuity and discontinuity was on display when Julia "Butterfly" Hill sent messages of support to Gibson on Facebook on 15 February 2012 and again on 15 December 2012. In an act of civil disobedience fifteen years earlier, Hill occupied a 1,500-year-old redwood tree (dubbed "Luna") in California in the US for a record 738 days, using solar-powered cell phones to conduct radio and television interviews with reporters (Hill, 2001).

Much of the hype associated with the "revolutionary" potential of the World Wide Web and digital media for environmental activism has been disproven—or at least muted—by the last two decades of lived experience. The empirical realities of the prevailing media landscape instead display a mixture of innovation and reliance on established strategies and techniques. Long-standing mechanisms of media production, transmission and consumption have been re-embedded into a range of new consumer technologies that foster distributed networks, increasing amounts of content, accelerating rates of communication, and new ways of organizing data and information. This process of relentless remediation is captured by Marshall McLuhan's notion of the "rear-view mirror" and the fact "we march backwards into the future" (McLuhan and Fiore, 1967; Levinson, 1999; Bolter and Grusin, 1999).[3] The result is a conflicting and occasionally confusing array of outcomes that are difficult to parse in terms of old, new and hybrid media formations. This task is particularly arduous when analyzing the entanglement of print, analogue, broadcast, digital and mobile media in a large and messy communications ecology (Goggin, 2011: 4). The demands of researching these developments are made more formidable by a kaleidoscope of representations, counter-representations and self-representations produced in the course of fiercely fought environmental disputes. Conflicts manifest in open public disagreement and debate, protest, legal proceedings, government action, political lobbying, and even physical violence, all refracted through modes of mediation (Lievrouw, 2011; Couldry, 2012).

Using a range of related disciplinary perspectives from the social sciences and humanities, the contributors to this book have accepted the challenge of analyzing and explaining the complicated relationship between environmental conflict and the media.[4] Through empirical investigation and critical analysis, they shine light on why media are central to historical and contemporary conceptions of power and politics in the context of environmental issues, and examine the role of media in helping to structure collective discussion, debate and conflict. Their insights reveal much about the forces, processes and practices that are observable through (i) specific media contexts, formats and technologies, (ii) activism and campaigns, (iii) the communication of crises, and (iv) news reporting and media frames that ventilate (and stifle) contested environmental claims. Discussing different sections of the globe—Australia, the UK, the US, Europe, Latin America, China, Japan, the Pacific Islands, Africa—the chapters demonstrate how conflicts emanate from and flow across multiple sites, regions and media platforms. Symbols and messages generated during conflicts by political challengers, elites and mediators are deployed to reach multiple audiences, which encompass like-minded allies, hostile opponents, political representatives, skeptical and sympathetic journalists, and the broader citizenry. The stories presented in this collection add much needed evidence to an observation made by sociology's master analyst of the network society, Manuel Castells:

> The uses of the Internet are integrated in a *broader multimedia strategy* that characterizes the actions of the environmental movement...*In sum*, the versatility of digital communication networks has allowed environmental activists to evolve from their previous focus on attracting attention from the mainstream media to using different media channels depending on their messages and the interlocutors they aim to engage. *From its original emphasis on reaching out to a mass audience, the movement has shifted to stimulate mass citizen participation by making the best of the interactive capacity offered by the Internet.* Thus, environmental organizations act on the public and on decision-makers by bringing issues to their attention in the communication realm, both in the mainstream media and on the Internet. (Castells, 2009: 327; emphasis in original)

Castells (2009: 303–339) highlights how social movements, the "long march of environmentalism" and networked political action are reprogramming cultural codes and political values across the globe. The following essays endorse his position, and proceed to identify the precise features of these *multi*-media practices through approaches and case studies that draw upon newspapers, magazines, broadcast television and radio, documentary and film, photography, websites, Internet video and YouTube, and social networking services such as Twitter and Facebook.

Returning to the case of the Observer Tree, Gibson's activities exemplify multifaceted media strategies targeted at fellow activists and environmentalists, as well as mass audiences accessing information through newspapers, radio and television current affairs. Examples include her use of Skype to talk with the leader of the Australian Greens, Christine Milne (see the Afterword to this book), before an audience of female environmental leaders; posting on YouTube a video of a visit to the Observer Tree by Australian Greens political icon Bob Brown; and uploading on Facebook a photograph of fellow activists who journeyed to the tree. The photo depicts crew members from the MY *Bob Barker* of the Sea Shepherd Conservation Society (see Crouch and Damjanov, 2011; Lester, 2010b). These messages and images targeted at fellow environmentalists are complemented by nationally broadcast reports about Gibson on Australian public service radio, commercial current affairs television programs, and a video question asked by Gibson via Skype from the tree's platform during an episode of *Q&A*, the Australian Broadcasting Corporation's (ABC) high-profile current affairs panel show. The creation of deliberately appealing visual images for the news media is also apparent. A visit to the tree by "Santa" was arranged in December 2011, and a photograph of a smiling Gibson and her festive visitor was published by Hobart's daily newspaper, *The Mercury*, on Christmas Eve (Hoggett, 2011b). This interweaving of actions, messages, events, reports and images is a recurring feature in the essays that follow.

The next chapter, by Simon Cottle, adds to our introductory analysis by helping to set the broader social and media context for the chapters that follow. Interdisciplinary in outlook and substance, Cottle's chapter situates mediated environmental conflict within the "politics of connectivity" and the "politics of representation". It then recommends research strategies that contextualize environmental affairs within broader "forces of change and global crises". Both forward-looking and agenda-setting, his essay captures the increasingly transnational character of environmental conflicts, and allows the interconnections between these conflicts and other forms of global crisis and contention to emerge. Approached as a whole, this introduction and Chapter One offer the reader a conceptual and analytical "road map" that connects the case studies and arguments presented throughout this volume.

PART ONE: "OLD" AND "NEW" TECHNOLOGIES

Categorical distinctions between old and new media are difficult to discern, as established technologies and formats are subject to digitization and remediation processes (Bolter and Grusin, 1999). As Daniel Palmer makes clear in Chapter Five, the power of wilderness photography to influence collective thinking about nature and the environment has been evident since the 1800s. Yet, it cannot be denied that networked digital distribution has, to paraphrase Palmer, dramatically transformed the meaning and value of photographic images (and the commercial and institutional relations that relate to this and other mediums). Similarly, the notion of occupying a tree to prevent loggers cutting it down is a strategy that predates laptop computers, the web and consumer mobile technologies by many years. But when a protester carries portable and affordable "do-it-yourself" tactical media tools (Garcia and Lovink, 1997;

Meikle, 2002) up a tree today, the parameters of communication and political action are *much more expansive* than 15 years ago.

Miranda Gibson's media practices are, for instance, significantly greater than the man dubbed "Hector the Forest Protector" in 1998. Hector was an electrical engineer named Neil Smith who scaled a tree on the slopes of Mother Cummings Peak in Tasmania's north to stop the progress of a proposed logging road (Lester and Hutchins, 2009).[5] He used a 2G mobile phone, computer and Internet connection, powering his computer with a car-sized 12-volt battery charged by a solar panel. His primary mechanism for drawing attention to his protest was email, which was used to contact politicians and journalists. Smith lasted 10 days before police removed him from the tree. Many years after Smith's forced eviction, the Observer Tree is constructed around the same strategy, but a series of powerful new tactical media and social software tools have since become available to Gibson—mobile broadband Internet, blogs, social networking services, user-generated video platforms, 3G and 4G phones and networks, easily accessed video-conferencing services, and cheap digital cameras—that allow her to connect speedily and reliably with the world from her elevated vantage point. These technologies, in addition to email, support a range of written, verbal and audio-visual communication techniques utilized by Gibson to prosecute her case for environmental justice in the forests. Her geographical isolation is overcome by her use of a 3G mobile telecommunications tower located in Maydena that has been in service since only 2009. This experience has, in Gibson's mind, only added to her interest "in using technology and using the Internet more for campaigning" (Gibson, 2012a), which is likely to spur additional innovations in the near future. The long march of environmentalism is now accompanied by the quickstep of networked digital communications, media and infrastructure, although, as is evident in later chapters, this advance proceeds unevenly in different regions around the globe.

The four chapters in Part One deal with the capacity of established and emerging media forms to help effect environmental change, examining the press, social networking media, online video and photography. Each essay analyzes the use of media in constructing environmental messages, images and discourses that are designed to elicit affective responses from their audiences. National parks, forestry, and anthropogenic climate change supply the topical focus of the contributions. The stories presented reveal a concern about the variable levels of *connectivity* enabled by media technologies across time and space.[6] *Connectivity* here refers to the density of "interconnections and interdependencies" produced by different media networks and technologies (Tomlinson, 1999). In Chapter Two, Michael Meadows and Robert Thomson show how the efforts of campaigning print journalists in late nineteenth- and early twentieth-century Australia continue to resonate across time, manifest in the national parks that citizens visit to this day. Displaying parallels with the national parks movement in North America, the circulation of printed news reports in these early years may have been slow, but journalistic advocacy for the conservation of natural habitats delivered lasting legislative and environmental outcomes. In contrast, Alex Lockwood (Chapter Three) analyzes the intersection of emotion, England's public forests and Twitter—the micro-blogging platform that exemplifies the emergence of an accelerated, real-time information order (cf. Tomlinson, 2007). His chapter explains how the "#SaveOurForests" campaign of 2010 and 2011 mobilized an intensity of media activity and emotive citizen expression that proved impossible for nationally elected representatives to ignore. Lockwood is followed in Chapter Four by Catherine Collins, who unpacks the narrative strategies evident in logging videos posted on YouTube. These YouTube videos tap into rich environmental and forest industry histories that are embedded in a user-generated video platform that features much ephemeral content and rapid-fire user responses (cf. Burgess and Green, 2009; Snickars and Vonderau, 2009). Collins systematically compares a series of videos about old-growth logging in the Pacific Northwest of the US with those produced in the context of the Tasmanian forests conflict. She outlines the narrative strategies of videos that both celebrate and condemn logging in both locations, as well as the user comments that respond to them. A key contribution of her chapter is to explain how competing narratives interact to limit the likelihood of dialogue between those in favor of and against logging. In an interesting coincidence, the most watched Tasmanian video discussed by the US-based Collins features (an unseen) Miranda Gibson, who was physically attacked by forestry contractors while protesting in the Upper Florentine Valley in 2008 (see Hutchins and Lester, 2011). Part One finishes with Palmer's chapter on photography, ecological criticism and climate change. Read together, these four chapters gesture toward the interplay of time, media and environmental disputes, offering analyses that project deep into the past, move through an accelerating present, and travel towards risk-laden environmental futures. In battles over the definition of existing and looming risks, we are all witnesses to collisions between competing concepts of time—"cosmological", "glacial", and "timeless" time (McAfee, 1999; Lash and Urry, 1994; Castells, 1996)—thereby helping to explain the intensity of environmental conflicts and crises at this point in global history.

PART TWO: ACTIVISM AND CAMPAIGNS

The third section focuses on the most publicly visible and often spectacular evidence of environmental conflict—activism and campaigns. Reflecting broader developments in social movement activities and research, environmental protests and efforts at collective action provide evidence of "mediated mobilization" (Lievrouw, 2011: 25–26). As outlined by Leah Lievrouw, this term describes the use of web-based media and social software to build interpersonal networks that are then mobilized on a larger scale to raise awareness of particular issues and achieve political objectives. Here we find familiar social movement strategies dating from the era of mass media transferred to the setting of the network society, which has transformed the capacity and scope of activism. Supported by physical protests and coordinated campaign actions, activist messages and symbols are now carried by digital communications networks far beyond the local, landing wherever vulnerability in government and corporate power can be found. Unsurprisingly, mediated mobilization is a "hit-and-miss" affair, met by stiff and well-resourced opposition from the state, developers and even the news media. For instance, analyzing the mobilization of citizens around environmental issues in Latin America, Silvio Waisbord (Chapter Seven) writes of the "well-known cases of social movements situated in a disadvantageous position vis-à-vis mainstream news organizations" and the subsequent lack of publicity that they achieve. But, drawing on a wide range of examples, Waisbord also maintains that this relationship is not entirely predictable. On those occasions when the commercial and ideological affinity between news media outlets, governments and resource extraction industries is weak, local interests are able to achieve favorable news coverage and influence policy making. As the five chapters in this part of the book remind us, a key objective of many campaigns remains the attraction of sustained commercial and public service news media coverage, an objective that endures even in an epoch of do-it-yourself media and user-generated content. Indeed, the novel application of these technologies is now a method used by activists to draw the attention of journalists and mass audiences to environmental protests, as news reports about the Observer Tree's role in a "hi-tech forest battle" (Hoggett, 2011a) ably demonstrate.

A notable feature of activism detailed in Part Two is the ease and speed with which both grassroots and internationally focused protests and campaigns link with actors and networks situated at the local, national, regional and global levels. The magnitude and scale of these increasingly routine "multi-scalar" relations (Flew, 2009) are historically unique, creating a world in which "the space of our existence has become both global and local at the same time" (Castells, 2009: 337). Even highly localized campaigns draw on a repertoire of strategies and resources that are shared online by activists living on different continents. Dan Brockington's analysis of celebrity intervention into environmental causes in Africa and other locations (Chapter Nine) describes the media and social conditions underpinning this existence. He outlines how the "images, messages and ideas" conveyed by Hollywood stars, such as

Harrison Ford about sustainable development in Africa, are embedded in the interaction between global capitalism and conservation movements.

It is noteworthy that Miranda Gibson's activities in southwest Tasmania are consistent with Brockington's arguments despite her different location and context, helping to confirm the relevance of his case. The Observer Tree's Facebook profile presents stories about celebrities that convey a sense of shared environmental awareness and consciousness. Examples include Leonardo DiCaprio's advocacy for the establishment of an Antarctic Ocean marine sanctuary and Daryl Hannah's protest against the building of an oil pipeline between Alberta, Canada and the Gulf Coast of Texas. Placed alongside these actors are "likes" and messages of support from popular musicians such as John Butler, Blue King Brown and Urthboy, and from Australian television gardener Peter Cundall, as well as an endorsement from environmental icon and political celebrity Bob Brown (cf. Street, 2004).[7] Underlining the transnational scope of environmental conflict, in addition to detailed coverage from news providers on the Australian mainland, Gibson's protest has also received attention from international news media outlets including Al Jazeera (English), *The Guardian* and the BBC in the UK, CNN in the US and a group of visiting South Korean journalists. A distinguishing feature of this reportage is the actuality generated by the high-quality digital photographs taken by Gibson in the tree that are used in stories, conveying a tangible sense of life on an elevated platform in a remote forest. This political and visual movement from the Observer Tree to screens viewed in Europe, Asia and North America is a regular occurrence, the success of which has surprised even Gibson:

> One of the big measures of success was early on in February [2012] when I organized the international action. I didn't have that many contacts around the world and I just put it out on the [Observer Tree] blog. We ended up getting responses from 15 countries...I was actually quite surprised at how many people and how many different countries we got in that action. I had a number of people contacting me from France at one stage. Someone in France had put up a story about me on a website over there. I then launched the campaign in Japan. I had a focus on spreading the word in Japan because that's where a lot of the [Tasmanian forest] products are going, and I started getting quite a lot of correspondence from people in Japan. I've had people contacting me from a lot of other different countries as well. Just the other day I was contacted by some people in America. They were doing a stall about the Tasmanian forests at their local library. I "Skyped" in to talk to people at the stall and I've had a few people contacting me from that. (Gibson, 2012a)

This statement hints at the emergence of a transnational public sphere (Fraser, 2007) supported by digital communications infrastructures, with environmental concerns a key feature of social movements and formal political systems. (The notion of a "green public sphere" and an "issue-specific public" centered on the environment is discussed by Guobin Yang and Craig Calhoun in Chapter Eighteen.) Gibson's reference to Japan is indicative of this complex situation, as her protest is also a highly visible component in a sophisticated campaign of "market-based activism". Coordinated by a coalition of environmental groups, targeted protest messages and lobbying are directed at Japanese manufacturers and retailers of flooring products who sell veneer sourced from Tasmanian regrowth and plantation eucalypt forests.[8]

The analyses of Waisbord and Brockington, as well as the evidence presented above, capture a variety of contemporary environmental protest and campaign "logics" (Cammaerts, 2012). The other three chapters in this section add valuable evidence and texture to this picture, showing how local campaigns are bound to transnational issues and media platforms in ways that are both explicitly and implicitly evident. These essays, all of which have an Australian focus, emanate from a major environmental media research symposium held in Hobart, attended by delegates from 13 countries. In these chapters, the structure of media practices and networks sees discussion connect the local to the global through: (i) travel journalism produced in and about conflict-ridden environmental landscapes for international magazines, newspapers and online publications (Lyn McGaurr in Chapter Six); (ii) the use of the web's most popular social networking service, Facebook, to distribute propaganda prejudicial to a Greens candidate during a formal political election campaign (Kitty van Vuuren in Chapter Eight); and (iii) fraught campaigns to conserve marine environments that are conducted in media, public and government forums, which attempt to translate "global conservation messages for a local audience" (Michelle Voyer, Tanja Dreher, William Gladstone and Heather Goodall in Chapter Ten). McGaurr, van Vuuren, and Voyer et al. also highlight the value of different approaches when investigating the logic and significance of environmental campaigns, drawing on judiciously selected case studies, meticulous content analysis, rich first-hand experience and wide-ranging interviews with environmental actors.

PART THREE: COMMUNICATING CRISES

Environmental conflicts are filtered through media frames, modes of communication, and the reception and use of content by audiences. The crucial political and social functions performed during these processes are particularly evident in reports about global warming, extreme weather events, habitat destruction and large-scale industrial accidents. Part Three examines such phenomena, analyzing the communication of crises that are, as identified by Simon Cottle in Chapter One, embedded into the conditions of "late modernity and globalized (dis)order" (also see Pantti, Wahl-Jorgenson and Cottle, 2012). Ecological catastrophes are both calamitous and politically charged events at a time when governments throughout the developed and developing world are prone to disagreement over appropriate responses to "natural" and "man-made" disasters. It is the collapse of the boundaries between these two categories—nature and culture—in a risk society (Beck, 1999: 145), and the unavoidable conflation of them in media representation and policy development, that underpins these disagreements. In this context, nuclear disasters, oil spills, toxic chemical leaks, tsunamis, earthquakes, "super-storms", floods, bushfires, extreme heat waves and famine function as fodder for spectacular "disruptive media events" and disaster "marathons" transmitted to distant audiences consuming a range of online, broadcast and print media (Katz and Liebes, 2007; Liebes, 1998). These sources now include compelling live Twitter feeds that aggregate images taken by citizens using mobile devices during floods and fires, information distributed by government, emergency service and news organizations, and sporadic examples of deliberately mischievous and inaccurate tweets (cf. Bruns, 2011). At the same time, these disruptive events are heart-breaking tragedies that cost the lives of countless humans and animals, inflict trauma and death on those who are personally affected, and lay waste to built and natural environments.

The chapters in Part Three show that much is at stake materially, emotionally, economically and politically in the communication of environmental crises. These four essays indicate that, depending on who is speaking, the intention may be to: (i) present evidence and/or warn of an impending crisis that can be avoided; (ii) relay the awfulness of an unfolding disaster; (iii) communicate the actions taken to alleviate suffering, and restore conditions of social stability and physical safety; (iv) reveal, obscure or deny the causes of a recent catastrophe; or (v) outline strategies aimed at reducing existing vulnerabilities and prevent future disasters. The trust embodied by the public figures delivering the message—journalists, politicians, emergency service workers, police, activists, documentarians, scientists—is related to the knowledge claims attached to their professional status. Their public reputations, the authoritativeness of their performances and the timing and form of their communications also play major roles in the reception of information. For instance, in the face of a coming or slowly unfolding crisis, the power of language is to the fore, with the Observer Tree blog providing a handy example. Repeated reference to the "protection" of the "precious" and "beautiful" forests from the "threats" and "destruction" caused by old-growth logging constructs a narrative designed to convince the reader of the need to act before the forests are "lost". Arresting photographs of pristine forests and/or scarred and barren landscapes bolster this narrative. The claims carried by these narratives are met by industry and government responses that frequently downplay, reject or are outright hostile to the arguments of environmentalists.

Chapter Eleven pinpoints the power of narratives in the generation of popular understandings of the environment. Morgan Richards offers an illuminating historical perspective on the evolution of wildlife documentaries and their role (or lack thereof) in creating widespread awareness about potentially catastrophic environmental futures. In particular, she looks at the "iconic visions of wildlife and nature" produced by the BBC Wildlife Unit and David Attenborough for the international television market over many decades. These productions are used to support her argument about a contested transition that has taken place, from spectacle-driven "blue chip" productions that fail to engage with ecological politics to "green chip" documentaries. This more recent form of documentary confronts the overwhelming consensus about climate change in the scientific community, portraying "catastrophic landscapes" and stories of animal extinction to audiences worldwide. Richards is followed by Myra Gurney (Chapter Twelve), who analyzes the strategic role of political speeches that address the crisis of climate change. Major speeches delivered by national and international leaders, including figures such as Christine Milne, generate extensive news and current affairs coverage. Yet we have a comparatively limited understanding of their role in setting the "agenda and vocabulary for debate" and the formulation of actions to avoid or alleviate the consequences of climate change. Focusing on speeches delivered by Australian political leaders about "the great moral challenge of a generation",[9] Gurney dissects the rhetorical tropes employed in the course of speechmaking, and the continuing conflict between moral considerations and economic interests in efforts to reduce human-induced carbon dioxide emissions.

In the "calculus of risk" (Beck, 1999: 50) there are few more ominous equations than a nuclear accident, although the story of Bhopal in India is a reminder that other forms of industrial accident can be just as deadly (Bogard, 1989). News reports starkly inform readers, listeners and viewers that nuclear contamination emitted after a reactor failure can be fatal for those living in the immediate surrounding areas, severely damages the ecosystem, and adversely affects the health of subsequent generations. Radiation is, therefore, a lethal airborne toxin that also symbolizes a fundamental clash between contending visions of modernity. For many people in favor of the "nuclear option", clean and efficient nuclear power is proffered as a solution to global warming, offering a vision of modernity where technological innovation, scientific risk management and market mechanisms combine to produce acceptable safety levels and better outcomes for citizens and the planet. For those against the building of reactors, nuclear power is a source of nightmarish scenarios that underline the ineradicable danger at the core of nuclear science and a flawed modernity where the limits of science and nature are not respected (cf. Lowe, 2007). The latter position recalls a case posited by French cultural theorist Paul Virilio (2007;

Redhead, 2006) about the role of technology in an accelerated modernity. Each new invention produces its own negative by-product, "the accident". The advent of the automobile and road systems created the violence of the "car crash" and "road toll", just as the airplane gave birth to the "air disaster". The building of nuclear reactors in populated communities has set loose the monstrous consequences of the "major nuclear accident". In Chapter Thirteen Kumi Kato accounts for such a disaster, describing the cascading destruction of the earthquake, tsunami and nuclear accident that hit the east coast of Japan in March 2011. Exhibiting empathy for the victims and a restrained frustration at the actions of selected media organizations and the Japanese government, Kato identifies the lamentable *lack* of "critical media debate" about nuclear power generation, especially given the terrible events that took place at the Fukushima Daiichi Nuclear Power Plant complex. She also observes a rising awareness of the need for open discussion and exchange among members of the public who no longer wish to live with "nuclear threats", with online sites used to find alternatives to official sources and information. In Chapter Fourteen Clio Kenterelidou applies a public communication perspective to compare three environmental "mega-crises"—the Fukushima nuclear crisis of 2011, the 1986 Chernobyl disaster in the Ukraine, and the 1979 Three Mile Island accident in the US. Comparative analysis demands painstaking judgments requiring, in this case, the development of criteria to compare different national contexts over time and across nations. Kenterelidou produces evidence-based suggestions that may inform more effective crisis communication strategies.

PART FOUR: CONTESTED CLAIMS

Climate change is *the* environmental challenge of our time. It is an issue that encapsulates the perplexing interface between human life, the natural world and economic activity, serving as a "lightning rod" for contrasting ideologies and approaches to the interaction of these entwined spheres. As three of the four chapters in this part of the book suggest, the complexity of this issue produces innumerable opposing claims related to climate science and climate modeling, serving as a site where media power and counter-power, political communication and counter-communication, and economic risk and opportunity collide (Castells, 2007, 2009; Beck, 2009). Systematic scientific analysis of weather conditions, greenhouse gases and sea levels is peculiarly susceptible to savage public attacks by skeptics and reactionary populists for the following reason: "Giddens's paradox" (Giddens, 2009).[10] Coined by sociologist Anthony Giddens, this term refers to a phenomenon that aggravates arguments over the state of the environment globally:

> ...since the dangers posed by global warming aren't tangible, immediate or visible in the course of day-to-day life, however awesome they appear, many will sit on their hands and do nothing of a concrete nature about them. Yet waiting until they become visible and acute before being stirred to serious action will, by definition, be too late. (Giddens, 2009: 2).

This paradox means that the media, media spaces and journalism are the primary symbolic and information battlegrounds in which the epistemology of global environmental conditions is debated, even as the physical world changes.

At one level, the scale and intermittently hysterical tone of arguments about climate change in news and online media is unprecedented. The ferocity of these exchanges relates to the fragmentation of the cultural authority possessed by qualified experts, legislators and intellectuals in late capitalism, leading to disputes over who has the right to name and control sources of uncertainty in social systems (Bauman, 1987). At another level, these disagreements speak to the pattern of continuity and discontinuity detailed throughout the pages of this book. For many decades environmentalists have known that any public claim made for the conservation or preservation of natural habitats, irrespective of who is making it, is likely to be sharply contested by political opponents, business and/or governments. From this perspective, power and counter-power are mutually constitutive features of environmental conflict, locked in a struggle to dominate the cultural and political codes flowing through media networks and into the minds of citizens (Castells, 2009). Miranda Gibson's tree-sit provides ready evidence of this broader dynamic. In July 2012 stories appeared in the national news media about a "counter-vigil"—or a protest against the protest—set up at the base of the Observer Tree by a newly formed pro-logging group, Give It Back (ABC, 2012a). Monitored by Tasmanian police, a farmer named Michael Hirst had decided to camp on the forest floor beneath Gibson's platform to "get another side of the story across to the public" and "fight for Tasmania's resources" (ABC, 2012a). This action was followed in September by an alleged (and unrelated) arson attack on Camp Florentine, the long-running forest blockade run by Still Wild, Still Threatened in the Upper Florentine Valley (ABC, 2012b).[11] As a spokesperson for Still Wild, Still Threatened, Gibson was upset by the malicious destruction of huts and equipment housed at the campsite. After receiving news of the attack, she stated on her blog, "This week, we have seen yet another act of violence against the non-violent protest movement" (Gibson, 2012b). As the following chapters in this volume emphasize, these types of events reflect the heightened tensions in many contemporary environmental conflicts and disputes over climate change.

Chapter Fifteen focuses on the seemingly inevitable public uncertainty produced by the multitude of claims and counter-claims about the reality of climate change in the US. Robert Cox connects the collapsing belief in climate change among large sections of the American public to a decline in professional science journalism funded by print and broadcast news organizations. This deficit is being filled by non-journalistic online sources and "self-interested information providers" dedicated to exploiting the consequences of this disinvestment, who are then capitalizing on the opportunities presented by web-based media to peddle misinformation and manufacture uncertainty about climate science. Chapter Sixteen relays the potentially disastrous implications of this uncertainty for those populations who are largely unnoticed by people living in metropoles around the globe. Chris Nash and Wendy Bacon stress the disproportionate impact of climate change on small-island states in the Pacific Ocean threatened by rising sea levels. Drawing on the sociology of Pierre Bourdieu, they analyze news media coverage of the challenges faced by these islands. Nash and Bacon show how tiny islands such as Tuvalu and Kiribati have been reduced to pawns in a "geo-political power play" over competing approaches to climate change. Their essay is complemented by Chapter Seventeen, in which Alanna Myers critiques the oppositional framing of climate change science by leading news media organizations. Concentrating on Australia, her essay sits well alongside the chapters by Cox and Bacon and Nash, suggesting that the "reduction of a tangible environmental problem to a question of social and political values" is a widespread problem. The influence wielded by leading news outlets in the shaping of public and political opinion is also made apparent by Myers. Her approach outlines how the operation of media power subtly shapes the parameters of inclusion and exclusion in debates.

The final chapter in Part Four offers a welcome picture of progress and change in environmental claims making. Accounting for the rise of a green public sphere in China, Guobin Yang and Craig Calhoun describe new forms of public engagement in this increasingly powerful nation and global powerbroker. A number of communications technologies, including the Internet and alternative media forms, are now being used in China to warn about "the dangers of irresponsible human behavior toward nature" and to call for "public action to protect the environment". This development signals a noticeable widening of the terms of media discussion and claims about the environment, which contrasts with the situations described in the other chapters that identify concerted efforts to limit dialogue or "shout down" opposing points of view. Yang and Calhoun's notion of a green public sphere also tables a conceptual resource that deserves further attention and development. For example, the avoidance of ideological extremes during debates about climate change requires an established and reliable framework that can help conflicting parties enter into meaningful dialogue. In the hands of Yang and Calhoun, the adaption of Jürgen Habermas's theory (1989a) holds this promise. The growth of a green public sphere presents possibilities for talking about the environment in ways that deliver meaningful citizen involvement and productive political participation.

The book finishes with an Afterword by the national leader of the Australian Greens, Senator Christine Milne. Drawn from the keynote address she delivered at the environmental media research symposium in Hobart in November 2011, she offers a welcome personal perspective on many of the issues discussed in the preceding chapters. Her extensive experience as an environmental activist and influential political leader shows how the strategic and tactical use of media has evolved over the past 25 years. Whether Milne is talking about fax machines, newspapers or Twitter, her account highlights the ceaseless interaction between activism, media and political systems during environmental disputes, thereby underlining the relevance of the chapters presented in Parts One to Four.

In advancing the study of environmental media and conflict in the network society, we hope that readers carefully consider the arguments that have been presented here, and are responsive to the issues, concepts, approaches and evidence contained in the eighteen chapters and Afterword that follow. This wide-ranging collection moves across continents, shifts between media platforms and technologies, considers activism and campaigns in multiple contexts, travels between the local and global, and traces the mediation of slow-moving crises and unexpected disasters. The ambition of this scope is informed by the knowledge that the future of the planet is unfolding before our eyes, witnessed in and through media whether we live in cities, towns, remote areas or even at the top of a tree.

ENDNOTES

1 368 days at the time of writing.

2 http://www.flickr.com/photos/76527681@N03/sets/72157632238538588/

3 A statement that recalls Walter Benjamin's *Theses on the Philosophy of History* and the figure of the "Angel of History": A storm "irresistibly propels him into the future to which his back is turned, while the pile of ruins before him grows skyward. What we call progress is this storm" (Arendt, 1983: 164).

4 Perspectives include media studies, journalism studies, cultural studies, communication studies, sociology, environmental studies, anthropology and history.

5 The "Global Rescue Station" then followed this, in summer 2003–2004. Located in the Styx Valley, this 5-month protest supported by Greenpeace and the Wilderness Society cost around AUD$200,000. The Station involved a relatively sophisticated online presence, a base camp and a platform in a giant *Eucalyptus regnans*. Activists from countries including Japan, Germany, Canada and Australia supported the protest, which attracted international news media attention (Lester and Hutchins, 2009).

6 This point connects with a lesson delivered by Harold Innis (2007 [1950]) well over half a century ago about time-biased and space-biased forms of media.

7 In addition to having her campaign aided by Brown's visit to the Observer Tree in July 2012, Gibson was the recipient of the 2012 Bob Brown Foundation Environmental Courage Award (see http://bobbrown.org.au/content/index.php/foundation/home/).

8 These products are sold by Ta Ann Tasmania via a complex supply chain. This company is a subsidiary of a Malaysian parent company, Ta Ann Holdings—one of six major forest companies in Sarawak.

9 As Gurney explains, the former Australian prime minister Kevin Rudd invoked this phrase in a 2007 speech about climate change.

10 Our thinking on this issue is informed by reports produced by expert and institutionally endorsed bodies such as the Australian Academy of Science (2010), *The Garnaut Review 2011* (Garnaut, 2011) and the Intergovernmental Panel on Climate Change (2007a).

11 This arson attack in September 2012 is entirely unrelated to the activities of Give It Back and its members, who profess a belief in peaceful protest only. There is no suggestion here that any of it members took part in or approve of the alleged attack on the campsite in the Upper Florentine Valley.

▪ CHAPTER ONE

Environmental Conflict in a Global, Media Age

Beyond Dualisms

SIMON COTTLE

Environmental conflicts today increasingly need to be conceptualized and theorized in relation to endemic forces of change and global crises. This includes the ecological catastrophes and calamities generated by late modernity and globalized (dis)order. Conflicts centred on environment and ecology, even when taking place in local or national contexts, are often best conceived in terms of "world society" and what Ulrich Beck discerns as its "interdependency crises" (Beck, 2009). They are also invariably waged in and through available media and communications (Beck, 2009; Cottle, 2009a, 2011c). It is by these means that images and ideas circulate around the globe; identities and solidarities become invoked across space and place; and the legitimacy of environmental issues and political aims are variously elaborated and contested. It is also by such means that environmental issues can also be "scaled-up" or "scaled-down" from the local to the global (Cottle and Lester, 2011; Pickerill et al., 2011).

The ecology of media and communications today is also fast transforming (and globalizing) with the rise of new media—the Internet, mobile telephony and social media—now all communicating alongside and, increasingly, infused with mainstream media. The "politics of representation", long associated with the critique of mainstream media systems, their corporate structuration, market determinations and democratic deficits, is now having to recognize the emergence of a new politics, a "politics of connectivity" and one often conceived in more celebratory terms. This centers on new media and their enabling connectivity within everyday life and civil society and facilitation of democratizing participation, including the coordination of new social movements and mobilization of ecological and environmental issues worldwide. In networked societies "communication power" is becoming radically reconfigured and dispersed (Castells, 2009).

This short essay sets out, first, to provide a brief reprise of what we already know from the field of media and communication research that helps to explain the saliencies and silences as well as the sometimes spectacular nature of environmental conflicts when reported in the media. I then argue, in more interventionist mode, for two necessary departures following upon the changing ecologies of both environment and media just alluded to. Therefore, I suggest that scholars and students of communications and environmental conflict now need to take cognizance of not only the increasingly global nature of environmental conflicts, but also how these are often deeply enmeshed with other forms of global crisis and contention. Third, I then argue, if we are to make better sense of how environmental conflicts and crises of ecology are represented and mobilized in today's media ecology, we need to move beyond a series of dualisms in the field of media and communications research, including current theorizing centered around "old" and "new" media. Too often, critics of "old" and celebrants of "new" media alike overlook the real game in town, namely, the dynamic fusions and interplays *between* them. Both traditional media systems *and* new communication networks, I contend, overlap, interpenetrate and condition each other within today's fast-evolving media ecology, and it is this dynamic and complex interplay that demands examination. Today's increasingly complex communications field is characterized by *both* conglomerate media formations and new communication flows, and it is one in which environmental conflicts can become powerfully mobilized as well as conditioned. Paradigmatic preferences for "new" media or critiques of "old" media ultimately prove unproductive when applied in isolation to today's complex media ecology. The following elaborates on these three principal aims and claims.[1]

ENVIRONMENTAL CONFLICT IN THE MEDIA: A BRIEF REPRISE

Environmental conflict in the media has received considerable research interest in the last two decades or so, and some of the fundamental drivers of this coverage are well known (Hansen, 1993, 2010; Anderson, 1997; Chapman et al., 1997; Allan, Adam and Carter, 2000; Allan, 2002; Cottle, 2006: 120–142, 2009a: 71–91; Lester, 2007,

2010b; Cox, 2010). Deep-seated news values such as "deviance", "negativity", "conflict", "drama", "violence" and, in the case of visual media, "spectacular" or "arresting images" can predispose journalists to identify some environmental issues and concerns as more newsworthy than others (Lowe and Morrison, 1984). The upshot of the operation of such general news values in this context is clear: dramatic environmental events such as environmental disasters are likely to find news coverage, but not longer-term processes of incremental environmental deterioration or invisible hazards. Environmental protests, especially those involving conflict or violence, are also likely to attract news cameras, but not the more mundane processes of political lobbying or behind-the-scenes bargaining and negotiations. And environmental issues that can be rendered into simple binary oppositions (protestors versus police; local residents versus property developers; industrial polluters versus defenders of local wildlife) are also likely to shape the news agenda and not the multiple perspectives or detailed arguments of complex disputes. Those environmental concerns that are perceived to be culturally proximate and "closer to home" rather than those geographically distanced or "culturally remote," are also more likely to be selected as more newsworthy by nationally oriented news organizations and journalists.

The adequacy of news values as an explanation for the extent and forms of news coverage of environmental coverage remains at best, however, a *generalizing* approximation of the complexities and dynamics involved in environmental news reporting. Other factors and dimensions also need to be taken into account. The model of agenda-setting, for example, with its concern with how the news media signal some issues as relatively more or less important than others (McCoombs and Shaw, 1972), prompts a more contingent, less universalizing approach to environmental agendas, and can help us to map the rise and fall of environmental issues in the news over time. This, in turn, invites closer examination of *how* news issues are actually represented, and not simply where they are positioned in the news agenda. Studies of news "frames" and "framing", for example, have attended to processes of news selection and salience and how news representations are structured to promote a "particular problem definition, causal interpretation, moral evaluation and/or treatment recommendation" (Entman, 1993: 52; see also Gamson and Modigliani, 1989; Miller and Riechert, 2000; Reese, 2003; Hannigan, 2006; Hansen, 2010).

If the study of news frames encourages a more detailed appreciation of how news texts organize the representation of environmental issues in ways that invite particular understandings (and promote possible actions), the political economy of news organizations continues to provide a necessary backdrop for understanding the general determinants (commercial, competitive and conglomerate) that shape both national and transnational environmental conflict coverage based on the corporate pursuit of ratings, readers and revenue. And sociological studies of news production and professional journalist practices, for their part, help to provide an insider view on how environmental coverage becomes practically mediated in and through the daily practices and performance of journalism. Studies of gatekeeping and the professional enactment of norms of objectivity and balance as well as routine source dependencies also help us to account for the belated and biased reporting of climate change.

Here, for example, the professional enactment of the norm of balance is widely said to have produced a situation where a handful of climate change skeptics could secure prominent news access and cast doubt on the scientific claims about climate change, and long after the world's scientists had arrived at a near consensus on such matters. Maxwell Boykoff and Jules Boykoff, in this regard, have stated: "Rather than relying on external constrictions—such as overt censorship and editorial spiking of stories—the mass-media depend on internal constructions, disciplinary practices that produce the patterned communicative geography of the public sphere" (Boykoff and Boykoff, 2007: 1201–1202).

Sociological studies of environmental "source fields" have found them to be populated by competing interests and identities and structured by unequal opportunities to influence the news agenda and processes of news framing, though here, communication tactics and other resources—financial, organizational, cultural and symbolic—can all enter the frame and, sometimes at least, offset disadvantages of institutional power between, say, environmental protesters and vested corporate interests (Anderson, 2003; Hansen, 2010; Lester, 2007, 2010b; Hutchins and Lester, 2011). The empirical exploration of news sources and communication strategies also prompts investigation of processes of agenda building, or how environmental interest groups and others can maximize their influence by, for example, forming coalitions to collectively advance their preferred environmental agenda in the news media over time. Tactically, this can involve the provision by news sources of "information subsidies" (Gandy, 1980) and, more recently, "video subsidies", by campaigning organizations such as Greenpeace and citizen journalists (Allan and Thorsen, 2009). By such means, protest groups and others aim to get their message across by providing an attractive "subsidy" of packaged information and broadcast-quality film sequences to cash-strapped, time-poor, competitive news organizations (Davis, 2003), and, on occasion, it works.

Studies of "claims makers" and "claims making" in the news media, when combined with inquiry into the structured patterns of news access as well as the different discursive opportunities enabled (or disabled) by different news formats (Cottle, 2000b), also open up new insights into the nature and possibilities of environmental news access. Studies of "primary definers" and "primary definition" (Hall et al., 1978), for their part, continue to emphasize the operation of hegemonic power in the privileged news access granted by the news media to political, economic and social elites who, it is said, set the parameters of public discourse and define the nature of public issues. Certainly, whoever manages to define public threats, issues and conflicts in the media and to prescribe courses of action (or inaction) in relation to them occupies a commanding position and one of considerable communicative power. In the field of environmental news discourse, this communicative power encompasses more than just the "hierarchy of credibility" (Becker, 1967) or news-sanctioned patterns of hierarchical access. It also refers to the different epistemological underpinnings to different news stories, knowledge claims and forms of speech; whether, for example, they involve the advancement of scientific knowledge or lay knowledge. Much may hang on whether accessed news voices advance their point of view and arguments in the dispassionate, impersonal and statistically probabilistic claims of scientific rationality, or in the emotionally felt, communally lived and culturally experienced views and values of social rationality (Beck, 1992; Wynne, 1996; Cottle, 2000b).

Research into the media and the environment also tells us that we need to extend our sights beyond media-centrism and the exclusive focus on media practices and processes to examine the *interactions* between different institutional arenas and how these influence the representations and discourses of environmental news. Here, the earlier "public arenas model" (Hilgartner and Bosk, 1988) remains as relevant as ever for understanding the underpinning of a media constituted "public sphere" (Habermas, 1989b). It invites us to examine how social problems (including environmental issues) compete with others for attention within different public arenas (e.g., politics, science or the media) and how these and other institutional arenas have differing (i) "carrying capacities" (or available space and time to allocate to different issues), (ii) "principles of selection" (such as "news values") and (iii) "networks of operatives". This institutional model, clearly, holds relevance for the fields of environment and risk communications today, and it invites further examination of the contiguities and interactions between the public arenas of media, government and science—to name only the most obvious. It also encourages closer examination of the practices and professional cultures of their respective "networks of operatives", including, perhaps, specialist environment correspondents (Peters, 1995; Anderson, 1997).

It is not only established institutional arenas and their interactions with the news media that inform and shape environmental reporting today, but also protest groups, new social movements and, more recently, transnational activist networks that come together to form loose coalitions and mobilize both online and in the real world (Opel and Pompper, 2003; Jong, Stammers and Shaw, 2005; Castells, 2007; Cottle and Lester, 2011). Here, researchers have begun to examine the complex networks and communication strategies, tactics and celebrities deployed by increasingly media-savvy environmental groups and protestors (Anderson, 1997; DeLuca, 1999; Allan, Adams and Carter, 2000; Lester, 2006, 2007).

This work helps to ground the analysis of environmental coverage within the wider and moving terrain of environmental politics. The deep-seated and historically forged "cultural resonance" of the environment as a cipher for feelings and values about nature also finds expression in and through the news media, in its conventionalized "rhetoric of environmental images" (Cottle, 2006: 130–137) and in the image politics of environmental protests (Deluca, 1999; Lester, 2007; Doyle, 2011). Whether depicted in terms of the *spectacularization* of nature as pristine, timeless and outside of human history and society, or through symbolic images of nature as humanly despoiled, exploited and *under threat*, such culturally resonant media images provide an affective charge to environmental discourses and the politics of risk now circulating throughout societies—including via the news visualization of climate change (Cottle, 2009a; Lester and Cottle, 2009).

Though this rhetoric of environmental images is potentially universalizing in today's global media, we also know that the meanings and politics of such scenes can be interpreted quite widely and differently when seen through the prism of different cultures, indigenous practices and traditional identities. The predominant discourses on whaling found in the British press, for example, concern the protection of endangered species and the immorality (cruelty) of killing whales or cetaceans in general. In the Japanese press further discourses come to the fore, however, including the economic consequences of whaling for involved communities and industries and the cultural impact a ban on whaling has on existing whaling communities (Murata, 2007). Different communities of interest and identity can read the same images of whaling quite differently, it seems, and especially when they are linguistically "anchored" (Barthes, 1977) in the "vise" of directing headlines, captions and news language.

Audience studies tell us that people make sense of media representations of the environment through multiple and often overlapping interpretative frames, whether based on personal outlooks, political orientation or appeals to reason, evidence and/or affect, as well as audience expectations of media balance and reporting fairness (Corner et al., 1990; Macnaghten and Urry, 1998; Buckingham, 2000; Etkin and Ho, 2007; Lorenzoni et al., 2006). In this sense, more is going on than the rationally conceived information-based processes often assumed in approaches to the public understanding of science. For many, environmental risks are experienced "close to home" in settings and through identities embedded in local communities and familial relationships (Macnaghten and Urry, 1998), and it is here that feelings of "ontological insecurity" (Giddens, 1990) in the face of the perceived failure of officialdom to deal with environmental threats are often most keenly felt. Confronting potentially devastating environmental change and proliferating environmental conflicts, the media have a responsibility not only to illuminate the "bads" of global risk society (Beck, 1992) but also to democratize them by enfranchising all those who are affected by them, communicating across geographical frontiers and the world's cultural and other divides, and enabling participation in collective decision-making processes about how such threats should be managed or curtailed in the future.

ENVIRONMENTAL CRISES IN GLOBAL CONTEXT: DEEPENING THE FRAME

As studies of climate change and references to "world society" have begun to indicate, many environmental conflicts today are not confined within national territorial borders, their habitats, sovereign seas or surrounding atmosphere. In a globalizing and de-territorializing world, environmental conflicts can extend beyond countries and even continents to encompass the globe. Climate change, probably *the* global crisis *par excellence*, according to Ulrich Beck, expresses both the condition and consequences of "global risk society" (Beck, 2009), but it can thereby also summon into being cooperative, concerted responses at national and transnational levels of governance and civil society. In its wake, climate change also spawns a host of new environmental challenges (Intergovernmental Panel on Climate Change, 2007a) that serve to illuminate how global ecological crises today both generate and interlock with other environmental conflicts variously played out at local, national and transnational levels.

As governments and economies seek to adapt to and/or mitigate the effects of climate change they must ratchet up their responses both nationally and internationally and commit to environmental policies and treatises. Conflicts and contention inevitably surround carbon-trading schemes and the commitment to emission reductions, fossil fuel energy efficiencies and the development of renewable technologies. New conflicts and new disagreements inevitably emerge and erupt on this fractious terrain of environmental governance. These involve, for example, the production of bio-fuels that leads to the clearing of (carbon-sink) forests and substitution of bio-fuel crops for food crops in poor countries; the "carbon miles" involved in air freighting food exports from developing countries, and the economic and social costs to those societies and communities of not doing so; the return of nuclear energy as a preferred alternative to high-carbon fossil-fuels; the expansion of wind farms and the environmental costs of renewable energy schemes; the inconsistencies and contradictions of government taxation systems and differing systems of transport subsidies; the bureaucratic chicanery and corporate abuses of international carbon trading and offsetting schemes; the rise of China, India and Brazil as major carbon emitters alongside the US and the West; and the failure of major powers to ratify international protocols and/or to meaningfully support processes of climate change adaptation as well as mitigation in developing countries.

Though climate change has received significant research attention by media and communication scholars in recent years (for example, Carvalho, 2007; Boyce and Lewis, 2009; Eide et al., 2010; Doyle, 2011), we need to better map and analyze how it generates and interacts with different environmental issues and conflicts. We also need to rise to the challenge of how the global crisis of climate change both spawns and interacts with other global threats produced by late modernity, including impending energy, food and water shortages, population growth, enforced migrations and civil strife, and how these become played out and mobilized in and through today's media and communication ecology. In the words of the expert panel of the Rio+20 sustainable development summit in 2012: "The combined effects of climate change, resource scarcity, loss of biodiversity and ecosystem resilience at a time of increased demand, poses a real threat to humanity's welfare" (United Nations Conference on Sustainable Development Rio+20 Communiqué, 2012).

Major disasters around the world are now also on the increase, driven by four principal factors: climate change, rapid urbanization, poverty and environmental degradation (United Nations International Strategy for Disaster Reduction, 2012). Classifying "environmental conflict" or "disasters" too narrowly for the purposes of research, as with news agendas and program running orders based on traditional news subject demarcations, runs the risk of dissimulating the complex interpenetration of disasters with ecology and other global dynamics, ultimately under-playing their complex, interlocking and, frankly, more disturbing nature (Pantti et al., 2012; Cottle, 2011c). The Japanese disaster of 2011 involved an unfolding complex of an earthquake followed by a devastating tsunami that, in turn, unleashed a nuclear meltdown and economic crisis. These events contributed to a world oil price rise as well as contamination of marine species in the world's oceans, and increased nuclear distrust around the globe. A prominent UK newspaper, *The Independent*, emblazoned on its front page: "Four explosions, one fire, and a cloud of nuclear mistrust spreads around the world" (Greenslade, 2011). In the wake of Fukushima (2011), and before that, Chernobyl (1986) and Three Mile Island (1979), public concerns about the risks associated with nuclear power have seeped into national debates about energy policy and, more recently, the desired combination of fossil fuels and sustainable energy sources in the worsening context of climate change. Environmental conflicts, then, are often deeply enmeshed within encompassing global crises of climate, food cultivation and energy shortages and human (in)security, and interlock with major unfolding disasters. This renders increasingly problematic notions of environmental conflicts as clearly identifiable, relatively discrete, and much less nationally bounded and nationally originated problems.

Violent conflicts around the world, we also know, are no less over-determined, and often lead to gross environmental despoliation and degradation. This can occur directly through the intentional destruction of habitats and means of community survival by military forces or insurgents, or more indirectly and incrementally as local populations in conflict zones or refugees fleeing from them over-exploit the surrounding fauna and flora in their bid for survival. The environmental impacts of war, civil conflict and insurgency too often remain out of media sight, however, and seemingly also fall off the academic radar of most Western media and communication researchers interested in "the environment" or "environmental conflicts".

The relationship between the environment and political conflict and violence can be even stronger and even more direct than the military wasting of habitats to intimidate local populations and render their means of survival untenable. Over 40 percent of all intrastate conflicts in the last 60 years, according to a report by the United Nations Environment Programme (2009), have a strong link to natural resources. Civil wars in countries such as Liberia, Angola and the Democratic Republic of the Congo have all centered on "high-value" natural resources such as timber, diamonds, gold, minerals and oil. Conflicts in Darfur and the Middle East also involve the control of scarce resources such as fertile land and water (United Nations Environment Programme, 2009: 5). When approached through a wider perspective on intrastate violence, "new wars" (Kaldor, 2006) and the breakdown of civil infrastructure in "failing" and "failed states", as well as in respect of the expropriation of communities and exploitation of natural resources by states and powerful corporate interests, environmental resources are positioned at the core of many of today's worst and most intractable forms of collective violence.

However, if environmental issues and interests often underpin violent conflicts, so too can they become the means for striving for peace: "The recognition that environmental issues can contribute to violent conflict underscores their potential significance as pathways for cooperation, transformation and the consolidation of peace in war-torn societies" (United Nations Environment Programme, 2009: 5). In other words, environmental conflicts may also contain the seeds for increased cooperation, stability and human security in the future insofar as issues of natural resources, ecology and environments can be deliberately built into processes of peace building and subsequent peace management. How media and communications can perform a more developmental and supportive role within these processes of peace building and in relation to potential conflicts centered on environmental competition has yet to be clearly recognized and concertedly researched. Here, at least, useful precedents exist in the work on peace journalism, media and peace building, and in media involvement in processes of post-conflict civil society reconciliation and reconstruction (Lynch and McGoldrik, 2005; Price and Thompson, 2002; Wolfsfeld, 2004).

From the foregoing, a number of points can be highlighted that should inform our thinking and approach to mediated environmental conflicts in a global age. Though many environmental conflicts continue to be played out in particular local-national settings, increasingly, these cannot be assumed to emanate solely from these contexts or to be entirely accounted for by them. In an increasingly globalized, interconnected, interdependent and inegalitarian world, they give expression to wider crises of ecology, economics and energy, as well as processes of collective violence fuelled by disputes over natural resources and their commoditization in the international marketplace. Whether environmental conflicts are locally originated and confined or globally endemic and enmeshed in the ways indicated above, they have all become increasingly dependent on media and communications for how they become more widely known and perceived, and how they are mobilized and responded to (or globally dissimulated and disappeared). We turn now to consider how today's changing media ecology both signals and conditions ecological crises and environmental conflicts in a global, media age, whilst also challenging anachronistic dualisms in the research field.

ECOLOGICAL CRISES IN A DIGITAL MEDIA AGE: BEYOND DUALISMS

Global crises, including ecological crises, following Beck (1992, 2009), can be conceived as endemic, often enduring critical events and threats that emanate from within today's global (dis)order. They can range across and interpenetrate other processes of global interdependency and may also spawn myriad conflicts and contentions. They constitute material and discursive sites of action and response that can themselves extend, exacerbate or intensify processes of globalization and globality. Importantly, in today's mediated and mediatized world, the nature of their elaboration within the global formations and flows of media and communications shapes their public constitution, and this, in turn, variously enters into their course and conduct (Cottle, 2009b). This invites a deeper appreciation of how ecological crises and environmental conflicts, though *ontologically* rooted in the complex forces of globalizing late modernity, are nonetheless also *epistemologically* dependent on how they become discursively framed, visualized and dramatized in media and communications (Cottle, 2011c).

Earlier sociological, less globally encompassing formulations of "crisis" have tended to marginalize and/or underestimate the active definitional, performative and cultural sense-making processes involved in their *epistemological* constitution. In more recent social constructionist accounts their preceding *ontology* is bracketed or even denied when emphasizing processes of discursive definition and cultural framing in language, discourse and media. Today, however, it is imperative that we seek to bring these ontological and epistemological outlooks closer together. In a world based on incessant (unsustainable) growth, the inexorable expansion of capitalism and applied technocratic rationality, crises of economics, ecology and energy are endemic within the contemporary world (dis)order. How these become signaled and symbolized, silenced or spectacularized around the planet, how they become "known" by different publics, and how this conditions cultural perceptions and political processes, positions media and communications in a pivotal role in their unfolding trajectory. It is this media and communications entry *inside* crises today that blurs the former dualism (and use value) of ontological and epistemological approaches to the roles and representations of the media in ecological crises and environmental conflicts.

In today's rapidly evolving communication environment a further dualism in the field also now needs to be challenged, namely, the tendency to conceive and theorize media and communications in paradigmatic terms of "old" and "new" media. The study of traditional mass communications across recent decades has encouraged a critical orientation to the "politics of representation", broadly conceived, and theorized in terms of ideology-critique (or variants of discourse and framing analysis). When approached through the theoretical optics of political economy, cultural studies or the sociology of media organizations, for example, representations of environmental conflict are found to be shaped by corporate media and the determinants of the marketplace, prevailing cultural codes and hegemonic discourses, and/or bureaucratic routines and professional norms, but all are critiqued on the basis of their representational democratic deficits. The study of new media, however, appears to invite a theoretical orientation more attuned to the participatory "politics of connectivity" (Deuze, 2003; Castells, 1996, 2009; Dahlberg and Siapera, 2007), an orientation that has obvious appeal and possible explanatory relevance in the field of ecological movements and environmental protests.

The Internet exhibits a synergy well suited to the new wave of transnational protests and global activism (Bennett, 2003; Castells 1996, della Porta and Tarrow, 2005;

Van de Donk et al., 2004; Dahlberg and Siapera, 2007; Dahlgren, 2009). Researchers have suggested, for example, that the "the fluid, non-hierarchical structure of the Internet and that of the international protest coalition prove to be a good match" and "it is no coincidence that both can be labelled as a "network of networks" (Van Aelst and Walgrave, 2004: 121). The Internet, evidently, contains a socially activated potential to unsettle and, on occasion, disrupt the vertical flows of institutionally controlled "top-down" communications, and it does so by inserting bottom-up, interactive and horizontal communicative networks into the wider communications environment. This possibly helps to account for the sometimes inflated, not to say socially euphoric claims about the democratizing nature and impact of the Internet (Curran et al., 2012), as well as its theorization as if separate from and operating outside of the wider communications environment.

This dualistic tendency to theorize the nature of media and communications based on either critical views of "old" media or more celebratory views of "new" media is not the most productive way to conceive of mediated environmental conflicts and ecological crises in today's more complex media ecology. A similar argument can be been made, for example, in respect of the dynamic and interpenetrating roles performed by *both* "old" and "new" media in the Arab Spring, events too quickly and simplistically dubbed the "Facebook revolution" or "Twitter uprisings" (Cottle, 2011a, 2011b).

Something of the complexity of today's communications ecology involving both "old" and "new" media in dynamic interaction is usefully encapsulated in a study by David Crouch and Katarina Damjanov (2011) that explores the media-savvy deployment of both by eco-activists aboard the *Sea Shepherd* deep in the Southern Ocean as they seek to disrupt the Japanese whaling fleet:

> ...almost instantaneously images are spread across the globe via satellite up-links, webcams, and around-the-clock internet blogging. *Sea Shepherd* takes their environmental protest in this remote and unforgiving location, impossibly beyond everyday reach, and broadcasts it back to the world, disseminating it through the Internet and blogosphere; they constantly send out fresh broadcast-quality images and a barrage of news releases, twitters, and updates of events.... Furthermore, it is not only the tools and tactics derived from contemporary communications technologies that they use to garner support for their cause; by staging their anti-whaling protest as spectacular "pirate" attacks, they also exploit elements of popular culture, tapping into the social imagination of a potentially transnational public sphere. (Crouch and Damjanov, 2011: 186)

As this extract helps to illuminate, the distinction between "old" and "new" media is not always helpful when we grapple with the complex of communication channels and different mediums that together infuse today's contemporary media and communications ecology and which, as in this case, can sometimes be creatively harnessed and put to work. This same study also draws attention to the dramaturgical power of some environmental actions, where drama, spectacle and culturally resonant images are manufactured not only to attract mass media attention but also to communicate potent environmental meanings and messages. In such instances, in such ways, the politics of representation and politics of connectivity are clearly *both* at work and infused inside the production and circulation of environmental meanings simultaneously communicated to mass audiences and networked supporters via old and new media.

This is not, therefore, an *either/or* question, a question of "public sphere" or "public screens", or a debate about environmental communications approached either, on the one hand, through an academic prism of rationally conceived deliberation and public opinion formation, or, on the other, as affective image events dispersed via multiple screens and platforms to fragmented audiences. Rather, in a world of proliferating screen-based technologies, the forms and features of environmental communications need to be approached in ways that are sensitized to the politics of representation and connectivity, to communicative dimensions of "deliberation" *and* "display" (Cottle and Rai, 2006), to the play of "strategic action" *and* "cultural symbolisation" (Wolfsfeld, 1997; Lester and Hutchins, 2012a) and to the "cultural pragmatics" (Alexander, 2006a) of strategic "mediated public performances" that help constitute today's "civil sphere" (Alexander, 2006b).

Today's media ecology undoubtedly contains more political opportunities for dissenting views and voices than in the past, and these are increasingly in evidence through *interpenetrating* alternative networks and mainstream news media. Indeed, given their dynamic overlap and interpenetrations, including the rise of citizen journalism and increasing dependence of mainstream media on incoming images and social media (Allan and Thorsen, 2009; Hänska-Ahy and Shapour, 2012), the conceptual dualism of "alternative" and "mainstream" media, rooted in relatively medium-centric views, has also become less clear and more permeable in practice than in the past (Cottle, 2009a). Lance Bennett commented, for example, that

> …impressive numbers of activists have followed the trail of world power into the relatively uncharted international arenas and found creative ways to communicate their concerns and to contest the power of corporations and transnational economic arrangements. In the process, many specific messages about corporate abuses, sweatshop labour, genetically modified organisms, rainforest destruction, and the rise of small resistance movements, from East Timor to southern Mexico, have made it into the mass media on their own terms. (Bennett, 2003: 18–19)

The field of contemporary media and communications also performs a powerful role in transnationalizing environmental issues and conflicts: scaling them up by signaling "local" concerns as "global" issues; or scaling them down by rendering "global" concerns as local issues (Cottle and Lester, 2011; Pickerill et al., 2011). This need not always be conceived in terms of *either/or*, but rather *both/and* (Beck, 2006). Local and national reporting, for example, can situate conflicts and crises internationally and globally, and international and global media reporting and communication flows can draw out their embodiment in local and/or national contexts. Representational processes of scaling up and scaling down in the media, therefore, can either reveal the processes and forms of local-global interconnection and interdependency in environmental conflicts and ecological crises or, alternatively, conceal and dissimulate them. In conditions of globality and endemic global crises, a new journalism outlook is required: a "global journalism" (Berglez, 2008). This, however, can potentially be conducted and discharged through *any* spatial realm of reporting, and in respect of environmental conflicts and ecological crises from the local and national to the international and transnational. How the evolving ecology of media and communications can reveal or conceal, deepen or dissimulate the complex local-global relations of interconnection and dependency, effectively scaling them up or down in relation to the politics of space and place, will demand increased research attention in the years ahead.

CONCLUSION

This brief discussion has reviewed some of the key findings in the field of media and environmental conflict studies and recognized their continuing relevance for the contemporary research field. Such studies help to point to the multiple determinants and dimensions of media organizations and their cultural forms, and to the professional practices of journalists and the strategic operations of power by corporate interests and the engagements with the same waged by increasingly media-savvy activists and social movements on the environmental terrain. Representations of environmental conflicts are invariably an outcome of this force field of strategic and cultural power in dynamic movement.

The discussion has also sought to deliberately deepen the contemporary frame of reference and research by situating environmental conflicts in relation to wider processes of globalization and the production of global crises. The latter includes profound ecological crises such as climate change that in turn spawn myriad interconnected environmental conflicts. Environmental conflicts, in globalizing context, are frequently enmeshed with other forms of global crises and major conflicts, including major disasters, wars and political violence that interlock in complex and environmentally destructive ways. Though these may not always be the forms of "environmental conflict" most obvious to Western media academics, they demand increased recognition nonetheless, as do the roles of media and communication within associated processes of civil society reconstruction and environmental conflict resolution around the world.

Finally, the discussion has sought to challenge a number of dualisms in the media and communication research field that threaten to delimit our understanding of the complex ways in which media and communications, both "old" and "new", "mainstream" and "alternative", are now deeply etched into the very constitution of environmental conflicts and ecological crises: ontologically and epistemologically, strategically and symbolically, deliberatively and dramaturgically, cognitively and culturally, and in processes of scaling up and scaling down from the local to the global. The politics of representation, based on the critical interrogation of traditional mainstream media, cannot be displaced on the basis of a more celebratory politics of connectivity enabled by new media. In today's media and communications ecology, with its media formations and relatively fluid communication networks, both the politics of representation and politics of connectivity are at play, and both now need to be more systematically studied in their mutual interactions. Along with other dualisms discussed above, they now need to be brought closer together, conceptually and theoretically, when exploring how today's complex media ecology enters into and shapes environmental conflicts and ecological crises in a global, media age.

ACKNOWLEDGMENTS

1 This chapter was first presented as a keynote lecture at the Environmental Politics and Conflict in a Digital Media Age Symposium organized by Libby Lester and Brett Hutchins and hosted in Hobart by the University of Tasmania (17–18 November 2011), and draws, in parts, on the author's recent publications on global crisis reporting (Cottle 2009a: 74–79, 2011c), the Arab Spring (Cottle 2011a, 2011b) and transnational protests (Cottle and Lester, 2011). I'd like to thank the symposium organizers/editors of this book for their generous support and collegiality during my stay in Tasmania and since.

▪ PART 1

“Old” and “New” Technologies

▪ CHAPTER TWO

Campaigning Journalism

The Early Press, Environmental Advocacy and National Parks

MICHAEL MEADOWS AND ROBERT THOMSON

Ideas about the environment, mountains and their varied relationships with people began to creep into colonial Australia through stories published in the local press as the settlements expanded. A marked increase in the frequency of mountain wilderness imagery—writings, drawings and increasingly, photographs—is an important characteristic of the development of the 20th-century Australian press, at least until World War II. Something had drawn people's attention to high places as never before, the result of a range of often competing and contradictory discourses, including Aboriginal creation myths, a unique landscape, the influence of European ideas of landscape and leisure, and charismatic local individuals with a passion for the environment.

In this chapter, we will consider this process through two case studies. The first, drawn from the 1890s, investigates the appearance of arguably the first comprehensive collection of journalism focussed on the Australian environment. Written by explorer-journalist-politician Archibald Meston, a series of feature articles published in the iconic weekly magazine the *Queenslander* reveal ways of framing landscape and the environment that challenged the then predominant images of a "wide brown land" and prevailing ideas of economic development. The second example is from the 1920s and 1930s and has parallels with the North American experience of the founding of national parks. It explores the role of the *Queenslander* and its successors in southeast Queensland in publishing images and stories of wilderness that, we will argue, played an important role in the national parks movement in Queensland.

IMAGINING THE COUNTRY

The language we adopt and access as part of our everyday lives reveals and influences our perceptions of the environment, reflects our objectives and interests, and affects our actions. Cultural groups transform the physical environment into landscapes through the use of symbols—like words, stories and images—bestowing different meanings on the same physical objects (Abrahamsson, 1999: 51). As Seddon (1997: xi) concludes:

> The ways in which we perceive, imagine, conceptualise, image, verbalise, relate to, behave towards the natural world are the product of cultural conditioning and individual variation.

This "imagining" process creates, maintains and articulates ideas and assumptions about the world and our place in it with popular media like newspapers, radio, television and, increasingly, online publications, at the centre of this struggle over meaning. News media play a crucial "sense-making" role in society, particularly in relation to ideas and events which are beyond the everyday experiences of most people. This is especially relevant because the genre of news claims for itself a privileged position as the bearer of truth (Mercer, 1989; Meadows, 2012).

By their very nature, mainstream news media tend to support the status quo, leading almost inevitably to conflict, which remains one of the most prominent news values in determining the importance and relevance of a story. This is clearly evident in contemporary news coverage of climate change, with a handful of powerful, vested interests claiming a right of reply despite overwhelming global evidence of impending environmental disaster (Evans, 2011; Boykoff and Timmons, 2007). Media and journalism are thus important agents in this process of "cultural resource management" but have the potential to be "used strategically" to lead a population from "simple common sense to coherent and systematic thought" (Gramsci, 1988: 385). The examples we consider here of the colonial press in Queensland provide representations of mountain landscapes and the environment to audiences who otherwise would have remained ignorant of their presence and nature. This extensive body of stories and images set up a framework for thinking about such places in terms of cultural heritage rather than the prevailing "common sense" notion of economic gain.

Australia's mountain landscape environments have been effectively "written out" of popular conceptions of the country. The predominant, stereotypical image of Australia centres around ideas of a generic notion of "the bush" and/or "a wide brown land". It has ignored the influence of significant areas of mountain landscapes in the meaning-making process, the end result of a limited scholarly focus on literary and visual representations of place (Bonyhady, 2000; Whitlock and Carter, 1992; Bonyhady and Griffiths, 2002; Horne, 2005). The overwhelming body of this scholarly work on landscapes and the environment thus far has essentially ignored the role of popular culture, particularly the stories, illustrations and photographs published in local newspapers. We argue that these popular cultural resources, primarily because of their accessibility and pervasiveness, played—and continue to play—a critical role in shaping ideas of the environment and landscape and the activities associated with them. We will argue that it is inconceivable that such a widespread "structure of attitude and reference" (Said, 1993) evident in representations of landscape in the popular press has *not* contributed significantly to our understandings of place and, by association, to the broader process of imagining the environment.

AUSTRALIA'S FIRST ENVIRONMENTAL JOURNALISM

In 1863 a shift in thinking about mountain landscapes in Europe manifested itself in the conservative *British Alpine Journal,* which began to describe mountains in terms of sporting and scientific destinations rather than the romantic idyll which had dominated representation for centuries, paralleling changing notions of framing mountains through photography (Ellis, 1990; Bensen, 1998). This influence became evident in Queensland when, towards the end of the 19th century, colonial newspapers published the first extensive body of mountain literature (Meston, 1889a, 1889b). This arguably set up a framework for thinking about mountain landscapes—and thus the environment—preceding the first mass rock climbing movement in Australia from the late 1920s, based in the mountains of southeast Queensland, with smaller cohorts active in the Blue Mountains and Tasmania (Meadows, 2001).

In a career seldom free of controversy, Archibald Meston was among the first to see Australian mountain landscapes in ways that until this point in time had been the preserve of the writings of English travellers in the Alps or the Lake District. Queensland had its own "peaks and passes" with a unique local flavour, and despite Meston's often emotive language, the descriptions he made of the wild mountain landscape in north Queensland at the close of the 19th century are unparalleled in other early Australian literature. Perhaps his most significant contribution was to present Australian colonial newspaper audiences with spectacular images of their own country along with an acknowledgment, despite the inevitable contradictions, of Aboriginal concepts of place.

Born in Scotland, Archibald Meston migrated to Australia in 1859 and grew up on the Clarence River in northern New South Wales. He worked as a journalist between 1876 and 1881 and was later appointed as leader of a Queensland Government scientific expedition to the Bellenden Ker Range in north Queensland. In 1878, at age 25, Archibald Meston became the youngest man to be elected to any Australian Parliament. Described as "ambitious, dashing, irresponsible and the vacillating young Meston" during his three-year term, he was given the nickname the "Sacred Ibis" for his habit of making classical references, regardless of the topic at hand. He also regularly quoted from the Romantics—Coleridge, Shelley, Byron, Shakespeare, and Wordsworth—and often composed his own verse when so inspired. This "Mestonian" style became the hallmark of his later writings (Gall, 1924: 14). He was described as "muscular and very fit", and he vowed he could take on anyone in an athletic contest. In 1897 he was appointed Protector of Aborigines for south Queensland. He always maintained that Indigenous people were mistreated and claimed that if it were not for firearms and poison, they would have driven Europeans into the sea (Gall, 1924: 14).

His enthusiastic engagement with the rainforest-clad mountains in north Queensland began early in the 1880s, as he explored the area on multiple excursions between 1881 and 1904. While the main purpose of his northern expeditions was scientific, Meston's desire to claim first ascents of local summits emerges as a clear driving force in his writings. As with the summits of the Australian Alps, the high points in north Queensland's Bellenden Ker Range are elevated sections of ridges rather than stand-alone peaks. In 1889, as Meston and his six companions toiled upwards towards one such summit, Mt Sophia, they emerged from "severe climbing over rough rocks to a clear area near the summit". As Meston noted, they had more than the physical act of climbing to contend with:

> In cutting through the thick vegetation so as to reach the outer edge of the peak, a long strange gray-looking snake glided suddenly from a bush overhead, passed over my right arm, and disappeared. In making an involuntary leap back, I collided with a small tree covered with sharp thorns, which tore away my right sleeve and furrowed my arm in rare and fanciful patterns. (Meston, 1889a: 7)

A few months later, Meston returned to the north, this time climbing the second-highest mountain in the state, Mt Bellenden Ker. Meston and his companion, government botanist F. M. Bailey, were well rewarded when they reached the summit following a three-day climb—even the loquacious "Sacred Ibis" was momentarily dumbfounded:

> And what a view! For some time not one of us could find a voice. All was distinctly visible, in the perfectly clear atmosphere, in a radius of, at least, 100 miles in all directions. We were silent in the awful presence of that tremendous picture that had laid there unaltered since Chaos and the Earthquake painted it in smoke and flame and terror in the dark morning of the world! It was a hall of the Genii of the Universe, the Odeon of the eternal

> gods with its immortal floor paved with the green mosaic of land and ocean, and overhead the arched blue roof flashing in diamonds and prismatic radiance to the far skyline on the edge of the dim horizon. Eastward rolled the calm Pacific, visible from the Palm Islands in the south to Cooktown in the north. The white surf breaking on the Barrier Reef was a long white line on the slumbering azure of the slumbering ocean. (Meston, 1889b: 7)

In the published account of his second 1889 expedition to Bellenden-Ker, Meston made a strong plea for the retention of local Aboriginal (*Kuku Yalanji*) names for flora, fauna and landscape:

> …these grand, sonorous, and euphonious aboriginal names ought to supersede the meaningless Bartle Frere and Bellenden-Ker, given presumably in honour of two gentlemen who had as much connection with these two mountains as with the building of the Pyramids. The native names of mountains, rivers, streams and lakes ought to be jealously preserved in all cases. They are far more appropriate than any we could possibly bestow, and as a rule, much more pleasant on the ear. (Meston, 1889a: 693)

Meston was amongst the first (and few) public figures to urge the adoption of local Aboriginal place names, although his attitude towards Indigenous people was highly problematic. His persona veered uneasily between that of champion against Indigenous oppression and that of an overt racist, from acknowledging the impact of dispossession to boasting that each notch on his rifle represented "a dead nigger" (McKay, 1998). He was probably the first to publicly use mountain names drawn from the *Noongyanbudda Ngadjon* people's vocabulary, including *Wooroonooran* (Mt Bellenden-Ker) and *Chooreechillum* (Mt Bartle Frere). It is only in recent times that *Wooroonooran* has been re-appropriated as the name of one of North Queensland's major wet tropics' national parks (Wet Tropics Aboriginal Plan Project Team, 2005). Meston made constant reference to Aboriginal presence in the mountains in all of his writings about the north. He and his companions often came across "myall" camps and followed their extensive network of tracks through the scrub, but rarely, if ever, sighted any people. He concluded (1889a: 7):

> …the wild children of the mountains had received intimation of the approach of the expedition, for such a mighty host walked not in silence through the lonely scrubs any more than Lucifer's army through the affrighted Deep.

Within a few years, development of the half-tone process saw the first photographs appearing in Australian newspapers. But Meston had to wait almost a decade before he was able to return to re-live his Bellenden Ker experiences through a camera lens, when an extensive range of photographs taken by A. A. White on his 1904 expedition to the area was published. This was the first extensive collection of mountain photography to appear in the Queensland popular press, and probably the first in Australia. The collection of Meston's long, descriptive feature articles published during this time offer insights into the nature of the landscape and the environment—as well as the man. This collection is arguably the first example of journalism published in Australia that focussed specifically on the environment. Meston's vivid descriptions of flora, fauna and landscape were framed in terms of a collective cultural heritage and proposed ways of thinking about such places in opposition to prevailing ideas centred primarily on economic development.

ENVIRONMENTAL ADVOCACY AND THE NATIONAL PARKS MOVEMENT

From the beginning of European occupation, ideas of landscape in Australia have been associated closely with the notion of leisure. The foundations for the idea of Australian leisure were laid by the second half of the 19th century—essentially they were the cultural traditions of England, albeit influenced in part by the introduction of the eight-hour day (Lynch and Veal, 1997; Hamilton-Smith, 1998). This is evident in the emergence of activities like bushwalking, rock climbing and cycling. At the turn of the 19th century, tourism emerged as a key cultural activity in Australia, with mountainous terrain a key destination.

By the later decades of the 19th century the idea of national parks had spread across much of the English-speaking world. The world's first national parks were established at Yellowstone in 1872 and at Port Hacking, south of Sydney, in 1878, and by the 1880s and 1890s there were calls throughout the Australian colonies for the creation of similar reserves in areas considered to have a particular aesthetic significance. Throughout the history of the emergence of Australian notions of landscape and leisure, there is a strong link to the natural environment, with Australia becoming the first place in the world to proclaim wildlife reserves (Hamilton-Smith, 1998; Westcott, 1991).

The role of media in national parks movements elsewhere is difficult to assess accurately due to the dearth of scholarly work focusing on this relationship. The argument for national parks in the United States drew upon romantic conceptions of "landscapes, wilderness, and island paradise" (Warner, 2008:19). However, it was the influential writings of conservationist John Muir—the "Father of American National Parks"—that provoked a cultural shift from considering these spaces as opportunities for progress to understanding them in terms of their intrinsic cultural value (White, 1999: xiv). Muir emigrated from Scotland to the United States in 1849, and by 1875 was aware of the degradation of mountain environments in his adopted country. By 1892 he had formed the Sierra Club, an advocacy organization which inspired national parks movements in Canada, New Zealand and Australia

(Warner, 2008: 29). His solo climbing and walking forays into the wilderness of the Sierra Nevada with minimal food and clothing were soon legendary. He later became known as the "founding father" of "clean" climbing, an approach in climbing mountains which uses no artificial aids and causes no damage to the environment. It influences rock climbing culture in the Yosemite Valley in California to this day (White, 1999: 14).

Muir gained a national reputation as a wilderness writer and conservation advocate through a series of articles in the journals *Overland Monthly* and *Century Magazine* (Warner, 2008: 19–20). Muir and prominent literary figures like Walt Whitman, Ralph Waldo Emerson and Henry David Thoreau—along with landscape photographers like Carleton Watkins—reflected the "idea of divinity in nature" which underpinned the national parks movement in North America, although access to hunting and recreation were inherent in the early popular conceptions of nature reserves (Warner, 2008: 17; MacFarlane, 2003: 210; DeLuca and Demo, 2000: 241; Seddon, 1997: 8–9). Muir's descriptions of the Yosemite landscape in 1869 reflect those of Queenslander Archibald Meston two decades later. Gazing upon the 3,500-metre Cathedral Peak, Muir wrote:

> This I may say is the first time I have been at church in California, led here at last, every door graciously opened for the poor lonely worshipper. In our best times everything turns into religion, all the world seems a church and the mountains altars. (in White, 1999: 51)

As in the United States, the national parks movement in Canada enlisted artistic and photographic resources to depict wilderness as "healthy and pleasurable to experience" (Warner, 2008: 28). But perhaps it was Canada's first Commissioner of National Parks in 1911, James B. Harkin—who regularly quoted Muir's writings—who was most influential in that country's parks movement. Canada may have trailed the United States in terms of its advocacy for national parks, but it nevertheless set up the world's first national parks service under Harkin.

The formation of national parks in Australia has been very different from the process in North America. For a start, land in Australia—and thus national parks—are controlled at the state level rather than by the federal government, enabling local movements to play a more important advocacy role (Frost, 2004: 505). By the 1920s and 1930s in Australia, political players were not always enthusiastic in embracing a growing national parks movement, perhaps because they were more interested in rural development than in "locking away" public lands. But it is only in the rural press that we find dissenting voices—and even then, only occasionally. For the most part, during the inter-war years the national parks movement remained largely unproblematic. Not only did it have a distinguished patronage and support base both in the metropolis and in the regions, but also the land involved was generally regarded as being unviable, hence unsuited to primary production. However, this was not always the case, as several major national park declarations in Queensland suggest (Frost, 2004: 505). In the case of the timber industry, it was generally agreed that there were ample and more easily accessible timber resources elsewhere within the state forest network. That meant that during this period, the idea that national parks might stifle economic development was never strong. The contested agendas pushed by a later generation of national parks advocates involving ecology, environmentalism, and the reservation of larger areas of land for national parks were yet to emerge.

Throughout the 1920s and 1930s the impetus from influential, sympathetic journalists and passionate adventurers with engaging writing skills placed national parks stories firmly on the news agenda. This mélange created ideal conditions for favourable local media coverage presenting a compelling anthology of "coherent and systematic thought" that privileged the environment.

Between 1919 and 1940, the national parks cause in Queensland received a significant level of coverage in the Brisbane and regional southeast Queensland press. Apart from a handful of exceptions, the various reports and articles that appeared in the newspapers were remarkably sympathetic, and throughout this inter-war period all major Brisbane newspapers were at various times prominent advocates for the establishment of national parks. Whether this coverage actually influenced government policy on national parks is, of course, difficult to say. However, it is fair to assume that at the very least, the press helped to shape public opinion on national parks and promoted a way of thinking about landscape that was favourable to their creation and expansion.

Queensland's first national parks advocate of any standing was the grazier Robert Martin Collins (1843–1913), who was inspired by the idea during his trip across the United States in 1878 when he spent three days visiting Yosemite. On returning to Queensland, Collins began a low-key campaign lobbying government to have parts of the McPherson Range, on the border of Queensland and New South Wales, reserved as a national park. Although Collins did not live to see his plans reach fruition, the cause was taken up by young Brisbane engineer Romeo Lahey, whose family had extensive timber interests around Canungra in the vicinity of the proposed national park. Writing in the *Queenslander* under the pseudonym "Wanderer", Lahey described one journey in 1911 with a companion, W. E. Potts, through the dense rainforests of the Upper Coomera River in the proposed national park. There was a hint of Archibald Meston's descriptive enthusiasm and zeal in his account:

> The creek rose steadily at a grade of 1 in 6, and waterfalls were numerous, and as we had to climb around them travelling was pretty slow. Soon after starting we captured a splendid specimen of the China-blue lobster, about a foot long (300 mm), so we decided to keep him for Mr Tyson, the Government Entomologist, to examine. There were a great

> number of them in the creek, so we caught a dozen or so and cooked them for dinner.... While we were having dinner a storm gathered just overhead; the clouds were so low that it seemed as if we had only to put up our hands to touch them. They came from all points of the compass, turned and twisted for while, then collected and sailed off down the valley, thundering all the way. (Lahey, 1911: 8)

Interestingly, the Queensland Museum now lists the unique Lamington blue crayfish as endangered, with a usual length of between 100 and 130 millimetres. This suggests that the value of such observations made by adventurers and conservationists such as Lahey, and later Arthur Groom—and published in the popular press—extended beyond the aesthetic into the scientific domain. With hindsight, could such evidence be interpreted as a portent of climate change, perhaps?

Lahey made a number of tours of the district, both in and around the proposed national park, between 1913 and 1915, promoting among local landholders and townspeople the idea of establishing a park. He also spent a good deal of time making a detailed survey of the proposed park, which even then remained largely unknown to Europeans. In August 1915, following a change in government, Lahey's campaign finally proved successful, and the Lamington National Park (initially comprising some 19,000 hectares) was officially gazetted. In the meantime, a number of other smaller national parks had been established. Queensland's first national park, a parcel of land at Witches Falls on Tamborine Mountain, was proclaimed in 1908. The Bunya Mountains National Park (initially comprising some 9,000 hectares) was reserved in the same year, and the Cunningham's Gap National Park (initially comprising some 1,200 ha) was gazetted in 1909.

During the late 19th and early 20th centuries Australian newspapers more broadly were replete with news items and feature articles which celebrated the aesthetic credentials of Australian landscapes. This was, of course, not a uniquely Australian phenomenon—by the late 19th century writing about landscape had become an established genre within journalism, at least in the English-speaking world, reflecting a broader cultural preoccupation with ideas of landscape which had emerged in the West from the 18th century. In Queensland, Brisbane newspapers had been praising the merits of the remnant bush areas around the metropolis since at least the 1870s, and to a lesser extent, this was also the case with the regional press. By the 1880s, with the introduction of new printing technologies which allowed the production of halftones, the *Queenslander*—a pictorial weekly—included numerous illustrations of picturesque bush scenes and landscapes, and with the arrival of photography in the later 1890s, this continued.

At the end of World War I Brisbane-based newspapers once again began filling their pages with articles and photographs of scenic locations across southeast Queensland and sometimes further afield. One particular topic of interest in this immediate post-war period was the Lamington National Park, which during the war years had been almost completely ignored. By the end of 1918 a succession of writers began visiting the park and writing feature articles on this "paradise" at the southern border. Significant contributions during this period came from a group of journalists with strong environmental credentials.

Alex Chisolm, a noted amateur ornithologist, worked as a columnist with the *Daily Mail* from 1915 to 1922. He was a prominent figure in the Brisbane-based Field Naturalists' Club, a mix of scientists and amateurs. In the inter-war years Chisolm was at the forefront of the national parks movement in Queensland and used his connections with government to promote the idea of Lamington National Park. He popularised the ideas of conservation and national parks in his regular newspaper nature column.

Winifred Moore started as a journalist at the *Daily Mail* in 1916 and became interested in national parks a few years later when she began to visit existing parks and reserves near Brisbane—Lamington, Tamborine and Springbrook amongst them. She became an enormously influential figure as "social editress" of the *Brisbane Courier* (later *The Courier-Mail*) from 1921 until 1952. Her highly popular daily columns undoubtedly influenced generations of female readers (and, no doubt, more than an occasional male reader). She was a founding member of the National Parks Association of Queensland, formed in 1930 to lobby government and to promote the parks cause. It seems highly likely that she penned many of the regular articles on national parks which appeared in the newspaper during her extensive journalistic career.

Firmin McKinnon was another supportive journalist with the *Brisbane Courier* (and later *The Courier-Mail*) from 1913 to 1946. Although his primary interests centred on literature and the arts, during the 1930s he was also a member of the National Parks Association of Queensland. His later influence as a senior editorial executive undoubtedly played a role.

A fourth influential journalist during this period was George Harrison, who worked with the Ipswich newspaper the *Queensland Times* from 1890 to 1916, when he moved to *The Courier-Mail*. Whilst at Ipswich, Harrison came into contact with some of Queensland's earliest bushwalkers who were active in the region in the first decade of the 20th century, and this may have encouraged his interest in the national parks movement (*Queensland Times*, 1909: 3). For example, he wrote a series of articles on a geology summer school for secondary school teachers, during which a proposal was raised to create a national park to include the nearby Mount Greville (*Queensland Times*, 1910: 4). Along with Chisolm, Harrison was one of the first journalists to visit and write about Lamington National Park in the period following World War

I (*Queenslander*, 1919: 41; *Brisbane Courier*, 1921: 12; *Brisbane Courier*, 1929: 15). He was also a founding member of the National Parks Association of Queensland when it formed in 1930.

The work of these journalists was complemented by engaging first-hand accounts by Romeo Lahey and Arthur Groom of their often solo and challenging scrambles in southeast Queensland wilderness areas, in the spirit of John Muir (A. G., 1929: 27). In 1933 Groom and Lahey established Binna Burra mountain resort on the edge of Lamington National Park. One of Groom's articles in 1929 particularly captures the essence of the era and harks back to the romantic writing about the landscape by Archibald Meston and John Muir. Groom was renowned for walking vast distances, often alone, and he described for readers of the *Brisbane Courier* one solo exploration of the deep gorges in Lamington National Park (A. G., 1929: 27):

> The bed of the East Canungra Creek was weird by day; it was more than weird by night. The darkness was so dense and as though in contrast hundreds of little spots of phosphorus gleamed queerly from the steep, damp banks. The trees rose high and almost joined overhead. They almost shut out the sky. Clouds were racing. It was going to rain.

Through publication of articles like this during the 1920s and 1930s, often accompanied by spectacular photographs, the press played a significant role in promoting the idea of national parks. There are perhaps two main reasons why the campaign received such favorable coverage. Firstly, the national parks concept fitted admirably within a broader progressive town planning and public health agenda (then ascendant) which by and large had the support of the educated classes and reformist politicians. Secondly, the campaign emerged at a time when regional tourism was becoming more popular and being promoted by government as a new industry. Locations such as the Lamington National Park, Springbrook and the North Coast hinterland were frequently touted in the Brisbane and regional press as places of undiscovered beauty, as recreational playgrounds, and as Queensland's answer to the Blue Mountains. Of course, other more long-standing arguments for national parks involving notions of national identity and preservationist arguments could also be accommodated within the broader public health and regional tourism agenda.

By the late 1920s the debates over the future of the Lamington National Park which had taken place in the immediate post-war years had been more or less settled. National parks advocates had convinced the state government that the Lamington National Park should remain largely undeveloped rather than be turned into a resort-style tourist attraction criss-crossed with roads and dotted with hotels. Advocates had moved on to tackle new issues such as public access, park management and the formation of new national parks—topics that would feature prominently in local newspapers throughout the next decade.

CONCLUSION

Journalism, as "the primary method of framing experience and forming public consciousness of the here and now" (Adam, 1993: 45), is an important element of the multifarious ways in which we "imagine" social spaces, including landscape. Through the medium of the colonial press in Australia, discourses around mountain landscapes and the broader concept of the environment were able to spread their influence well beyond a narrow, activist audience. These early writings about the Australian environment reveal as much about the colonial landscape and how it has been imagined as they do about the nature of the landscape itself. The analysis of the role of the local press lends further support to the significance of local campaigns as an important element in understanding the extent and effectiveness of environmental advocacy (Lester, 2007, 2010a; Lester and Hutchins, 2012b). The contributions by Archibald Meston and environmentalists like Romeo Lahey and Arthur Groom, along with the works of a small but influential cohort of journalists who supported the Queensland national parks movement, represent vivid examples of the strategic role played by the press in managing ways of thinking about Queensland's natural heritage.

Perhaps surprisingly, there was little if any conflict involved in these examples, apart from some criticism of Meston's use of hyperbole. This suggests that these two cases of journalistic intervention—forty years apart—were able to lead the population from the "simple, common sense" idea of associating landscape with economic development to accepting the "coherent and systematic thought" inherent in the emerging concept of the environment. In each case, the media played a strategic role in winning consent for a particular set of ideas and assumptions around landscape and our relationship with it that were at odds with prevailing ideologies.

▪ CHAPTER THREE

Affecting Environments

Mobilizing Emotion and Twitter in the UK Save Our Forests Campaign

ALEX LOCKWOOD

A number of globally recognized environmental protests over the past thirty years have focused on the protection of trees in various locations (individual trees, woodland and forest), or have employed trees and forested spaces in their campaigns for the symbolic power they provide in the mobilization of public interest and action (see Anderson, 2004; Flam and King, 2005; Rival, 1998; Rossiter, 2004; Zelter, 1998). In the UK there is a strong symbolic connection with trees and woodland and, in particular, publicly owned woodland (Tsouvalis, 2000). In the twelve months to February 2010, the English adult population made over 317 million visits to forests and woodland (Natural England, 2010). As Jones and Cloke state in their work on the place of trees in the construction of nature-society relations: "Trees in Britain and elsewhere have become carriers of some people's environmental anxiety and love for nature, cropping up in various discourses on environmental crises, countryside change and habitat loss, and quality of urban life" (2002: 6). Ancient and heritage forests are the homes of veteran or old growth trees not to be "trenched around, tarmacked, parked under or urinated on [...] threatened by matches or assaulted by tree climbers" (White, 1997: 222). Indeed, in the UK in 2006, the right-of-centre Conservative Party changed its logo from a torch to a British oak in party colours in recognition that "certain trees such as veteran oaks [...] build up layers of positive association which make them key cultural icons" (Jones and Cloke, 2002: 30; see also Sullivan, 1985).

The associations that trees and woodland evoke in the public imagination are not all positive, however. Trees and woodland can also be places that instil fear or feelings of isolation or exclusion. What is not in doubt is that trees and woodland often have a strong affective dimension in both their physical appearance and symbolic use (Jones and Cloke, 2002: 36) that makes them an important object of disaggregated study when exploring the construction of nature-society relations and the spaces of activism.

For Rival, there is a specific brand of environmental protest that can be formulated as a "Western Tree Activism" where trees "stand not only for life, but also for social justice and public space" (1998: 16). Central to this Western Tree Activism is the concept of biophilia, first introduced by Wilson as "the innate tendency to focus on life and life-like processes" (Wilson, 1984: 1). It has become "a throb in the heart-beat of the ecopsychology movement" (Hegarty, 2010: 65), an etymology that expresses an affective love of life in the construction of nature-society relations, a negotiation Rossiter summarizes as "the social contingency of all our understandings of what counts as 'nature'" (2004: 140). Since Wilson first published on biophilia, the fields of environmental psychology and ecopsychology, as well as media and cultural studies, have examined the emotional content of nature-society relations (see Hinds and Sparks, 2008; Kovel, 2008). Johnsen (2011) has provided a conceptual framework for the use of nature for emotion regulation that examines the ways in which environments affect emotional processes; for example, by making it easier to reflect on one's feelings (175). Among others, Hegarty (2010) has surveyed individuals on nature-connectedness and found a positive emotional tone running through all the contributions. Participants mentioned "explicitly a relaxing effect, and a personal value gained from being part of natural surroundings" (Hegarty, 2010: 69). Hegarty found that respondents agreed that "being in nature felt good, or was associated with pleasant feelings" (2010: 70). Such feeling for nature or "place attachment" (Kelly and Hosking, 2008: 578) has been shown to facilitate pro-environmental behaviours. Such evidence supports Brown and Pickerill's argument for the recognition of feelings in activism, which they see as "a hopeful and reparative engagement with the role of emotions in activist spaces" (2009: 25).

Where then does emotion sit in the study of environmental protests and media processes? A useful starting point for widening the scope of such studies is to incorporate the effect of emotions on both activists and on publics in the area of environmental conflict. For Hutchins and Lester, "[e]nvironmentalism represents a unique conception of place in the network society, valuing the physical environment for its symbolic and physical qualities, not for its potential economic use value" (2006: 436).

But environmentalism is not just reported on in traditional media; it also finds new modes of expression and facilitation via emerging media platforms. As Hutchins and Lester propose: "The media is more than a site for environmental action; it plays a significant role in shaping debate and influencing outcomes. It is here that representations are determined, images softened or distorted, and power granted or denied" (2006: 438). They draw extensively on the work of the sociologist Manuel Castells, who places environmentalism at the centre of his analysis of the information society and communication powers. Lester and Hutchins provide a robust consideration of Castells' work that explores the concept of flows: that is, the power of flows and the flows of power in a networked society (see Hutchins and Lester, 2006, 2011). They explain, for example, that for Castells:

> Reservoirs of power and resources are contained within business, media, political and social networks, but it is the capacity to control and/or influence how these networks connect and interact that determines which actors possess most power. [...] This ability is conceptualised [by Castells] as "switching power"—"the ability to control connection points between different networks" (e.g. business, media and economic networks). (Hutchins and Lester, 2011: 161)

Hutchins and Lester argue that "the environmental movement desires, but conspicuously lacks, switching power. [...] They cannot *control* connection points between business (e.g., the forestry and fossil fuel industries), political networks (governments at various levels), and news media (established broadcast, print and online outlets). Rather, through the use of mobile and social networking media, actions by protest groups [...] aim to temporarily *destabilize* or, optimally, *disrupt* the smooth functioning of capital and government" (Hutchins and Lester, 2011: 161; emphasis in original).

It is notable that Hutchins and Lester refer to the emotions or feelings of activists—their "desires"—although they do not take further this line of enquiry into the role played by affect in the negotiation between media organizations and environmental campaigners over the control of connection points. Castells has in fact begun to describe emotional and affective content within these flows, particularly empathy (Castells, 2009). This chapter takes off from such an exploration of affect as it shapes and influences the processes by which environmental issues develop into crises as they are carried through established and emerging media platforms and networks. This affective tone is central to the strategies of activism and its own sustainability (Brown and Pickerill, 2009), to environmental organizations as they campaign (Jasper, 2011) and to environmental journalists as they seek to influence their audiences (Lockwood, 2012). However, forms of media, both established and emerging, act as more than simply carriers for those affective and emotion-laden messages with the aim of reaching a public (Lockwood, 2010, 2012). In particular, emerging media platforms and their use have structures that amplify affect in new, unexpected and powerful ways.

Here I focus on the affective content of an environmental protest that took place in the UK between October 2010 and February 2011, in opposition to the UK government's plans to sell off England's public forests. I examine the ways that emerging forms of social media, in particular the microblogging platform Twitter, were central to the campaign's success by facilitating the circulation and amplification of affect from protesters and members of the public opposing the government's plans. This circulation of affect was both an explicit element of the campaign and a product of the mechanisms of the media platforms it employed. I explore these complications between environmental protest and media processes in nature-society relations through the concept of an "intimate public" (Berlant, 1997: 5) revolving around both a real and symbolized love of trees. In taking this approach, I hope to add to the understanding of experiences of affect in the making of public and private worlds. Scholars such as Lauren Berlant (1997, 2008), Ann Cvetkovich and Anne Pellegrini (2003) and Kathleen Stewart (2007) have put into circulation new concepts of public feelings to "challenge the idea that feelings, emotions, or affects properly and only belong to the domain of private life and to the intimacies of family, love, and friendship" (Cvetkovich and Pellegrini, 2003: 1). Where private or, increasingly, privatized feelings are allowed into political protest, they are stage-managed and redacted of political agency. For public feelings scholars, a critical program is one that destabilizes this worldview of politics as the proper mechanism for managing and privatizing public feelings, in a similar way to how environmental protest aims to destabilize or disrupt the flows of power. The mediation/mediatization of public feelings is a vital component of study for understanding environmental conflict. What became known as the #SaveOurForests campaign is then a useful case study for the way that an environmental conflict was mediated via a public feelings culture to impact on British political life. Due to the speed and scale of the success of the campaign, the period between October 2010 and February 2011 characterizes the emergence of a "structure of feeling [bringing] into being alternative cultures" (Cvetkovich, 2003: 12): in this case, how a walk in woods can be connected through informed mediatized spaces to mobilize a political force.

THE PROTEST: AN OVERVIEW

In the UK in October 2010, Forests Minister Jim Paice MP announced plans to sell up to 100 percent of England's Public Forestry Estate (PFE), over 258,000 hectares of woodland, including ancient heritage forests and dozens of Sites of Special Scientific Interest. New legislation written into a Public Bodies Bill would go through Parliament and give government powers to dispose of the Estate, which accounts for

18 percent of all woodland in the UK. The public and media outcry against these plans crossed the political spectrum, engaging hundreds of thousands of people fearful of the loss of public goods such as access rights, well-being and the biodiversity benefits of England's woods.

The Telegraph newspaper, politically a right-of-centre paper, first reported the story. Forest campaign groups, in many cases formed in the 1980s to resist the Conservative prime minister Margaret Thatcher's earlier attempts to sell off Britain's forests, organized public meetings and action. Campaigning organization 38 Degrees asked its followers through digital communication (email, Facebook and Twitter) if it should prioritize the issue in its campaign action; it then crowd-sourced money to pay for a YouGov poll that found 84 percent of people opposed the sell-off.[1] An activist, Tamsin Omond, joined forces with Rachel Johnson, editor of *The Lady* magazine and sister of London mayor Boris Johnson, who opened her address book; Omond coordinated a public letter to *The Telegraph*, with 84 signatories of public figures including the archbishop of Canterbury labelling the sell-off as "indefensible".[2]

After a number of smaller steps to ameliorate the situation, on 17 February 2011 the government bowed to public pressure and announced the closure of its consultation period (Department for Environment, Food and Rural Affairs, 2011a) over the future of the PFE, as well as the removal of the relevant clauses from the Public Bodies Bill (Department for Environment, Food and Rural Affairs, 2011b). Environment Minister Caroline Spelman MP apologized to the House of Commons in what was affectionately labelled a "yew-turn" in government policy.[3] The initial stage of the #SaveOurForests campaign was over, although the sale of England's PFE remains a very real threat.

SAVE OUR WOODS

Not long after the government's proposal to sell off England's PFE was announced, three individuals unaffiliated with any charity or campaigning group, animated by a need to act and a lack of transparency around the proposals, set up the Save Our Woods[4] website as a forum for gathering together and centralizing access to information and debate on the forest sell-off. Other websites were also launched by individuals,[5] as was a petition by 38 Degrees. Local campaigns and online actions, mainly driven by Save Our Woods, 38 Degrees and individual social media users including politicians such as Labour's shadow environment minister Mary Creagh, led to over half a million people signing the petition.[6] Across local and national media, public resistance to the forest sell-off made front-page news. In the Save Our Woods forum, peers sitting in the House of Lords including Liberal Democrat peer Lord Greaves, forestry experts including retired head of policy for the UK's Forestry Commission, Rod Leslie, and civil servants leaking information anonymously began an epistemological response to the government's plans. Lord Greaves told the House of Lords that: "The young people who run [Save Our Woods] have done a very good job [...] creating a forum where people could exchange information. I believe that all this has contributed to the amount of knowledge and understanding in the campaign groups being much greater than it was at the beginning. [...] The involvement of the Internet, Twitter, Facebook and all these realms that I do not know much about has been a complete eye-opener to me" (Greaves, 2011). As Hen Anderson, one of the founders of the Save Our Woods website, posted on Twitter:

> @hen4: Blows my mind that, we are able to communicate directly, publicly, on a forum with Westminster #saveourforests, —cool innit? [10 Feb]

The non-affiliated position of the Save Our Woods website was central to its success in fostering credible debate and action (Anderson, 2011, personal communication). Following Freelon, the Save Our Woods website can be seen as an example of "communitarian political spaces [that] are thought to offer the most conducive atmosphere for the furtherance of collective political objectives" (2010: 1180).

The three founding individuals of Save Our Woods met through the microblogging service Twitter. Launched in 2006, Twitter is a social media platform that allows individuals to message their followers (via "tweets") and to follow other users' messages, each of which must be 140 characters or less. Such microblogging (Twitter is not the only company offering the service) broadcasts an individual's thoughts or actions to a connected network. Java et al. (2007) found three types of users (friends, information seekers and information sources) using Twitter for four main reasons: daily chatter, conversations, sharing information (mainly in the form of links) and reporting news. In their study of the news reception of Twitter between March 2006 and 2009, Arceneaux and Schmitz Weiss (2010) found a favourable reception and categorized this positive acceptance into three sub-themes: New Sensibility, Commercial Use and Civic Use (Arceneaux and Schmitz Weiss, 2010: 1268). Within the Civic Use sub-theme, Arceneaux and Schmitz Weiss identified as a key utilization of the microblogging tool "the way various groups could use Twitter to mobilize around specific causes" (2010: 1270).

Not all responses have been so positive to the potential of social media in general, and Twitter in particular, as a democratic tool. According to Biggar: "'The capacity for web 2.0 applications like Twitter to resemble a dyadic, peer-to-peer communication network has yet to be fully realised [...] the exponential expansion of networked communications technologies has potential to be more redundant than useful at times" (Biggar, 2010: 16). However, as Arceneaux and Schmitz Weiss argue, as far back as the telegraph, new technologies have initially inspired negative responses because they disrupted established modes of communication and the ideas about space, time and public/private spheres those established ideas represented (Arceneaux and Schmitz Weiss, 2010: 1265).

TWITTER AND THE CIRCULATION OF EMOTION

As Stein reminds us, it is important to recognize that "while technology may influence social movement activities, movement groups can also influence technology use" (Stein, 2011: 364). Such disruption is what Hutchins and Lester identify as the *aim* of environmental protest groups in their use of emerging media forms. Critically, the Save Our Woods founders began to publicize their resource via Twitter and to encourage protest via the #saveourforests hash tag (a hash tag acts as a search link to all other "tweets" posted with the same tag). I became aware of the @SaveOurWoods user on Twitter (all users are prefixed with the @ sign) when I spotted one of his or her tweets:

> @SaveOurWoods Do you #LoveTrees? Send us your pictures & stories to hello@saveourwoods.co.uk #SaveOurForests [20 Jan]

This message was re-tweeted by another user into my stream, as I was not following @SaveOurWoods messages at that time. This re-tweeting is a critical function of Twitter's social mechanism; it allows immediate sharing of the messages that any user sees with her own followers. In this way, users of Twitter become "message multipliers"; such multiplication is registered in "trending" topics which, according to Twitter, "help people discover the 'most breaking' news stories from across the world".[7]

For Alfred Hermida, Twitter is a new and important tool for both professional and citizen journalists. It is one of a set of new "always-on communication systems [that] are creating new kinds of interactions around the news, and are enabling citizens to maintain a mental model of news and events around them" (Hermida, 2010: 298). For Hermida, such "awareness systems" have given rise to what he calls "ambient journalism" (2010: 298). The tool's usefulness for environmental campaigning and activism in the dissemination of information is obvious here. However, what has accompanied this multiplication and amplification of messages across individuals' broadcast networks is its affective component. Writing in the *New York Times Magazine*, Clive Thompson (2008) labelled this not "ambient journalism" but an "ambient intimacy" which, as Arceneaux and Schmitz Weiss suggest, "we acquire [with] a greater awareness of many individuals, a group far larger than what we could keep up with through personal contact" (2010: 1269). Or, as Murthy puts it: "Even if one does not post on a particular day, [one feels] an aura of other users through the feed of microblog posts" (2011: 782). The affective nature or intimacy of people's usage of

Twitter and other mobile, always-on communications systems suggests that Twitter and other social media should be assessed, as Hermida (2010) has argued (drawing on computer science modelling), as connectedness-oriented communication systems where the informational content of the message is of secondary importance to the emotional, relational content (Baren et al., 2003). Or as Ruth Rettie (2003) argues, communication can create a sense of connectedness or feeling of being in touch; in "awareness systems" this may be more important than the content of the communication. Indeed, Rettie found the need for connectedness was the most important factor in making a choice between communication channels for families and friends staying in touch over distances.

Any analysis of this campaign must attend to its feelings and how these were integral to both the content and form in its success. The emotional force of the Save Our Forests campaign, explicitly elicited from the Save Our Woods campaigners and 38 Degrees as well as generated spontaneously from social media users, is, I argue, an example of where a movement group has influenced technology use—in engendering "nature-connectedness" for pro-environmental behaviour in opposing a government-proposed sell-off of England's forests. The emotions generated and expressed by those involved were central to the campaign's momentum. See, for example, the Valentine's cards campaign organized by Save Our Woods, in which cards carrying explicit messages of emotional connection with trees and woodland were sent to members of the House of Lords before an important vote on the Public Bodies Bill; many cards were made of leaves, or in the shapes of trees.[8] Twitter was explicitly used to spread positive emotionally laden protest messages, links to blog posts and pictures:

> @tentspitch (Sarah Read) Read this David Cameron. Nothing you ever said has been this eloquent or from the heart. From @yasminehamid http://tinyurl.com/4qgjpak [13 Feb]

> @sylvanmuse (Sylvan Muse) @Savebritforests My love of trees & woods and how strongly I feel against forest sell-off http://sylvanmuse.blogspot.com #SaveOurForests

Through Twitter, other emotions such as disconnected anger were harnessed and turned into connected pressure:

> @Dogcatchicken (Tim) Well this save our forests thing has just take off... I was angry... just typed in #saveourforests and found other people just as angry! [4 Feb]

Connectedness was then used to multiply and amplify messages. Monitoring the #saveourforests hash tag revealed thousands of people re-tweeting the campaign every hour, driving people to sign the 38 Degrees petition and to learn more via the Save Our Woods website. People came onto social networks who had not been there before:

> @sylvanmuse (Sylvan Muse) @WildLives @alexlockwood @hen4 @wildelycreative Lotsa folk I've spoken to are thinking of joining Twitter specially for #SaveOurForests! [31 Jan]

Leading up to 17 February the Save Our Woods campaign used its website and Twitter in "making nature visible" (Rossiter, 2004: 142) by creating space for individuals to share emotional and affective connections. Following the search term/tag #saveourforests, a political network (Buchanan, 2003) of hundreds of thousands of loosely affiliated individuals whose "interests may be specific, local, and disconnected from a conscious membership in a broader global movement" (Rojecki, 2011: 96) self-identified through social media to articulate a shared yet non-prescriptive pro-environmental position. In this way, individuals used social media for a political goal, while Save Our Woods refused to commandeer the platforms for personal or organizational benefit (in fact, many environmental organizations were criticized for their inaction[9]) but instead curated an affective protest that found its own momentum:

> @Dogcatchicken (Tim) Now #saveourforests is everywhere in the media... and the Gov is running scared... it really is people power!!!
>
> @jennyclairedunn (jenny dunn) @jobeyG Me too! I'd never been bothered by politics before! REVOLUTION!!! Lets get a posse together for #SaveFristonForest #saveourforests

The emotional content of tweets was also made explicit by campaigners and non-affiliated individuals in the days following the government u-turn:

> @Dame_de_Lotus (Sarah Noone) I know I sound mushy, but I'm really proud of the #Nocton and #SaveOurForests campaigns. They've reassured me that we can fight back! [17 Feb]
>
> @wildelycreative (Karen Wilde) Seems love is everywhere... to all you helped #SaveOurForests, from me *grin* http://twitpic.com/40t936 [17 Feb]

By forcing the government to cancel its plans to sell off England's Public Forestry Estate, the Save Our Woods group, a non-affiliated trio of activists, showed how it was possible to utilize emerging media platforms including Twitter to mobilize a massive political force in affecting legislative acts. As such, the campaign has been claimed a victory for social media. Writing in *The Guardian*, environmental journalist Fiona Harvey wrote: "The mobilisation of such a large opposition to a policy that ministers thought would pass under the radar, and the strength of the response, shocked David Cameron's government. 'It was a genuine cock-up on our part,' said one Downing Street insider" (Harvey, 2011).

PHENOMENOLOGIES OF SOCIAL MEDIA, COMMUNICATIVE CAPITALISM

For Jodi Dean, the phenomenologies of social media (tweeting, posting, communicating) are affective experiences. She says a tweet "marks the mundane by expressing it, by breaking it out of one flow of experience and introducing it into another. Now part of a shifting screen of comments and images, the mundane moment tags a mood or sensibility" (Dean, 2010: 98). These social media sensibilities are affective because they are enjoyable:

> Affect [...] is what accrues from reflexive communication, from communication for its own sake, from the endless circular movement of commenting, adding notes and links, bringing in new friends and followers, layering and interconnecting myriad communications platforms and devices. Every little tweet or comment, every forwarded image or petition, accrues a tiny affective nugget, a little surplus enjoyment, a smidgen of attention that attaches to it, making it stand out from the larger flow before it blends back in. (Dean, 2010: 95)

However, Dean is pessimistic about mediation as a political tool so long as the media remains in ownership of and manipulated by powerful elites. Enjoyment is a trap: "We are captured because we enjoy" (Dean, 2010: 121). This may even explain the disappointment many campaigners felt at the news that the government was performing its u-turn, as it meant the end to an intense period of being part of something, or what Rettie (2003) identifies as the "connectedness" that accrues by using such connectedness-oriented systems. Such anti-climaxes are recognized as the emotional fallout of activism (Brown and Pickerill, 2009). However, for Dean, the circulatory procedures of social media are the workings of a "Communicative capitalism [...] that economic-ideological form wherein reflexivity captures creativity and resistance so as to enrich the few as it placates and diverts the many" (Dean, 2010: 4). The intensity is used against political efficacy: "the affective charges we transmit and confront reinforce and extend affective networks without encouraging—and, indeed, by displacing—their consolidation into organized political networks" (Dean, 2010: 119).

Others are not convinced by Dean's pessimism. Scholars such as Mark Deuze offer a more positive view, where digital media aid a political vision of public goods in public hands (Deuze, 2011). In the #SaveOurForests campaign, the democratic potential of mediated spaces to protect environmental places gained traction. For Dean, the faith placed in the ideologies of networks and publicities by activists using new media is an idiosyncratic aspect of communicative capitalism. Activists continue to emphasize the democratic potential of the Internet, even while experiencing increases in economic inequality and the consolidation of neoliberal forms of capitalism in and through the uses of globally networked communication (Dean, 2010: 31). The iterative re-tweeting, circulation and amplification of media systems such as Twitter *are* affective, but this "deluge of images and announcements enjoining us to react, to feel, to forward them to our friends, erodes critical-theoretical capacities" (Dean, 2010: 2–3). Yet the #SaveOurForests campaign is an example of these emerging media platforms, and their affective mechanisms, contributing to a pro-environmental (and anti-capital) outcome, an example of Rival's Western Tree Activism in opposition to "the domination of the non-human world [by] the Western industrial network of knowledge and power" (Jacques, 2008: 10). My argument is that the campaign was successful *because of* the deployment of affect carried through, and amplified across, these emerging media platforms when viewed as integrally affective: that is, as connectedness systems, not communication media. Such a success cannot be constrained by Dean's communicative capitalism. In attending to the emotional or affective values of those mediums, we can come closer to an understanding of the phenomenological factors involved in making an environmental risk visible, and of that risk becoming understood, or *felt*, as a crisis, and then *responded to*. As Berlant suggests, such momentary mediations as tweets are pivotal to the possibilities of forging responsible public cultures. In suggesting this, she argues for:

> a mode of criticism and conceptualisation that reads the waste materials of everyday communication in the national public sphere as pivotal documents in the construction, experience, and rhetoric of quotidian citizenship. [...] The very improvisatory ephemerality of the archive makes it worth *reading*. Its very popularity, its effects on the law and on everyday life, makes it important. Its very ordinariness requires an intensified critical engagement with what had been merely undramatically explicit. (Berlant, 1997: 12)

The thousands of messages and pictures shared and shared via Twitter and Facebook as symbols of nature-connectedness were moments of everyday life under threat, gathered together and mediated into a public feelings agenda that has already gained political efficacy. The role played by connected individuals in different affective environments—both forests and online media spaces—channelled public feelings into a mobilized political force to halt legislative moves. The ways in which the campaigners used emerging media platforms to harness a "switching power", through the cumulative effects of affect in disrupting the flows of political processes, was unprecedented for its speed, force and dis-organizational mode.

CONCLUSION

Dean is not only pessimistic; she proposes the need for a "critical media theory [which] anchors its analyses of technologies, users, and practices in an avowedly political assessment of the present" (Dean, 2010: 3). I agree, although this chapter is not a counterclaim for the good of social media networks. In the case study presented, there are many other factors in the campaign's success including the lack of enemies (as in the case of climate skeptics; see Lockwood, 2010), national and local media coverage; and the already established symbolisms of the trees (Bennett, 2010; Jones and Cloke, 2002). Rather, I see this critical media theory as allying itself with work exploring the complicated and multifaceted role of affect in public and private cultures. There is in the #SaveOurForests campaign—led by the activities of the Save Our Woods campaigners but also enacted by at least half a million individuals empowered by an "ambient intimacy" or "connectedness" (Rettie, 2003) that amplified the individual's experiences of both nature and media—a way of conceptualizing potential new public cultures and a more environmentally responsible citizenry.

For Lauren Berlant, such a citizenry is an "intimate public" where participants "feel as though it expresses what is common among them, a subjective likeness that seems to emanate from their history and their ongoing attachments and actions" (Berlant, 2008: 5). Much of what was common among those engaged in the campaign were acts of everyday citizenship; for example, walking in the woods. England's Public Forest Estate has public goods because a public *creates them*. And public citizenship "is a status whose definitions are always in process. It is continually being produced out of a political, rhetorical, and economic struggle over who will count as 'the people' and how social membership will be measured and valued" (Berlant, 1997: 20). In a Westminster Hall Debate following the government u-turn, Sir Peter Soulsby, then MP for Leicester South, asked Forestry Minister Jim Paice MP, specifically, "who counts?"—that is, who will be represented on the independent panel that the government was to establish after its u-turn and debate the future of the Public Forestry Estate? In the context of the #SaveOurForests campaign, this question of citizenship—of "who counts" or, to put it another way, who "#LoveTrees"—remains central to the next stage of analysis.

ENDNOTES

1 See http://labs.yougov.co.uk/news/2011/01/27/keep-our-forests-public/.

2 See http://www.telegraph.co.uk/earth/earthcomment/8290596/These-are-our-forests-how-can-they-be-for-sale.html.

3 See for example http://www.guardian.co.uk/environment/2011/feb/18/forests-public-sector-cuts.

4 http://www.saveourwoods.co.uk.

5 For example, see http://savebritforests.blogspot.com/.

6 See http://www.38degrees.org.uk/page/s/save-our-forests.

7 See http://support.twitter.com/articles/101125.

8 See http://saveourwoods.co.uk/get-involved/love-trees/happy-love-trees-valentines-day/.

9 See for example Jonathan Porritt's attack on environmental NGOs http://www.guardian.co.uk/environment/2011/feb/06/jonathon-porritt-conservation-forests-sell-off.

▪ CHAPTER FOUR

Clear Cuts on Clearcutting

YouTube, Activist Videos and Narrative Strategies

CATHERINE COLLINS

In a seminal essay on the competing information campaigns about old growth forest extraction in the United States' Pacific Northwest, Jonathan Lange (1993) demonstrated that what was regarded as acceptable policy really depended on the values and worldviews of the competing interest groups. Environmentalists and timber workers were seldom engaged in face-to-face meetings; their information campaigns were carried out instead in the media. New technologies—especially social media—offer a way of reaching potentially large audiences. They also have the advantage of allowing opposing sides to control the message that reflects their interpretation of facts, values, and acceptable policy alternatives with less mediation by reporters and journalistic gatekeepers. One would reasonably expect that the messages disseminated through social media would reflect the iconic story of the authors of these messages. If one values a pristine wilderness over economic values, the audience would expect to discover in the narrative appeals to preservation and an articulation of the non-economic worth of the forest. I am interested in how the debate over old-growth extraction policy is engaged through the medium of YouTube, how this medium encourages groups to tell their stories through images and commentary, and how narrative strategies may compete against each other in such a way that there is little possibility of dialogue.

Debate is always about a clash, but the kinds of propositions that are debated change the arguments that are advanced and the relations between opposing sides. Propositions of fact—whether something is or is not true—are settled by turning to verifiable data, and may lead to disagreements when factual information is not made available. Propositions of value ask the audience to make judgments about the justness, ethicality, or worth of something. Conflict becomes heated when the opponents have different value hierarchies. Debates over propositions of policy—should something be legislated—may lead to conflicting interpretations of the *best* way to address problems, but debating policy propositions also leaves unarticulated an assumption that one's own values are appropriate and those who oppose one's judgments are intractable. The resulting conflict over policy *others* the opposition as it seeks ways of defeating rather than collaborating with those whose values and worldview are different. In environmental debates, especially those over the issue of cutting down old-growth forests, whether the proposition seems to be factual or policy based, the competing interest groups rely on propositions of value that leave little tolerance for competing value positions.

APPROACH

With the development of the Internet, scholars, organizations, and politicians saw the potential for extending the number and type of voices contributing to civil society. Kavada (2005) argues:

> the internet provides a space for political organizations to publish their opinions and gain a foothold in the public sphere...communicate directly with their audiences, bypassing the mechanisms and commercial bias of the mainstream media...[and serve as] a powerful networking tool...to foster solidarity within the organization and deepen the public's involvement with its cause. (Kavada, 2005: 208–209)

Social activists have discovered the potential of the Internet, and scholars have examined the production and consumption of Internet messages linking social activism to the diffusion of information (Earl, 2010), the relationship between online and mainstream media with respect to power and activism (Lester and Hutchins, 2009), and the way social movements create common identity (Eaton, 2010) and encourage collaboration (Allen, 2010), especially through an illusion of closeness (Pliskin and Romm, 1997; Nip, 2004).

Realizing the perceived potential of the Internet is complicated because its use is not controlled by the disseminators of messages, as one might assume, but by "its 'pull' nature which suggests that online it is the users who have control of the interaction and can easily choose to avoid political coverage" (Kavada, 2005: 210). Scholars of persuasion since the classical era have understood that effective messages are shaped to particular audiences. Messages that have fidelity with the audience's values and experiences and reinforce the narratives that the audience accepts will gain adherents; messages that fail to adapt to their audiences are unlikely to persuade. Organizations disseminating messages through the Internet need to find ways of making their messages attractive, because failing a good advertising campaign, they may well pull only those audiences who are already adherents of their cause. Burgess and Green argue that most Internet users "are far more likely to watch videos hosted on YouTube than they are to log into the website regularly" and that as a social network, "the video content itself is the main vehicle of communication and the main indicator of social clustering" (2009: 58).

As a social network, YouTube viewership depends on the recommendations of others. Referrals from related videos, organizations, and YouTube searches and links embedded on sites like Facebook bring additional audiences to these videos. One challenge to using YouTube as a platform for message dissemination is that it is not overtly designed to foster collaboration, but when the content inspires response, the YouTube community has found ways of creating interconnectivity. Easy to upload, free of gatekeepers censoring one's content, and having the potential of reaching a huge community of viewers, YouTube is an appealing site for activist social movements to exploit.

My case study includes narratives posted to YouTube from two locations: the Pacific Northwest (Oregon and Washington), because they have the largest remaining old-growth timber in the United States; and Tasmania, where more old-growth timber is harvested than in the rest of Australia combined. Both of these geographical locales have experienced symbolic and real conflict over policy decisions that allowed continuation of old-growth extraction. Environmentalists and, to a lesser extent, those involved in the timber industry have chosen to exploit the new media and tell their story through videos posted to YouTube.

Two types of YouTube searches were conducted: one using the primary term "old-growth forests" along with one of three secondary terms—"Oregon," "Washington," or "Pacific Northwest"; and a second search again using the primary term "old-growth forests," along with the secondary term "Tasmania." I selected for study those videos that had received at least 1,500 hits because I was interested in message campaigns that had a substantial viewership. Tasmania and the Pacific Northwest each posted 10 videos that met this criterion. The lowest number of hits from Tasmania was 1,818; the highest number was 16,146, which was the only video with more than 10,000 hits. The lowest number of hits from the Pacific Northwest was 2,141 and the highest was 20,042, with two videos receiving more than 10,000 hits. The results are summarized in Table 1.

Table 1: Old-Growth Forests in Tasmania and the Pacific Northwest

Videos about Tasmania	**# of hits**	**Posted**
"Tasmania's Ancient Forests"	1,818	2007
"Activists Shut Down Gunns Ltd"	1,935	2009
"'Endangered' Tasmania's Wild Places"	2,231	2008
"The PM & the Frog"	2,353	2009
"The Upper Florentine Trilogy: Part One"	4,093	2007
"Come to Camp Weld"	4,675	2006
"Lower Weld Valley"	5,035	2007
"Tasmania's Forests: A Global Treasure"	7,649	2006
Wild Tasmania trailer	7,764	2007
"Tasmanian Forestry Contractors"	16,196	2008
Videos about the Pacific Northwest	**# of hits**	**Posted**
"Northwest Old Growth Forest Legacy Campaign"	2,141	2009
"Saving Agarikon"	2,441	2009
"HELLog" ("HELLog 2")*	3,594 (21,412)	2007 (2007)
"Old Growth Forest Defenders—Sisters"	3,528	2007
"Earth Tribe TV.org—Old Growth Forest"	5,393	2006
"Amazing Spotted Owl Mouse Grab"	7,131	2007
"Old Growth Timber of Washington State"	7,795	2009
"Winter Wren Bird Song"	8,774	2006
"Threats to Spotted Owls"	11,652	2007
"Opal Creek—Oregon Cascades"	20,042	2007

* The second version of this video was not identified with the search terms.

The most popular video from the Pacific Northwest is about Opal Creek, an area that was hotly contested as a potential site for old growth extraction. The most popular video from Tasmania was "Tasmanian Forestry Contractors", in which two forest workers attack environmental activists.

Rather than separating Tasmanian and Pacific Northwest videos, I will look at each of the videos from both groups as one of four types: narratives that explain old-growth forests and their ecological functions, narratives valuing old-growth forests, narratives that celebrate logging, and protest narratives. Following the discussion of these four categories, I will look at one video as a representative case of the kinds of issues scholars and activists need to consider in assessing the use of YouTube for disseminating their competing information campaigns.

NARRATIVES DEFINING OLD-GROWTH FORESTS

The videos in this category offer images of lush forests and glimpses of endangered flora and fauna under threat from logging. Aesthetically appealing footage is accompanied by information-loaded arguments. "Northwest Old Growth Forest Legacy Campaign", typical of this category, is more scientific in its orientation, emphasizing the ecological value of old growth forests and the need for preservation. Discussion of habitat conservation, watershed protection, climate control (carbon sinks), and the interdependence of living organisms structure the argument. To combat the economy versus environment frame, the video reminds the viewer, "Old growth forest recreation is an economic engine for the region" (2009). Although stopping old-growth logging hurts those in the timber industry, destroying the recreational value of the land economically disadvantages many small businesses in the region ($5 billion annually). Because both sides of the policy issue have constituencies that might be adversely affected by policy decisions, the message suggests some common ground. By appealing to a commonsense vision, the narrative offers hope for recovery from past errors in judgment.

Similar definitional arguments are apparent in "Earth Tribe TV.org—Old Growth Forest". The narrator directly tells the viewers that rather than lecturing about old growth forests she will take them on a walk to show them what makes a forest old growth. Pitched at a young audience, the definition of old-growth forests is easy to comprehend. The importance of and function of the forest seems to be its existence value. In several videos in this category, including "Saving Agarikon", species extinction is linked to logging. The scientist who discovers a forty-year-old fungus thriving in an old-growth area tells the viewers, "the loggers are going to totally destroy this situation here so we may as well just harvest it" (2009). "Threats to Spotted Owls" narrows the function of old-growth forests to habitat protection, especially the importance for spotted owls, and asserts that the only hope for the spotted owl is to preserve remaining, though largely diminished, stands of old-growth forests. The popular pull of the video is probably due to the close identification between spotted owls and deforestation in the psyche of the Pacific Northwest. Although short and visually interesting, the video would probably draw far fewer viewers if the spotted owl was not such a powerful emotional symbol. In "Amazing Spotted Owl Mouse

Grab" there is no argument. The implicit message is that the amazing feat of the owl will never again delight visitors to the forest once logging leads to species extinction.

NARRATIVES VALUING OLD-GROWTH FORESTS

The videos in this category represent the values of environmental activists as they celebrate the worth of pristine forests for current and future generations. Utilitarian values are generally eschewed for aesthetic, amenity, and existence measures of worth (Perlman et al., 1997: 44–45). Many of the videos emphasize contestation between these values and utilitarian—generally economic—demands. Sponsored by Bob Brown and the Greens, "Tasmania's Ancient Forests" tries to balance competing interests to appeal to a wider audience; for example, by weighing the direct use value for the timber industry against the worth of the forest for the tourist industry. The latter allows for economic gain for Tasmanian workers without destroying an old-growth forest. Balancing the utilitarian value subtly challenges the economy versus environment frame that characterizes much of the debate over old-growth extraction. The video also balances visual appeal with verbal arguments: scenes of a pristine forest are countered with a shot of a harvested area replete with blackened stumps. This is the backdrop for the verbal articulation of the shortsightedness of extraction policies: "a death sentence to these ancient trees and their wild life" and "surely this isn't Australia's future?" (2007). The verbal argument reinforces the jarring visual contrast between lush forests and harvested ruins.

Almost exclusively visual, "'Endangered' Tasmania's Wild Places" provides an example of a less successful argument. Exquisite nature photography and pleasant mixes of instrumental music and nature sounds appeal to the aesthetic value of pristine forests. Any narrative direction comes through limited use of subtitles identifying the location being shown or providing an occasional comment such as "in risk of being mined" (2008). The viewer is left to make an argument from the largely visual appeal. Few shots depict the landscape after extraction, so even visually there is limited direct argument. The bequest value or the transformative value of nature on human behavior could be developed with the images, but the claims are not made directly. Although pleasant, the video is unlikely to pull an audience beyond those who already agree with the visual narrative. Similar concerns are apparent in "Lower Weld Valley Threatened Forests from the Air". With little variation in the images and no verbal argument, the film is largely unpersuasive. The same can be said of "Winter Wren Bird Song in Old Growth Forest of North Cascades". Without having been presented an overt argument, the viewer is left to enthymematically link the images and sounds of birdsong to the existence value of nature and an aesthetic appreciation of the diverse flora and fauna that make up an old-growth forest ecosystem.

"Tasmania's Forests: A Global Treasure, a National Responsibility" offers more effective visual argument. Shots of pristine nature are juxtaposed with shots of wildlife that has been poisoned, trees that have been clearcut, and humans directly engaged in the destruction of the forest. Human agency is directly indicted, not passively assumed as is the case in "Endangered". One subtitle reads, "Tasmania for its size has one of the highest rates of landclearing in the world" and another reads, "Nothing survives" (2006). Images switch from the forest to woodchip mills and factual information on how much woodchip product comes out of Tasmania. Similar treatment is given to the issue of poisoning wildlife. A direct reference to the bequest value comes with the subtitle, "This destruction is denying us, our children and grandchildren the opportunity to experience these natural wonders" (2006). The film takes the argument a step further in arguing that momentum is building through the protests, whether they are protests against Gunns, public arts projects, or the establishment of the global rescue station. Visualizing a future, a role for the viewer adds depth to the argument. Individuals depicted as working for the future are clean-cut, well dressed, and "typical", not marginalized. Finally, the music is more upbeat, compelling the viewer's attention. In addition to the pull that results from sponsorship, the video has strong narrative appeal in plot, characters, and depiction of the scene before and after extraction, and in the future. As the second most frequently viewed video, "Tasmania's Forests" offers a stronger model of a persuasive appeal.

The trailer for *Wild Tasmania*, a 2007 documentary, has a much stronger narrative line. The narrator provides context and introduces the issues at stake, the photography is professional and rapidly intercut to create heightened interest, and the music creates a sense of drama. Unlike the other videos in this category, this film lays out the problem, the "fine balance between economics and environment" with "community expectations from both sides" that "are huge" (2007). At stake is species endangerment, the end of "the last great predators of the Tasmanian forest." Visually compelling, suspenseful, and contextualized as a community-based conflict, *Wild Tasmania* is understandably a more watched film than the others we have discussed. To pull an audience on the Internet requires a good narrative, one that is coherent and has fidelity for the audience.

The most successful video in terms of the largest number of hits, in both this category and of all of the Tasmanian and Pacific Northwest videos, is "Opal Creek—Oregon Cascades". The narrative is carried through visual appeals of a stroll through the forest and a family enjoying the recreational value of Opal Creek. It continues to pull viewers at a rate of nearly 2,000 hits per month at the time of writing. Most viewers associate Opal Creek, an easily accessible family recreation site, with a successful campaign in the 1990s to return the lands to public ownership and designate it as a wilderness and a scenic recreation site to preserve the area from clear cutting. Given its pull, it is disappointing that the narrative does not overtly engage in the policy debate over old-growth extraction.

NARRATIVES CELEBRATING LOGGING

The search revealed few videos in support of logging, and those that did meet the search criteria do not engage the debate over old-growth cutting. "Old Growth Timber of Washington State" shows a series of still photographs of log trucks interspersed with images of tress, mountains, and streams in Washington. Created by Tellefsen Trucking, the only overt argument is in a photograph of a billboard that reads, "We have a medical problem—we're sick of envirocrat do-gooders" (2009). The image is visible for less than five seconds. The audience for this video, like many of the protest videos, seems to be those members of the community described in the video. A similar approach is apparent in "HELLog" and "HELLog 2", two versions of a short videotape showing a helicopter logging trees. The second version with 20,000 hits adds the slug, "No tree left behind" (2009). Neither version contains a verbal message. The sound of the helicopter is the only accompaniment to visual footage. YouTube responses suggest that viewers focus on the pilot's amazing skill in a hazardous flying situation. Neither video does much to explicitly engage the viewer in the values or policy arguments over old-growth logging.

PROTEST NARRATIVES

Protest narratives show environmentalists actively engaged in efforts to stop old-growth logging. "Activists Shut Down Gunns Ltd Triabunna Woodchip Mill" is typical. The lyrics to a protest song serve as the verbal narrative. Lots of handheld camera work, shots of protest marches, and word-intensive subtitles accompany images of forest destruction. Narratively, in both the plot development and in stylistic choices, the video is the antithesis of the successful protest that it is designed to celebrate. Similar problems are evident in "The Upper Florentine Trilogy: Part One". Rather than plot-driven, this narrative largely focuses on character. A more extensive view of environmentalists does little to make them representative of a diverse audience. Their appearance reinforces the stereotype of environmentalists as left-over hippies. In contrast, "Come to Camp Weld" does a better job establishing the need for activists before focusing on the protest camp they have established in the Weld forest. While the activists are still depicted in stereotypic terms, their efforts are given credibility by a politician's statement: "I support the protestors. I want them to be here...seeing the animals going through the forests...protecting that for future generations, these are the people who should be the heroes" (2006).

"Old Growth Forest Defenders—Sisters of the Siskiyous 2" (no longer available on YouTube) visualizes the protest rally that is a tradition in and celebration of fighting old-growth logging. Old women link arms and block a logging truck from going up to the extraction site. The whole video typifies old-style political protest. The arguments are clearly designed for an audience that already accepts the anti-logging vision. Anyone who was not part of the group consciousness would likely dismiss the message as harmless and a bit eccentric. "The PM & the Frog" skillfully employs irony and animation to protest Australia's old-growth logging. After beginning with Prince Charles's campaign to save rainforests, the video shifts to a cartoon animation of Australian politicians, including former prime minister Kevin Rudd, mocking frog preservation. The script is reminiscent of the campaign in the Pacific Northwest where activists on both sides displayed bumper stickers reading, "I like my spotted owls fried" or "Kill a logger, save an owl", among others.

ASSESSING THE VIDEO MESSAGES: A REPRESENTATIVE CASE

In choosing a video to assess in more depth, I turned to one with significant audience pull. It is not surprising that "Tasmanian Forestry Contractors" (2008; no longer available on YouTube) has the highest viewership of the Tasmanian videos. A camera recorded the confrontation between protestors and forest workers, unbeknownst to the contractors and a representative from Forestry Tasmania (FT), the corporation that hired the contractors. Two forestry contractors are caught on film: one uses a sledgehammer to smash the windows of the car containing the protestors, and one screams obscenities throughout the confrontation. There is no commentary with the video; its mimetic presentation of contractor violence, while representing the frustration on the part of the contractors, functions to vilify the forest contractors and FT, and by extension those who support the extraction of old-growth trees.

The videotaped incident received significant media attention. Rather than advancing the debate over logging policy, the video polarized competing interest groups. Media coverage following the incident makes it clear that both sides identify the other as intransigent, and implied that collaborative efforts to resolve the conflict were unlikely. For example, Michelle Paine's story in *The Mercury* says: "A representative of forest contractors said they had been pushed to the limit and he predicted further violence in the forests" (2008: 9). She explains that the contractor had lost about $30,000 as a result of previous protests that interrupted his work. Given the way the media reports the loggers' narrative, one would logically assume that the cycle of violence will continue because a choice has to be made between the right to earn a living and the right to protest—a subset, I believe, of a broader and more prevalent narrative frame that pits the economy against the environment and assumes that economic interests should always prevail. Both the broader and the more narrow frames create a conflict that seemingly cannot be resolved by discursive means—assuming both sides maintain their argumentative stance—and suggest instead that winning demands resolution by any means possible to maintain one's own interests. When either side resorts to violence as a strategy, it will initiate a cycle of violence that overshadows discursive means of resolution.

The activists' narrative following the incident assumes the righteousness of "legitimate peaceful protest" and sees the attack as evidence of a victimage story. In this narrative the activists' agency is subsumed by the violent agency of the other side (the protestors become helpless victims), and by what Kenneth Burke would term a compelling scene: a pristine environment threatened with extinction as a result of greed. Framed in this way, the narrative encourages polarization of activists and loggers as it sets up a contest between the forces of good and the forces of evil whose greed threatens the existence of old-growth forests. The narrative does not encourage efforts to understand the other side, nor does it facilitate collaboration with opposing parties or compromise.

Evidence for how the narrative construction of both parties leads to intransigence is plentiful in the comments posted to YouTube. The majority of comments in the first three days after the video was posted reveal polarizing narratives employed by both sides. Vilifying the other "unifies individuals in movements, provides a clear target for movement action, and allows activists to define themselves and their positions in opposition to those of their adversaries" (Vanderford, 1989: 166). Environmentalist sympathizers characterize the forest workers as greedy, ignorant, and solely responsible for the violence. Consider these characterizations of the loggers in the YouTube comments section: "they carry on like fascist thugs"; "logging contractors truly are pond scum"; and "a bloody disgusting bunch of Neanderthals" ("Tasmanian Forestry Contractors," 2008). Those viewers who support the loggers are equally polarizing in their characterization of the environmentalist protestors: "These 'Benefits professionals' protestors have no understanding what its like to have your livelihood constantly out at risk"; "These self-indulgent, sanctimonious feral basket-weavers deserve everything they get"; and "go and get a job you lazy bastards, those contractors pay taxes that pay your dole" (2008). Most of the name-calling of environmental protestors is directly related to the economy versus environment frame.

There are few comments posted in response to the video from either side that seek to discover common ground. Evidence of polarized responses suggests that the message most likely spoke only to those who already agreed or disagreed with the environmentalists' position before seeing the video.

While the Internet is valued because it allows new voices to enter the public sphere, this video is less likely to extend dispassionate discussion of the issue of harvesting old growth and more likely to reinforce in-group solidarity. Bormann argues that groups frequently exchange dramatizing messages that he calls fantasy themes: "when participants have shared a fantasy theme they have come to symbolic convergence in terms of common meanings and emotions that can be set off by an agreed-upon cryptic symbolic cue" (1985: 131). When environmentalists post the footage to YouTube they offer a message that reinforces their sense of group identity and the importance of their protests. In naming the other as they do, in resorting to similar frames for retelling the story of the forest contractors incident, the opposing sides in the issue of old-growth extraction each demonstrate in-group symbolic convergence that casts the other side as intractable.

In-group identification, in theorist Kenneth Burke's terms, is a form of identification, but he cautions: "one need not scrutinize the concept of 'identification' very sharply to see, implied in it at every turn, its ironic counterpart: division" (1969: 23). In the process of establishing identification with some, others are excluded. Burke fears the "wavering line between identification and division...[is always] bringing rhetoric against the possibility of malice and the lie...[and] deliberate cunning" (1969: 45). A common enemy against which one might measure one's own values and actions may heighten the appeal of group belonging. Thus, while symbolic convergence strengthens in-group solidarity (an important function of the Internet) it does not necessarily lead to a greater understanding of the issues or the reasons for taking a particular stand on this policy debate.

The forest contractor incident also gained much needed news coverage. As Hutchins and Lester (2011) note, news media attention is crucial to sustaining environmental protest campaigns. Referencing the incident they conclude, "the desired public image of the forestry industry and Tasmanian State Government was, at least for a moment, punctured and destabilized, forcing both to defend their records and activities" (Hutchins and Lester, 2011: 168), and that response is a concession that environmental issues "are deserving of a response in the public domain" (171).

Although polarizing rhetoric is unlikely to lead to discursive resolution, protest strategy has long recognized the potential success in policy determinations of escalation/confrontation strategies. If one side to the dispute can push the right buttons and get the other side to over-react, sympathy goes to the victims and may move policy decisions toward their desired resolution. In the short term, the battle may be won by such divisive strategies, but conflict that is not properly managed tends to return, and those who have been the losers learn new strategies for arguing their ends. More importantly, the kinds of narratives we employ reflect and shape who we are as a community and how we will work with others. In writing about social ethics, Stanley Hauerwas argues, "trust is impossible in communities that always regard the other as a challenge and threat to their existence" (1981: 11). The kind of narratives we choose to employ, the characterization of self and other, the attribution of motives and values shape us as groups and communities. If the stories we tell and the symbolic convergences we create encourage trust and cooperation, then those narratives encourage the participants to trust and collaborate with one another; if they vilify they invoke symbolic warfare, minimizing our inclinations to listen to and seek identification: morals "are the linguistic projection of our bodily tools and weapons. Morals are fists...the moral elements in our vocabulary are symbolic warfare" (Burke, 1965: 192). Naming the other as a redneck, Neanderthal, basket-weaver or Benefits professional fosters division rather than cooperation based on identification of commonality. Choosing to air the video clip was symbolic warfare that carried the same ethical implications as the name-calling responses to its content and strategic use.

CONCLUSION

Although old-growth extraction is a policy issue, the weighing of competing values, the choice of narratives that foster identification versus division, the symbolic warfare and real conflict that surrounds these policy debates are as important as discussions of costs and benefits. Most videos discussed here do little to explain why one's values and interests should be accepted. Arguments to support one's position or refute one's opponent's claims are seldom presented; the protest videos celebrate or denigrate protest without arguing for the rationale behind their position. Although the Internet has the potential to dramatically increase the audience for one's message, if the narratives disseminated do not reflect a message crafted to appeal to people besides those already identified as sharing one's position, using the medium is unlikely to enhance one's cause: message construction demands an understanding of the social network of which they will be a part. To pull an audience, YouTube videos must articulate shared values and compelling reasons for cooperation and joint action, and they must offer well-crafted narratives that have coherence and fidelity for the viewers.

Like other Internet platforms, YouTube has a literacy that needs to be developed if one is to use this social network effectively. The challenge for activists is to immerse themselves in this channel of communication, taking advantage of the opportunities it affords while recognizing that these messages require careful crafting to pull a significant audience and convey a meaningful message. It is not sufficient to just post a video to YouTube: pretty pictures and clichéd messages have not drawn large audiences. None of the environmental messages posted from activists in these two venues went viral. Because of the pull nature of the medium, environmental activists and loggers need to develop messages that compel attention without creating a climate in which opposing sides cannot find a discursive space for negotiating policy decisions with respect to old-growth forests. Activists need to create messages that take us beyond the mirroring and matching of competing interest groups in the old growth debates that Lange found ineffective nearly twenty years ago in the Pacific Northwest.

▪ CHAPTER FIVE

Photography, Technology, and Ecological Criticism

Beyond the Sublime Image of Disaster

DANIEL PALMER

> *Its [man's] self-alienation has reached such a degree that it can experience its own destruction as an aesthetic pleasure of the first order (Benjamin, 1973: 242)*

Among the various ways in which knowledge about climate change is mediated, documentary photography has long been important as a means to visualize its environmental effects. Photographers operate as global witnesses to a complex problem that few people understand let alone experience directly, trading on photography's reputation as a conveyor of visual truth. Supplementing and updating the well-established role occupied by photojournalists in the twentieth century, their work provides memory icons for an uncertain age. Indeed, it can reasonably be claimed that over the past few decades an increasing number of professional photographers have shifted from a broadly humanist to ecological framework. Although the idea of witnessing remains central to the rhetoric of the photography of climate change, the subject poses particular challenges—not least, the complexity of the scientific and political issues, and even more obviously, the relative invisibility of CO_2 emissions and the effects of climate change itself. This chapter therefore explores the rhetoric of environmental images in order to consider some of the difficulties that accompany dominant genres of such imagery, and argues that the effectiveness of photography as a mediating process deserves more scrutiny. I briefly explore some emerging photographic approaches that take a more overtly participatory and collaborative approach to image making. I argue that these approaches signal a shift from a single photographer seeking to capture universal truths to photography as a more localized and indeed location-dependent activity—in short, as a form of tactical media.

Various motivating forces are at work in the photography of climate change, alongside photographers' own commitments to bring awareness to the issues. Consider, for instance, the various awards established in recent years. The Prix Pictet, established in 2008, claims to be the world's first art prize dedicated to artists who use photography to convey crucial messages regarding sustainability. According to its website, www.prixpictet.com, it has "a unique mandate" which is "to use the power of photography to communicate vital messages to a global audience" in order "to uncover art of the highest order, applied to confront the pressing social and environmental challenges of the new millennium". Each year a theme is selected; thus far these have been Water, Earth, Growth and Power. The finalists—predominantly, Euro-American artists—are nominated by a panel of experts who include directors and curators of major (northern hemisphere) museums and galleries. In contrast to "art of the highest order", Hewlett Packard's 2009 Capture Change competition addressed ordinary photographers and their desire to document the changing world. Its promotional website beckoned:

> Flashfloods, hurricanes, droughts, heat waves and other natural disasters are becoming commonplace. Each one of us is being affected. The impact of climate change is no longer something that will come in the future—it is happening right here… right now. www.hpcapturechange.in.com (accessed October 2009)

HP's competition encouraged amateur photographers, their customers, "to record these changes happening around you". While both the Prix Pictet and HP's Capture Change competition can be applauded for spotlighting the urgency of responding to climate change, unsurprisingly, neither addressed the fact that the causes and impact of climate change are geo-politically specific. It is well established that the world's poorest populations, who produce the least carbon emissions, will be disproportionately affected and are the least prepared to deal with the effects of climate change. Ironically, in spite of HP's India-based competition being open only to "users of Indian nationality", its promotional rhetoric reinforced a universal view of climate change by not drawing attention to specific issues facing the Asian subcontinent.

For obvious pictorial reasons, extreme weather events like floods and droughts have featured heavily in the photography of climate change. The North and South Poles also feature prominently. To be sure, the poles are the "canary in the coalmine" when it comes to the earth's warming temperatures. Scientists have concluded that the poles have heated more than the rest of the planet due to the cumulative effect of the lack of reflection of the sun's rays on the ice compared to dark water. The poles may seem remote, but what happens there may be more important than any other aspect of climate change, because of rising sea levels. In relation to their communication by photographs, it is worth noting that the poles have always been communicated to the public through media—very few people have the opportunity to visit them in person. Moreover, they offer symbolic proof that climate change is a global problem, quite literally in the sense that both Antarctica and the Arctic are the preserve of multiple nation-states. At the same time, the symbolic status of the poles has become a marketing opportunity (Hansen and Machin, 2008)—their whiteness, pristine quality and lack of human habitation lends them a spiritual quality. Yet, photographs of melting glaciers that once might have functioned as persuasive signs of the visible impact of climate change have inevitably become a cliché. Indeed, images of polar bears have become so familiar that droll versions—such as images of a bear staring at its reflection in a melting glacier pool—are now common in newspapers.

Images of the poles may nevertheless still reveal crucial evidence to scientists, and by extension, offer important information to broader publics. Thus, in September 2011 the Australian *National Geographic* photojournalist Jason Edwards was awarded the Eureka Prize for Science Photography—an award sponsored by *New Scientist* magazine for photographs that most effectively communicate an aspect of science—for an image titled *Receding Glacial Cap with Cryoalgae*. This remarkable image effectively records a specific effect of climate change recognizable by scientists. An evocative description of the image is offered on the website of the Australian Museum (2011):

> A beautiful photograph is even more powerful when it exposes hidden dangers and educates an unsuspecting viewer. Proof of that is a mesmerizing image of a storm front rising over a rugged and weathered glacial cap. It reveals the blue and pink stains of cryoalgae—snow algae that grows in ice—as the cap recedes because of global warming. The photo was taken on the Antarctic Peninsula, an area that has undergone one of the highest temperature increases in the world over the past 50 years, with the mean annual temperature rising by more than 3°C. The result is significant thinning of ice caps, recession of glaciers and the break-up of ice shelves

The language is emphatic: the image "exposes hidden dangers", "educates", offers "proof" and "reveals". But as Sean Cubitt (2005: 131) has observed, the "popular mediation of ecological anxiety is not an exact science". Unfortunately—and tellingly—when the image appeared on the museum's public website, no date was given. Ironically, the receding glacier was thus symbolically re-frozen by the photograph as an image of sublime majesty. As Cubitt has argued (2005: 46), the sublime always has the "pretence of a timeless, universal" order—beyond history or debate—and the image is always in danger of appearing as a diversionary spectacle.

NATURE AT RISK

Photography has long had a privileged role in environmental campaigning. In my own country of Australia, the role of photography in the successful 1976–1983 protest against the Franklin River Dam is the most famous example. The Franklin blockade—widely recognized as Australia's most famous environmental protest—was "a globally important moment for green politics" (Hutchins and Lester, 2006: 433). Peter Dombrovskis' *Morning Mist, Rock Island Bend, Franklin River* (1979), arguably the most celebrated landscape photograph in Australian history—actively used by conservationists and said to have helped determine the federal election of 1983—was built on a model of nature photography pioneered in America. In the nineteenth century the photographs of Carleton Watkins had influenced the unprecedented decision to set aside Yosemite Valley as a state park in 1864, and the photographs of William Henry Jackson had figured in the decision to create the first national park, Yellowstone, in 1872. In the twentieth century the tradition was carried on by greatly admired photographers like Ansel Adams, while Dombrovskis himself was influenced by his mentor Olegas Truchanas, whose landscape images had become especially popular following his accidental death in 1972. Today, such wilderness photography is frequently criticized for indirectly promoting a sense of wilderness in which humans have never existed. Given that the original inhabitants of Tasmania have a history of at least 40,000 years, such an idea is both potentially offensive and demonstrably untrue (Flanagan, 2010: 127). As Adrian Franklin has argued, the representational space of wilderness raises important questions about, for example, what is being represented and *to whom* (2006). Nevertheless, the philosophy of Truchanas and Dombrovskis was simple and remarkably effective—their sincere belief was that if people could see the beauty of Australia's "wild" places then they might be moved to save them.

Photography's relationship with climate change is about more than just the conservation of the natural environment, since it involves much more than simply preserving visible scenes of beauty. Indeed, as I suggested at the outset, climate change is largely out of sight or literally invisible to the extent that it is based on predictive statistical modeling. The effects of climate change, according to many climate scientists, cannot (yet) be seen directly. Consequently, in no small measure, visualizing climate change is a method for making the invisible visible (Doyle, 2011; Schmidt and Wolfe, 2009). As such, the proliferation of images of strange weather events is somewhat paradoxical. Photographs of floods and drought do not in themselves constitute evidence of climate change, even though they might be attached to legitimate arguments about their increasing frequency and intensity. Moreover, while the

Figure 1: *The Science of Climate Change: Questions and Answers*, Australian Academy of Science, 2010.

visual has always been important to science communication, scientists often seem to harbor a residual degree of suspicion about the affective dimension of images—undoubtedly because reason and accuracy are traditionally attached to words and data. All of this points to a further problem at the level of photography's status as evidence. The quality of the images used in the Australian Academy of Science brochure *The Science of Climate Change: Questions and Answers* (2010) appears to be a case in point. With the exception of an ominous cover image of a desert cloaked by dark clouds, the bulk of the photographs used in that document—intended for wide distribution to persuade Australian policy makers of the need to act—appear to be willfully bland and generic. The fact that many were sourced from global, commercial picture libraries such as Corbis is telling indeed.

Julie Doyle has written at length about the use of photography in climate change campaigning. She has been specifically critical of Greenpeace's philosophy of "bearing witness" to environmental damage (2007a; 2007b). Reflecting on a mass-circulated image produced by Greenpeace of a walrus stranded on a melting piece of ice, she suggests that:

> Whilst undoubtedly providing an important impetus to the politics of climate change, particularly in light of the formation of the Kyoto Protocol in 1997, at the same time such images produce a distancing effect, relegating climate change impacts to a remote and inaccessible place, where animals and habitats are affected rather than humans. (Doyle, 2007a: 142)

This distancing effect is fundamental and predominant. As Doyle put it more recently, "one of the difficulties in engaging people with climate change is due to its historical framing as an environmental issue, which has led to a separation of humans and culture from the environment" (2011: 3). This separation—unwittingly supported by the scientific community—also has the side effect of rendering climate change "as a distant and future event rather than a present and immediate reality" (2011: 3).

Furthermore, and by contrast, Doyle notes that the "photographs of glaciers represent temporally the *already seen* effects of climate change" (2007a: 129). As she argues, this is "a distinct problem when considered within the context of the historical efforts by environmental NGOs to bring attention to the global problem of climate change, *before its impacts could be seen*" (Doyle 2007a: 129). To be sure, the medium of photography—at least in its traditional, unmanipulated forms—is limited by its temporality of what is or "has been", rather than what *may be*. Indeed, images used as testimonies of the changing landscape have played a central role in communicating climate change to the public, epitomized by before-and-after projects such as Gary Braasch's best-selling book *Earth under Fire: How Global Warming Is Changing the World* (2007). The before-and-after technique was also deployed to powerful effect in the final episode of the BBC's *Frozen Planet* (2011), in which David Attenborough counterposed a photograph by Frank Hurley from the Ernest Shackleton expedition to the Antarctic in 1916 with recent footage of the same glaciers, revealing a dramatic retreat over the past century. However, while these comparisons are disturbing, we nevertheless cannot assume the positive effect of comparative before-and-after images; they arguably do little to enable the viewer to do anything about the loss presented (O'Neill and Nicholson-Cole, 2009).

NATURE "OUT THERE"

Problematically, the kind of images I have discussed so far can suggest that the effects of climate change occur "out there", in an abstract "nature"—distanced from people's everyday experience. In this sense, nature under threat is the environmental variant of the sublime image of disaster, with all its attendant problems. The discourse of climate change is understandably preoccupied with the imagination of possible futures, of "scenarios, narratives, and contingency plans that project toward or back from uncertain futures" (Yusoff and Gabrys, 2011: 517). But as the record of science fiction shows, our fascination with the aesthetics of disaster occurs at moments in history when it becomes easier to imagine the end of the world than alternative futures. The motivation for apocalyptic images is clear: to shock people into action. Thus Al Gore's vision of global warming in *An Inconvenient Truth* (2006)—a film that was crucial in establishing the mainstream cultural legitimacy of climate change—foregrounds the potential magnitude of impacts through a visual and textual language of danger and threat. The risk, of course, is that this language produces a sense of impotence in the face of an overwhelming threat (Miles, 2010). As Simon Cottle has argued in relation to the "news spectacle", to avoid becoming "voyeurs" of "our own impending peril" we need to feel engaged in the present (2009a: 88). Notably, *An Inconvenient Truth* ends with a section before the credits titled "Are you ready to change the way you live?" and the optimistic statement: "The climate crisis can be solved. Here's how to start". This is followed by a number of practical solutions on how to reduce one's carbon emissions and recommendations for climate activism. These suggestions appear in dissolving white text, interspersed with the credits. This sober coda section of the film is arguably one of the most powerful, not least due to an emotive pop song with the lyrics "I need to move, I need to wake up, I need to change, I need to shake up, I need to speak out, something got to break up, I've been asleep and I need to wake up, now". It is notably lacking in any imagery at all.

Taken in isolation, apocalyptic imagery suggests, as Walter Benjamin (1973) famously wrote, that our alienation is such that we can experience our own destruction as an aesthetic pleasure. Yates McKee coined the phrase "the misanthropic sublime" to describe the abstract pleasure to be taken in such images (Bos and Lam, 2009: 68). Others have likewise argued that apocalyptic imaginaries serve to disavow or displace social conflict and antagonisms. Erik Swyngedouw (2007; 2010) characterizes the contemporary consequences of the imagination of environmental disaster as "post-political" environmental politics because they operate by externalizing the nature of environmental threats rather than focusing attention on the inner workings of capitalist relations. McKee has similarly argued that what I am calling the sublime image of disaster acts to "displace attention from the unevenly distributed causes and effects of climate change with an appeal to a universal ethos of environmental preservation addressed to an equally generic 'humanity'" (Bos and Lam, 2009: 58).

As we have seen, the photography of climate change is thus implicated in larger philosophical and political debates within environmental advocacy. As Timothy Morton argues in *Ecology without Nature*, his study of "nature writing" as a rhetorical form, "nature always slips out of reach in the very act of grasping it" (2007: 19). What he says about writing—that it "overwhelms what it is depicting and makes it impossible to find anything behind its opaque texture" (2007: 19)—is also true of photography. Photo-*graphy* is indeed a form of "writing"—or what he calls *ecomimesis*—and its sublime and apocalyptic genres are two sides of the same thing; each installs us in a relation of otherness to what is being depicted: "Just when it brings us into proximity with the nonhuman 'other', nature re-establishes a comfortable distance between 'us' and 'them'" (2007: 19). While Morton acknowledges that "[a]ppealing to nature still has a powerful rhetorical effect" (2007: 24), he concludes that the chief stumbling block to environmental thinking is *the image of nature itself*, and he calls for a new form of ecological criticism.

EMERGING APPROACHES IN THE DIGITAL MEDIA

The birth of the modern environmental movements is intimately linked to images, and specifically to landscape and nature photography. My claim, however, is that the conventions of wilderness and other forms of sublime or apocalyptic photography that posit nature as a pure "outside" are effectively obsolete. I am not suggesting that *poetics* are redundant—as Emily Potter observes, the "demands of environmental change expose the limits of representation" (2009: 77), and poetic forms of representation can draw attention to the very evasiveness of the issue. What is clear, however, is that the age of climate change coincides with the need for a more complex understanding of "nature", and together with the rise of networked digital media produces a context in which new modes of photography are demanded. More generally, photographic images are currently undergoing a dramatic transformation in their meaning and value. In particular, the ubiquity and networked character of screen-based digital images means that in important respects, photographs are becoming less about memories of significant past events and more about the communication of often banal experiences in the present. As such, I want to conclude by considering some emerging forms of photography relevant to the activation of ecological citizenship in the age of climate change. What follows are three examples of photography that seek to move beyond the *representational* image of nature to a more *connected, performative* and *activist* orientation.

My first example is a project that utilizes Photovoice (www.photovoice.org), an increasingly widely used tool in community development and contemporary social research. Photovoice, "participatory photography for social change", involves participants—often marginalized groups—taking photographs in response to a particular issue. The ideas in the images are then drawn out and negotiated through group discussion and "clarified with captions" as a means to bring them to the attention of a broader audience and decision makers in particular (Chandler and Baldwin,

2010: 30). In 2009 Queensland academics Lisa Chandler and Claudia Baldwin utilized Photovoice "to elicit the values and concerns" of residents, artists and visitors to the small lakeside community of Boreen Point about the uncertain danger of rising sea levels and flooding due to climate change (2010: 30). Their experiment was

Figure 2: Ron Smith, *The Wedding Tree*, from the At the Water's Edge Photovoice project conducted by Claudia Baldwin and Lisa Chandler, Floating Land, 2009.

Figure 3: Annie Thurlow, *Our Livelihood Threatened,* from the At the Water's Edge Photovoice project conducted by Claudia Baldwin and Lisa Chandler, Floating Land, 2009.

conducted at Floating Land, a biennial site-specific environmental art event held in the coastal community of Noosa, Australia, but it also included participants from the seriously threatened Pacific Islands of Tuvalu. The images were first presented at Noosa Regional Gallery in 2009, and were later shown after the event on the festival's website, where they were open to further comment. The researchers argued that the resulting visual "stories"—formed from groupings of photographs—provide persuasive forms of "expressing shared perspectives of climate change" from communities "at the water's edge" (Chandler and Baldwin, 2010: 30).

Despite the acknowledged methodological limits of their research—notably, its rather biased sample of participants—this experiment with Photovoice is a refreshing example of a different use of photography in the communication of climate change. Rather than the medium of a single photographer seeking to convey a universal message, photography functioned here as a vehicle for multiple specific conversations within a diverse community. Chandler and Baldwin emphasize that the collaborative, participatory process was a valuable attempt at building consensus in a local community. They cite one respondent who noted that the project "reinforces our need to act now" (2010: 33). However, equally valuable are the differences and conflicts that inevitably result from the discussions—even among participants who

Figure 4: *Participatory Sensing: A Citizen-powered Approach to Illuminating the Patterns that Shape Our World*, Center for Embedded Networked Sensing, University of California, 2009.

might appear to be in agreement. In other words, this is a productive use of photography not to confirm what is already known and visible, but to raise questions and dialogue. The mainstream media could take note of this approach.

A second, related direction in emerging photographic practice involves the use of camera phones to engage citizens with the environment. As more and more people in the developed world use smartphones as their primary cameras, the practice of photography is rapidly evolving. Sites such as Flickr, with its vast user groups and sophisticated organization of images via metadata, also mean that online photo-sharing has become increasingly common. Accompanying these developments is so-called "participatory sensing", which uses embedded devices like the smartphone to capture data about oneself and one's community in order to extend a history of citizen science projects. Research underway by scholars at Los Angeles' Center for Embedded Networked Sensing at the University of California is at the forefront of this trend (Goldman et al., 2009). There, people are encouraged to use a mobile phone's geo-tagging camera to work together for specific purposes, from simple projects such as mapping the best bicycle routes to targeting the location of invasive plant species in national parks. Their work in identifying sites of local pollution is a brilliant example of decentralized eco-monitoring, in which emerging risks can be identified before they become full-blown crises.

Contrary to often-voiced fears that mobile technologies lead to forms of virtual habitation and placelessness, participatory sensing becomes a way in which locative media can deepen our sense of connection to place. The process of collecting imagery turns citizens into active visual agents rather than mere spectators of the sublime image of disaster, engaging them as environmental data gatherers in various forms of participatory mass observation. Participatory sensing is thus an emerging form of mass data collection that may, under appropriate conditions, be conceived as a form of "empowering" self-surveillance—an opportunity for individuals and groups to "provide possibilities for self-exploration, community discovery, and new knowledge creation" (Shilton, 2010: 132). One simple example in relation to climate change involves gardeners identifying the changing growing seasons of plants and trees. Such a project not only responds to Doyle's call to focus on "how climate change is made relevant to people's everyday lives" (2011: 8), but also seeks to aggregate that information in productive ways.

My final example, while not specifically related to climate change, stands as a further model for what can be achieved in activist networked photography. It also involves grassroots mapping, but this time in the form of home-made "satellites". It emerged in the wake of the BP Deepwater Horizon oil spill in the Gulf of Mexico in 2010, when a group of citizens and activist mappers developed a set of novel DIY tools for sending inexpensive digital cameras up in helium balloons and kites. By setting the cameras to automatically take pictures every five seconds, they were able to generate aerial photos from up to 300 metres in order to document the effects of the spill. The project was led by Jeffrey Warren, a fellow at the MIT Media Lab's Design Ecology Group who is also the founder of grassrootsmapping.org, which promotes public geographic information systems for cartographic areas in dispute. The facilitators encouraged people to participate and sought donations to pay for petrol for local fisherman to take people out to zones that were out of bounds for ordinary news crews. Their stated aspiration for the data gathered is that it would be useful "in both the environmental assessment and response, as well as in the years of litigation following the spill" (Grassroots Mapping, 2010). Importantly, all the imagery captured was released into the public domain and is free to use or redistribute.

Like participatory sensing, grassroots mapping belongs to the realm of what David Garcia and Geert Lovink have called "tactical media" (1997): "Tactical media do not just report events, as they are never impartial they always participate and it is this more than anything that separates them from mainstream media". By refusing the illusion of objectivity, tactical media projects can be conceived as an emerging

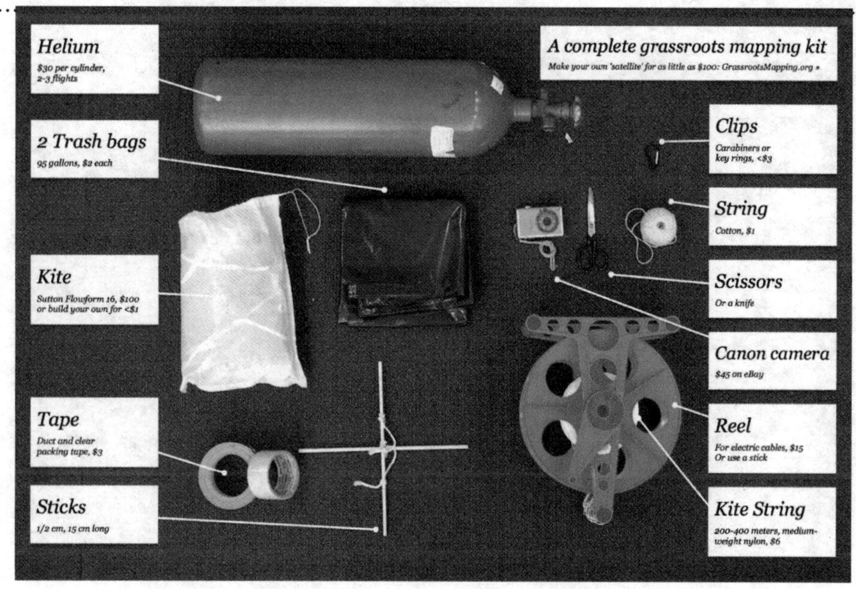

Figure 5: *$100 Satellite Poster*, 2010, Grassroots Mapping Community, an open source collaborative group. Photograph: Jeffrey Yoo Warren. Creative Commons Attribution license.

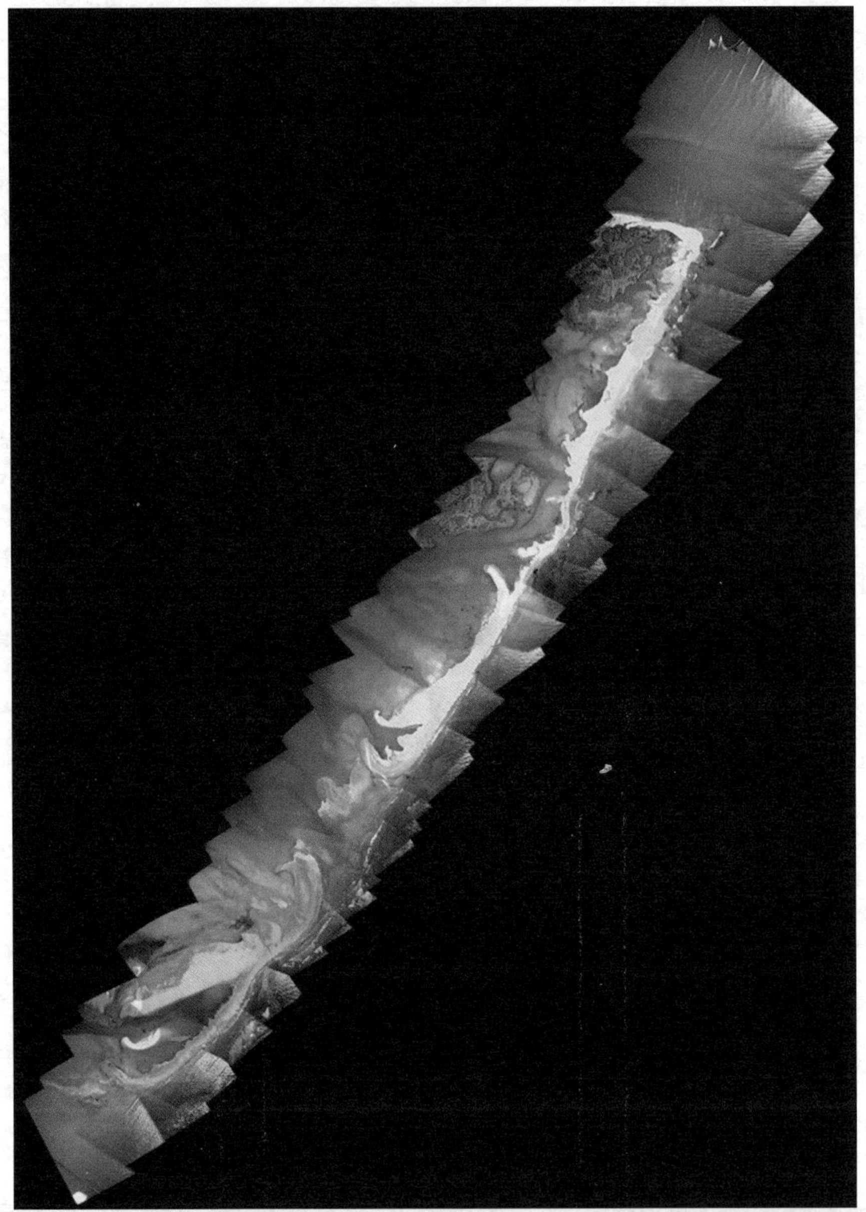

Figure 6: *Gulf Oil Map*, 2010, based on images captured by Adam Griffith as part of the Program for the Study of Developed Shorelines at Western Carolina University. The map was processed, prepared and published by Stewart Long.

platform for critical journalism, existing not just as a supplement to mass media but as part of a complex media ecology. What this example of grassroots mapping reveals, with its reliance on collaboration, is that the pervasiveness of networked digital media is well suited to assist in the formation of mobile and adaptable activist coalitions.

CONCLUSION

As we have seen, conventional ways of thinking about nature and the environment situate them as objects "externalized" from humans and culture. A more nuanced understanding of human-environmental relations is needed today. The three examples above illustrate how interactive and networked technologies can enable a more dynamic relationship with the photographic image of nature. Photography cannot visualize the future. However, emerging platforms for photography such as those discussed in this chapter should be recognized for their efforts to turn ordinary citizens into producers and participants of public imagery around the environment, rather than mere consumers. While I do not want to overestimate the contribution that creative photographic approaches might make to debates around climate change, it is apparent that photography can play a role in helping ordinary citizens to re-imagine environmental data and help us to ecologize our collective lives—not by producing universal narratives, but by introducing local, sharable techniques.

Notably, each of the examples above are collaborative, coinciding with a broader shift in art making that has seen the artist's role shift from a sole creator to a collaborator or facilitator. Moreover, each recognizes the need to do more than merely document the world "out there" in a single image. For the most part, environmental photography—even the most creative and critically celebrated—remains locked into conventional representational modes. But as Malcolm Miles has suggested, art must find new ways to represent nature. Art must go beyond the imagery familiar to us from mass media, "not by being more realistic or more beautiful—both outdated criteria—but by changing the conditions in which nature is produced" (2010: 24). More performative modes of photography are appropriate to this task, in which the camera is used as an instrument of analysis and discovery. Photographs are thus conceived "not as melancholic 'that-has-been,' but more as a future oriented and interrogative 'what-will-be?'" (Iversen, 2007: 105). Paradoxically, then, one mission for engaged photographers, as for "environmental art" in general, must now be to unsettle the self-evidence of nature itself, addressing the environment "as a contingent assemblage of biological, technological, economic and governmental concerns whose boundaries and agencies are perpetually exposed to conflict" (McKee, 2007: 557). There is much work to do.

ACKNOWLEDGMENTS

I am grateful to Shelley McSpedden for her invaluable research assistance in the preparation of this chapter.

▪ PART 2

Activism and Campaigns

▪ CHAPTER SIX

Not So Soft?

Travel Journalism, Environmental Protest, Power and the Internet

LYN MCGAURR[1]

> Tactical media emerge out of the margins, yet never fully make it into the mainstream. There is no linear career path through the world of newspapers and television channels. (Lovink, 2002: 258)

▪ ▪ ▪

> RESEARCHER: So you talk about power. Who has the power, then, in travel writing?
>
> TRAVEL JOURNALIST: I guess ultimately the editor of the paper. Then, obviously, the editors of each section have a certain amount of power. I mean, to be honest, maybe—. Yeah, it's a good question. Perhaps I should try that more. (2009, interview with author)

In these days of direct access to foreign publics via online media, appealing to journalists for publicity and legitimization can seem like an old-fashioned, even reckless strategy for environmental groups to pursue. Yet some of Australia's largest environmental organizations deploying networked technology strategically and tactically have failed to use it "to effect genuine transformation in the structural distribution of media and political power" (Lester and Hutchins, 2009: 590–591). Rather, they have used it in the manner of more traditional protest actions—that is, to attract the attention of journalists and gain access to mainstream print and broadcast news (Hutchins and Lester, 2011; Lester and Hutchins, 2009). In so doing, they have relied on elements of tactical media (Garcia and Lovink, 1997; Lovink, 2002, 2005; Meikle, 2002)—"cheap 'do-it-yourself' media, made possible by the revolution in consumer electronics and expanded forms of distribution (from public access cable to the Internet) ... exploited by groups and individuals who feel aggrieved by or excluded from the wider culture" (Garcia and Lovink, 1997). In its purest form the term "tactical media" is more restrictive than will be useful for the purposes of this chapter, particularly in regard to its emphasis on anonymity and rejection of the institutionalization that attends the evolution of many non-government organizations. However, when Geert Lovink coined the term in the 1990s, he made reference to tactical media practitioners working "both inside and outside the mainstream media" (Lovink in Meikle, 2002: 120). This, together with his views that investigative journalism is "the basis of all 'tactical' output" but editorial desks are servile and censorious (Lovink, 2002: 258), makes evidence of new-media tactics in the "soft" genre of travel journalism intriguing.

To date, news journalism has been the genre of most interest to scholars investigating the role of the media in environmental conflict (Allan et al., 2000; Anderson, 1997; Boyce and Lewis, 2009; Cottle and Lester, 2011; Hutchins and Lester, 2006; Lester, 2006, 2007, 2010b; Lester and Hutchins, 2009). In an effort to increase the salience of their arguments and "generate *dissensus* among the powerful" (Wolfsfeld, 1997: 29, original emphasis), environmental groups often attempt to use news coverage of their activities to highlight the broader economic impacts of environmental damage. This "scope enlargement" (Gamson and Wolfsfeld, 1993: 116) draws in other industries and, through them, other media genres. In this regard, the physically rooted, place-based character of many environmental threats that inspire protest actions, including actions that make tactical use of new media, means that some environmental groups have come to appreciate the value of wooing tourism operators to their cause or attempting to counter the ability of the tourism sector to divert attention away from environmental conflict or damage that might diminish a destination's reputation as an attractive place to visit. This has resulted in examples of environmental groups and elite sources engaging with travel journalists in ways that are politically instrumental (McGaurr, 2010, 2012). While local and national journalism in democracies addresses audiences who may eventually be given an opportunity to change the government policies at issue through the electoral process, transnational journalism can bring political and economic pressure to bear on governments and businesses by drawing the attention of international markets to the conflict, with the potential for associated economic repercussions. In the case of travel journalism, these repercussions can, conceivably, extend beyond the tourism industry, because the genre's audience of corporeal and armchair travellers may include consumers of the destination's exports and potential investors in its enterprises. Campaigns that attract overseas media attention to local environmental disputes have the added benefit for movements of demonstrating that physically rooted conflicts are part of a global environmental endeavour (Hutchins and Lester, 2011).

This chapter moves beyond news journalism to examine the intersection of activism and travel journalism, which is an underexplored area in the existing literature. My theoretical starting point is Lovink's concept of tactical media as adapted by Lester and Hutchins (2009), who see value in exploring the broader strategic and tactical uses of new media in environmental conflict and regard many environmental NGOs as "worthwhile, albeit imperfect organizations" (2009: 582). Although focusing primarily on the actions of travel journalists themselves, each of the two examples I present in the case study analyzed later in this chapter begins with the deployment of a new-media tactic by environmental activists. In the first example, the original web protest emerges in a form that readily corresponds with Lovink's description of tactical media as anonymous outsider do-it-yourself new-media activism. In the second, the initial online campaign seeks legitimacy by publicizing the involvement of large, well-established NGOs, thereby making a virtue of the very "professionalism inside the office culture of these networked organizations" with which Lovink takes issue (2002: 260). By tracing the outcomes of these examples of new-media environmental activism as manifested in (or absent from) the traditional or new-media practice or texts of two travel journalists, I attempt to determine whether either represents a sustained challenge to "the many influences, pressures and constraints that encourage the media to gravitate towards the central orbit of power" (Curran, 2002: 148). In so doing, I ask more broadly whether new media can enhance the agency of travel journalists and, through them, the power of environmental groups, in the face of structural forces that favour elite sources. By structure I mean "the basic 'rules of the game' as they are understood by the various players" (Wolfsfeld, 1997: 13). By agency I mean the freedom of journalists writing for mainstream media organizations to "accept, reject, interpret, and modify structural rules" (Wolfsfeld, 1997: 14).

Gadi Wolfsfeld (1997) has described the battle for media access and meaning fought between elite and non-elite sources as part of a larger battle for political control. In a media landscape in which elite sources have privileged access to publicity, non-elites often find they must create newsworthy disruptions if they wish to attract the attention of journalists. Once access is achieved, the creation of news is driven by information and frames[2] (Wolfsfeld, 1997: 34). Although most of the media's frames are provided by elites, the media do retain a supply of anti-elite frames that they are willing to apply to stories involving non-elite sources who are able to provide them with appropriate information and take advantage of elite vulnerabilities (Wolfsfeld, 1997: 55). Ultimately, in the battle over access and control of meanings, the media may assume one of three roles: servants of the elites, semi-honest brokers or *"advocates of the underdog"* (Wolfsfeld, 1997: 69, original emphasis).

Government tourism offices are elite travel journalism sources that attempt to reduce their relationship with travel journalists to a simple exchange of services in which free transport, accommodation, information subsidies and/or logistical support are traded for positive publicity. In relationships between government tourism offices and travel journalists, there are structural inducements for the journalists to avoid referring to political conflicts governments do not wish to see reported. Firstly, as reviewers of tourism destinations, travel magazines and travel sections of newspapers are not just tacit allies of the tourism industry (Fürsich and Kavoori, 2001) but also active members dependent on its continued good health to maintain their supply of readers and advertisers. Secondly, much of the advertising upon which travel media rely comes from government tourism offices (Fürsich and Kavoori, 2001). Finally, although some of the larger United States newspapers and travel magazines, and some individual travel journalists, refuse free transport and accommodation, many still accept it (Austin, 1999), while in Britain and Australia an expectation that free visits will be available appears to be built into travel journalism culture (Hanusch, 2012; Seaton, n.d.). A 2009 survey of Australian travel journalists conducted by Folker Hanusch (2012) found that a large majority believed free travel was essential for their jobs. Even so, an overwhelming majority believed they would tell the truth even if this upset those providing the free travel. To the extent that it is possible for any journalist to tell the "truth" of a subjective experience, it may sometimes be possible for a travel journalist to accurately describe an individual tourism experience without reference to non-elite sources. In the case of destinations that build their tourism reputations on natural environments that NGOs and other activists consider threatened, however, these groups and individuals are likely to feel their views deserve to be reported in travel journalism to balance messages from elite sources who seek to shield journalists from evidence of damage or controversy.

This chapter uses the qualitative method of the case study in an attempt to generate "concrete, practical and context-driven knowledge" (Flyvbjerg, 2001: 70) informed by "a nuanced view of reality, including the view that human behaviour cannot be meaningfully understood as simply a set of rule-governed acts found at the lowest levels of the learning process, and in much theory" (Flyvbjerg, 2001: 72). As will emerge from my analysis, however, a case study can also reveal structural forces that can prevent environmental campaigns from reverberating more audibly through the travel journalism genre or decisively challenging the power of media institutions and elite sources. The examples discussed are taken from Australia's island state of Tasmania—a tourism destination where large-scale environmental conflict has been mediated for decades (Lester, 2007). Even as wilderness and the island's other natural attractions became the cornerstone of Tasmania's tourism brand (Department of Premier and Cabinet, 2006), fierce battles over the fate of its old-growth forests have commanded media attention locally, nationally and internationally. Those battles have continued into the current decade and have occasionally drawn the attention of transnational travel journalists. My case study is informed by the published articles of two such international travel journalists who have written about Tasmania's environmental conflicts, but also by news reports, blogs, websites and government publications, and by semi-structured interviews with the journalists in question, other travel journalists, government tourism public relations practitioners, tourism operators and other significant players.

TRAVEL JOURNALISM MEETS ENVIRONMENTAL CONFLICT IN TASMANIA

In 1995 Tasmania adopted the logo "Discover your natural state" for all its interstate marketing (Tasmanian Department of Tourism, Sport and Recreation, 1995: 8). Although other logos have since featured in its marketing, "Australia's natural state" was still appearing on websites and brochures produced by Tourism Tasmania for its overseas markets well into the first decade of this century (Department of Premier and Cabinet, 2006; see, for example, Tourism Tasmania, 2008, and Tourism Tasmania North America, 2008). Both of the travel journalists in this case study noted the logo in articles they published in 2008 (Greenwald, 2008; Miles, 2008b), and research conducted in 2010 for the state government's tourism authority, Tourism Tasmania, found "the greatest trigger to influence intention to visit Tasmania was Wilderness" (Tourism Tasmania, 2011: 6). While tourism now surpasses forestry in terms of its contribution to Tasmania's gross state product and employment (Department of Economic Development, Tourism and the Arts, n. d.), forestry has traditionally been an important source of jobs in a community suffering from persistently high levels of poverty. The government business enterprise Forestry Tasmania manages just under 40 percent of the state's forests, of which it says approximately half can be harvested (Forestry Tasmania, n. d.).

As a government authority, Tourism Tasmania is required to "market Tasmania as a desirable tourist destination" (Tourism Tasmania Act 1996), and its Visiting Journalist Program has been a powerful tool by which the government has sought to gain media endorsement for its framing of Tasmania as "natural". International travel journalists hosted by the program are generally given free transport and accommodation, an itinerary and a Tourism Tasmania guide, and each visit is expected to "produce a piece that, in print or broadcast media, positively profiles Tasmania to potential visitors" (Tourism Tasmania, 2012). As successive Tourism Tasmania annual reports record, this program has enjoyed considerable success over many years in attracting travel journalists writing for prestigious British and United States travel publications. In the first decade of the 2000s, however, environmental groups and individual conservationists began to challenge Tourism Tasmania's framing of the island as "natural" in ways that captured the attention of some of the same hosted journalists.

THE LEGACY OF THE STYX

In November 2002 Australian NGO the Wilderness Society, together with the World Wildlife Fund and Planet Ark, bought billboard space in Sydney Airport and erected a billboard juxtaposing an image of Styx Valley forest, due to be logged in 2003–2004, with an image of forest that had been cleared and burnt. The billboard headline "Discover Tasmania before 2003" repurposed words that are part of the URL of Tourism Tasmania's holiday website www.discovertasmania.com.au. The airport advertising agency removed the billboard after one day, saying it had not been approved in advance by the airline Qantas. A report in Tasmania's biggest newspaper, the *Mercury* (Ribbon, 2002), suggested the state's premier and minister responsible for Tourism Tasmania, Jim Bacon, had pressured Qantas to remove the billboard. Qantas denied this, but Bacon publicly attacked the environmental groups who had erected it and claimed there was "no truth in the statement that Tasmania's forests are threatened in the way that the last of our forests are about to disappear" (Ribbon, 2002: 3). Two weeks later, an anti-forestry website with the URL www.discover-tasmania.com was posted anonymously. When the website owner was exposed a week later, the *Mercury* identified him as a former Tasmanian tourism operator, Gordon Craven, who said he was acting alone and had set up the site primarily to protest against removal of the airport billboard and to highlight the cruelty of the forestry industry's use of 1080 poison to kill animals that browsed on seedlings (Bailey, 2002: 3). Tourism Tasmania's chief executive at the time was reported by the

Mercury as labelling the website "a 'direct hit' on the state's reputation which threatened visitor numbers and could have a multi-million-dollar impact" (Bailey, 2002: 3). In the same article, Craven was quoted as saying that he "set up the site to shame the Government and Forestry Tasmania who use tourism, and the tourist operators who just stand by and let it happen" (Bailey, 2002: 3). In a related dispute about Craven's registration of the associated domain name www.discover-tasmania.com.au for a tourism website that linked through to his original site, the World Intellectual Property Organization (2003) ruled against Tourism Tasmania. In later years, one of the forestry conflicts publicized by Craven was a campaign against plans by major Forestry Tasmania customer Gunns Ltd to build a AUD $1.9 billion pulp mill in the state's premier wine tourism area, the Tamar Valley—a dispute that also captured the attention of British travel journalist Paul Miles.

Miles is an environmentally and socially conscious freelance journalist who lists the NGO Tourism Concern as one of his non-media clients (Miles, n.d.). Prior to visiting Tasmania late in 2007, he had been aware of its forestry conflicts—so much so that until then he had avoided writing about the island (Miles, 9 March 2009, interview with author). After learning of the pulp mill dispute, however, he decided to tour Tasmania as a hosted and guided guest of the Visiting Journalist Program, [3] thereby acknowledging the appeal of travel journalism as a vehicle for raising concerns about environmental issues. In his words, "I thought it might be quite good if I could try and write about the pulp mill and the controversy around that as well as writing about the good things that are happening in the state" (Miles, 9 March 2009, interview with author). The extent of the value tourism destinations place on coverage in prestigious, high-circulation publications—in this case, the British edition of *Conde Nast Traveller* and London's *Financial Times*—is evident in Tourism Tasmania's decision to host Miles in spite of his admission that he intended to investigate the pulp mill dispute. Nevertheless, the journalist felt his candour made the authority nervous. As he described it, "I think that because the tourism people were possibly anxious about what I might see or do they pretty much kept me as busy as possible in the time that I was there" (Miles, 9 March 2009, interview with author).

Having secured editorial agreement from *Conde Nast Traveller* to cover the dispute, Miles organized an alternative guide to show him around the proposed pulp mill site on the one free day his Tourism Tasmania itinerary allowed him (Miles, 9 March 2009, interview with author). The resulting article (2008b) was written in the third-person style of news journalism, favoured the frames of those opposed to the mill and included the counter-arguments of Gunns. Miles' feature for the *Financial Times* (2008a), by contrast, was in the first-person style of a traditional travel journalism narrative and covered much of his Tourism Tasmania itinerary. In the draft he submitted to the newspaper, he also described the process of clear-felling and burning associated with old-growth logging and explained the potential effects of the pulp mill on the island's wine and food industries.[4] The newspaper cut this part of the story before publication, which Miles saw as demonstrating that power in commercial travel journalism publishing resides with editors and publishers (Miles, 9 March 2009, interview with author). Miles, however, engaged in a form of "hit-and-run" (Meikle, 2002: 119) tactical media that briefly challenged such power. Beneath the URL of Tourism Tasmania's website for tourists noted after his article, he listed the URL of the tourism anti-forestry website www.discover-tasmania.com.au (Miles, 2008a). As Miles recalled in 2009: "I listed that and they included it in the feature.... That's as close as I've got really, in the *FT*. I think that was kind of a little bit subversive actually. I think they just assumed it was a link to the Tourism [Tasmania] website" (Miles, 9 March 2009, interview with author). Yet, although Miles was able to challenge the newspaper's power briefly by including this URL, on this occasion it does not appear to have shamed the Tasmanian government. Moreover, the structural forces that generally privilege the messages of elite sources in travel journalism were not altered by this gesture. Commenting almost a year after the article appeared, Miles remained despondent about the power imbalance in his relationship with editors:

> Sometimes I don't even know if/when a feature is going to appear. Subs and editors seem to go ahead and do their own thing. It is rare for them to bother to check with writers that we agree with any changes, but sometimes, thankfully, they do. Occasionally I give some feedback, but it doesn't achieve much and as far as the readers are concerned, it's already too late. (Miles, 10 March 2009, personal correspondence)

UNDER THE RADAR IN THE TARKINE

In February 2005, shortly after an Australian election in which Tasmania's forest disputes had featured prominently, the United States NGO Ethical Traveler mounted an online letter-writing campaign in association with the Wilderness Society (Ethical Traveler, 2012). The campaign asked members to urge the Australian Government to hold firm on all of its election promises in relation to the protection of Tasmania's old-growth forests. In reporting Ethical Traveler's actions, a related publication, *Earth Island Journal*, noted the importance of tourism to Tasmania's economy and observed that "the partner groups view their joint environmentalist-traveler campaign as a logical next step in protecting these extraordinary forests" (McColl, 2005). *Earth Island Journal* ended by stating that Ethical Traveler's executive director, Jeff Greenwald, had given assurances that Ethical Traveler would "monitor" the situation to ensure the Australian government held to its promise (McColl, 2005).

Jeff Greenwald is also a noted travel journalist and travel book author. In recounting a 2007 tour of Tasmania as a guest of Tourism Tasmania, he said he'd arrived with few preconceptions and had been only superficially aware of Ethical Traveler's 2005 campaign in conjunction with the Wilderness Society. While in Tasmania, however, Greenwald had also made contact with people to whom, as he explained, Tourism Tasmania would probably rather he had not spoken (Greenwald, 6 March 2009, interview with author). The resulting article published in *Islands* magazine in 2008 duly praised a variety of tourist attractions and described the destination as a "global treasure," but it also deployed Tasmanian conservationists' framing of old-growth logging as a warlike conflagration. In addition, the article called on the Australian Government to make "wise choices" for Tasmania's forests, but it did not mention Greenwald's association with Ethical Traveler.

Greenwald maintains a personal blog, but he did not use it while in Tasmania in 2007; nor did he blog for Ethical Traveler during this visit, although he had posted blogs on that site before and would do so again. His decision not to blog is especially interesting in view of his having created the first international blog (Greenwald, n.d.), which he published on what is claimed to be the world's first commercial online publication, *Global Network Navigator* (O'Reilly and Associates Inc., 1993), while circling the globe by land in 1994. He has described the experience of blogging at that time as giving him a sense of being "almost on fire, that the excitement and heat of my journey was something I could broadcast in no time at all. It was a very giddy feeling" (Greenwald in Shapiro, 2004: 250). Despite this initial intoxication with online media, however, more than a decade later he was apologizing on his blog (Greenwald, n.d.) for his tardiness in failing to blog while in Tasmania, and observing that he didn't like blogging while on assignment. At the very least, this comment privileges traditional media assignments over the journalist's commitment to readers of his blog.

Greenwald expressed no sense of obligation to Tourism Tasmania for hosting him in 2007 (Greenwald, 6 March 2009, interview with author). Rather, when discussing the kinds of articles he wrote, he spoke in terms of his publishers' expectations. *Islands*, he said, had wanted a traditional atmospheric travel narrative, whereas *Afar*, which funded a second trip to Tasmania inspired by the first, wanted something far more people-focussed (Greenwald, 6 March 2009, interview with author). The extent to which Greenwald was prepared to write to direction is evident in the following interview extract, which refers to a dispute over proposals by Forestry Tasmania to pave a section of road through an area of rainforest in the Tarkine to create a "tourist loop":

> [O]ne of the reasons I was so keen to come back was I really wanted to be able to focus my attention on the political conflict unfolding in western Tasmania. Now, though I was

> given that mandate by the magazine *Afar*, it was also made clear to me that it wasn't to be a good-versus-evil story. *Afar* magazine wanted to see the human side of the issue from both sides and as a result we spoke to people who were passionately committed to their point of view on both sides of the forestry argument and the Tarkine road argument. So, you know, there was no—... Though I may have my own prejudice in terms of what I'd like to see done to the area, the function of my story is not to editorialize but to just present how different people who grew up in the same part of Tasmania can have such different views about how to use the land and what the future of that forest should be. (Greenwald, 6 March 2009, interview with author)

Just as the value of international publicity had been sufficiently seductive to persuade Tourism Tasmania to host Miles in 2007 despite his stated interest in the pulp mill, it was attractive enough to elicit an unusual degree of cooperation with Greenwald from the Tarkine National Coalition environmental group in 2009, as its campaigner Scott Jordan explained:

> [W]e gave him some names of people that we perhaps under normal circumstances wouldn't have told the media to go off and get comment from these people. But his project was perhaps a bit different from what we'd normally do in media so we tried to assist him where we could. ... [H]opefully, if we generate some passion in people who are overseas reading that article [they might] then join the campaign in some way and contribute to letting our government know that this area does have international recognition and is worthy of protection. (Jordan, 18 June 2009, interview with author)

When Greenwald's *Afar* article was published in the United States in 2010, almost a year had passed since the journalist's second visit to the island. From Jordan's perspective, the article had some value in bringing the dispute to international attention but would have served the interests of his NGO better if it had been published in 2009 as originally promised, had devoted more space to the tourist loop issue, and had quoted Jordan himself (Jordan, 4 March 2010, personal correspondence). Apart from its high-impact headline, "Bedeviled Island", it was less overt in its political subjectivity than the *Islands* article had been. In November 2010 *Afar* won the top award in the magazine category in one of the United States' most prestigious travel writing competitions, the Lowell Thomas Travel Journalism Awards, run by the Society of American Travel Writers, and "Bedeviled Island" won the top award in the category of environmental tourism article (Society of American Travel Writers Foundation, n.d.). In this second Greenwald example, then, journalism took precedence over activism in relation to coverage of Tasmania's forestry debates.

CONCLUSION

In the Australian summer of 2007–2008 Miles and Greenwald had personal commitments to environmentalism and connections with groups whose campaigning for ethical tourism sometimes includes environmental issues. In both cases, past uses of new media in environmental protests contributed to their ability to mount insider challenges to the definitional advantages enjoyed by elite tourism sources. Indeed, there was an element of Lovink-style tactical media in Miles' hit-and-run inclusion of one of Craven's URLs in his *Financial Times* article. By contrast, Ethical Traveler's online campaign in 2005 was openly networked, but it nevertheless saw its message delivered in a mainstream travel article that did not acknowledge its author's association with the group, thereby turning the tables on government tourism organizations that thrive on positive publicity from hosted travel journalists who neglect to inform readers of their assistance (Gartner, 1993). However, neither the *Financial Times* nor the *Islands* example represents a sustained assault on media power, as evidenced by Miles' continued lack of control over his published texts and Greenwald's willingness to meet *Afar*'s editorial requirements. Moreover, although personal blogs offer travel journalists the ability to circumvent the editorial control exerted by media institutions, as yet they rarely provide an income or the symbolic capital afforded by mainstream publication. In these circumstances, it seems unlikely that the more direct access to readers and the tactical opportunities offered by online technology will be exploited sufficiently by travel journalists who sympathize with environmental groups in dispute with the governments of tourist destinations, thereby gaining sustained definitional power in the genre as a result. But Miles' article in *Conde Nast Traveller* and Greenwald's article in *Afar* suggest another possibility—a future, perhaps, in which mainstream travel publications more routinely put journalism ahead of their relationships with elite sources in order to establish a level of credibility and reader interest capable of commanding a print or online subscription price high enough to sustain the business in an era of falling advertising. In such travel journalism, the politics of the environment may one day command greater attention beside representations of its tourism appeal. As Curran observes, "[t]he media can persuade, change and mobilize. However, the principal way in which the media influence the public is not through campaigning and overt persuasion but through routine representations of reality" (Curran, 2002: 165).

ACKNOWLEDGMENT

The author worked for Tourism Tasmania from 2002–2008.

ENDNOTES

1 Small sections of this chapter have appeared in the following journal articles: McGaurr, L. (2010). Travel journalism and environmental conflict: A cosmopolitan perspective. *Journalism Studies*, 11(1), 50–67; McGaurr, L. (2012). The devil may care: Travel journalism, cosmopolitan concern, politics and the brand. *Journalism Practice*, 6(1), 42–58.

2 Wolfsfeld adopts Gamson's definition of a frame as "[a] central organizing idea for making sense of relevant events and suggesting what is at issue" (Gamson in Wolfsfeld, 1997: 31).

3 Miles travelled to Australia with Tourism Australia and paid for his own airfares from the mainland to Tasmania and back.

4 Un-subbed version of Miles, 2008a, provided by Miles in personal correspondence, 9 March 2009.

CHAPTER SEVEN

Contesting Extractivism

Media and Environmental Citizenship in Latin America

SILVIO WAISBORD

Since the rise of European colonialism, Latin America has been a major global source of natural resources. The voracious push for natural commodities resulted in the predatory exploitation of minerals such as gold, silver, tin, copper, and other resources over the centuries. The region's history cannot be understood outside the evolution of the extractive economy. The complex development of the extractive industries shaped political, economic, and social structures, particularly in the Andean countries (Chile, Bolivia and Peru). Aside from a legacy of violence and social exclusion, large-scale exploitation left untold environmental destruction.

This historical pattern of intensive exploitation of natural resources and environmental degradation continues, as Latin America remains a prime destination for extractive industries. During the past two decades, global demand for extractive goods, particularly metallic and non-metallic minerals, have led to a remarkable expansion of mining, oil and gas drilling, and forestry across the region. In 2011 the continent attracted 28 percent of total global mining investments and was the target of more large projects (above $1 billion) than any other region in the world (Ericsson and Larsson, 2012). The production of minerals such as gold and silver reached unprecedented volumes in the past decade. The export of extractive commodities is a significant percentage of the total of export earnings. In Peru, for example, mining exports totaled 59 percent of the country's foreign earnings (U.S.A. Department of Commerce, 2010).

This recent boom has been the result of the combination of three main factors. First, during the 1990s, free-market administrations passed policies to favor foreign investments in the extractive sectors. Premised on the notion that the extractive industries offered opportunities for development, they assiduously courted global companies by passing incentives, such as royalty schemes and tax breaks, to attract investments. These incentives have remained in place, even as several countries have shifted to the left since 2000. As many critics assert, the coming of left-center and populist administrations has not fundamentally changed the business climate in favor of global extractive industries (Svampa and Antonelli, 2010). Furthermore, "progressive" presidents and governors have enthusiastically championed the model of "extraction-as-development", leaving untouched conservative reforms to attract foreign investments and passing legislation to promote commodity extraction and export (Burdick, Oxhorn and Roberts, 2009). There have been significant continuities between the extractive policies of conservative and populist administrations. For example, President Cristina Fernandez de Kirchner, who has championed redistributive and progressive human rights policies, vetoed a law proposal to protect glaciers in the Andes mountain range that would have limited mining and oil production. Her decision undoubtedly favored the business plans of Canada-based Barrick Gold for a massive gold mine on the Argentine-Chilean border. Local communities that depend on glacier water for farming roundly criticized the decision.

Second, a tradition of weak environmental regulations and poor enforcement of existing laws has made the region attractive for global extractive companies. The extractive sector in the region has long been characterized by not only abysmal labor conditions, but also industrial practices that are banned in the United States, Canada, and other countries for their negative environmental impact. The region has hardly been a model of well-managed extractive resources and environmentally sound projects.

Third, Latin America's poor governance and political stability have also offered incentives for global extractive companies. The region's relative lack of internal conflicts that may endanger multimillion-dollar projects appeals to global capital's search for predictable political conditions. Chronic governance problems reflected in the weakness of accountability mechanisms, widespread corruption, and the absence of participatory mechanisms favor the rapid implementation of business projects. In some countries, large extractive permits have been granted without regard for "prior consultation" laws that require extensive citizen participation in such type of ventures. The fact that governments quickly awarded concessions to companies in past years has raised questions about the transparency of the process and the accountability of public officials.

As the number of domestic and global investments grew, environmental justice movements sprung up throughout Latin America (Dougherty, 2011; Giarracca, 2007). The potential and real negative impact of extractive industries on water sources and livelihoods has been a matter of concern across communities. Citizens' observatories and industry sources have documented a steady increase in citizens' actions against extractive projects through public rallies, road blockades, and formal requests. In many communities, citizens' assemblies have sprung up to make demands about environmental and other issues. Popular protests against extractive companies and governments have become common across the region.

Citizens' negative reactions to projects with massive environmental consequences are not unexpected if we consider that "where there is extraction there is contestation" (Bebbington, 2009). This is particularly true in the context of large-scale ventures rife with opacity, lack of public participation, and connivance between industrial and political interests. Constant promises of egalitarian development and a bright future filled with jobs and social services sound hollow in political-social contexts filled with public distrust of official voices and a history of extractive economies with a poor environmental record. Furthermore, citizens' activism is a likely response in contexts plagued by the absence of the State, as demonstrated by the weak enforcement of regulations and a tradition of patrimonialism and mismanagement.

Citizens' actions have crystallized in a budding environmental movement. Although the region has long had a history of environmental and conservationist groups, the massive scale of industrial projects has reinvigorated environmental activism. Citizen's mobilization has been central to the implementation of projects such as building hydropower plants, mining, gas and oil exploitation, and deforestation. One factor that contributed to popular resistance is that the recent extractive rush has expanded into agricultural communities that had no experience with mineral extraction and forestry. For communities unaccustomed to the consequences of extractive industries, new projects raised concerns about the environmental impact on traditional livelihoods and water sources. Local mobilization has been articulated as "environmental citizenship" (Latta and Wittman, 2012) pursuing rights and justice in three areas: the recognition of demands and interests that don't match the narrow interests of business and political officials; the participation of local citizens and institutions in a range of decisions affecting the fate of extractive industries; and the social distribution of the benefits of economic development (Schlosberg, 2007; Carruthers, 2008). Among other demands, citizens have raised concerns about the environmental burdens of extractive industries such as contamination, water pollution, deforestation, and land degradation (Boelens et al., 2011).

MEDIA AND ENVIRONMENTAL CITIZENSHIP

Citizens' mobilization around the environmental impact of extractive industries in Latin America brings up interesting questions for media scholarship concerning, for example, the interaction between local and global forces in environmental media politics; the politicization of the environment through communicative actions; the tension between left-wing populism's defense of progressive media policies and resolute defense of media politics in favor of extractive companies; the role of old and new media in environmental citizenship; and the clash between citizens' broad view of development and social good and the developmentalist rhetoric of extractive industries and governments.

My interest in this chapter is to discuss the uses of the media by environmental movements mobilized against extractive projects. As indicated by many studies, the media play a central role in the social definition of environmental risk (Allan, Adam, and Carter, 2000; Anderson, 1997). The media play two roles in the coverage of social movements and the environment: they frame the issues and actors, and provide a battleground for various interests and sources to compete over the social definition of problems (Gamson and Meyer, 1996; Hansen, 1991). Stakeholders rely on various sourcing strategies to get media attention, including public relations and the staging of news events (e.g., press conferences, public rallies, parades). While government and business usually resort to standard public relations routines (e.g., advertising, press releases), citizens' movements typically use strategies of "resource-poor" actors (e.g., street theater, low-budget media). Also, editorial positions as well as journalistic norms and practices shape news coverage.

Recent experiences of citizens' mobilization against extractive projects in Latin America show two patterns. One pattern shows citizens' opposition against projects supported by large corporations and state officials. This pattern corresponds to well-known cases of social movements situated in a disadvantageous position vis-à-vis mainstream news organizations (Gamson and Wolfsfeld, 1993; Gitlin, 1980). In these cases, the press consistently misrepresents mobilized groups that challenge powerful political and industrial interests. The news media frequently emphasizes violent and radical aspects, distorts demands, and focuses on stereotypes. While some analysts attribute such biases to the fact that media companies largely express the interests of a particular socio-economic order, others conclude that distortions are the result of the reliance on political and business sources coupled with the difficulties that social movements have in gaining access to newsrooms (Lester and Hutchins, 2009). Given these conditons, enviromental movements rely primarily on their own media to express demands and promote civic actions. The majority of environmental activism in Latin America fits this pattern.

The other pattern shows that the mainstream media is not firmly and consistently opposed to mobilized citizens. News coverage is not uniformly contrary to local interests, nor does it fall into the typical problems of stigmatization and exclusion of local voices. This pattern is demonstrated by the *asambleismo*, a social movement born in Gualeguaychú, Argentina that emerged in opposition to the establishment of pulp mill plants on the Uruguay River between Argentina and Uruguay.

What explains these two different patterns of news coverage in the mainstream media—one of opposition and distortion of citizens' movements, and the other of favorable coverage of popular protest? Because there's no single relationship between civic society, media, and the state around environmental conflict, there is no single pattern of media use and news coverage. When popular insurgency confronts political and corporate interests that are supported by the media, the coverage is, expectedly, critical of citizens' actions. In these cases, citizens rely primarily on their own media—so-called alternative media—to voice interests and make demands. By contrast, when the mainstream media lacks corporate links to extractive industries and opposes government officials who are supportive of extractive projects, or prominent officials support environmental activism, then the coverage is different. Therefore, we should not assume that mainstream media news is consistently negative. Instead, it is necessary to study news coverage by considering the relation between media companies and governments as well as the position of official newsmakers vis-à-vis environmental protests.

WATER OVER GOLD: MEDIA AND RESISTANCE AGAINST EXTRACTIVE PROJECTS

The first pattern—the uses of alternative media by citizens to express demands against corporations, governments, and the mainstream media—is demonstrated by numerous conflicts triggered by extractive projects throughout Latin America. In the past decade there has been a surge of citizens' resistance against extractive industries. In many instances, opposition resulted in violent clashes between citizens and the police. In Peru, for example, the government decreed a state of emergency after demonstrations ended with several deaths in 2011 and 2012. The demonstrations were organized against a controversial $5 billion project led by US-based Newmont Mining in the northern part of the country, with technical assistance from the World Bank.

Central to citizens' mobilization were concerns about the impact of gold mining on water supplies. Although conflicts have erupted around various projects, they have been particularly visible in cases such as open-pit mining, which relies on water, cyanide, and explosives. This type of extraction has major environmental implications, for it requires massive amounts of fresh water and electricity, the destruction of vast forests, and the use of toxic substances. The rallying cry "water is more precious than gold" (and other minerals) has been common from Peru to Argentina. Concerns about the impact of extractive projects on environmental health and livelihoods have also been at the forefront of local citizens' assemblies. Mineral and forest exploitation not only damage the soil and water supplies, they also create mosquito-breeding sites in large craters that increase malaria and dengue transmission. In some cases, strong resistance caused governments to backpedal and cancel extractive projects (Bebbington et al., 2008).

> Citizens' activism has given voice to alternative notions of development (Carruthers and Rodriguez, 2009). It has directly challenged the idea of development endorsed by presidents, governors, and companies for whom development means jobs and social improvement. This "developmentalist" vision is embedded in the modernist aspiration that narrowly equates economic/industrial growth with social progress. In contrast, citizens' assemblies have counter-offered a holistic model of human development that prioritizes participatory rights and community well-being. They have chipped away at common truisms in official narratives that portray extraction as the harbinger of "progress," "jobs," and "economic development" by questioning the actual number of jobs created, the unequal distribution of earnings, and the environmental devastation caused by specific types of exploitation (Urkidi, 2010). Environmental citizenship follows a long-standing Latin American tradition that contests the conventional meanings of development and proposes a comprehensive notion of human development as a local, participatory process that prioritizes community needs, including living in harmony with the environment (Walter and Martinez-Alier, 2010).

Local actions have used traditional and new forms of communication. In the cases of indigenous communities, several of whom are affected by the growing footprint of extractive industries, they have tapped into ancient communicative traditions linked to community work ("minga" or "minka" in Kichwa) to bring out concerns and identify collective solutions. Community radio stations owned by communities, local associations, and churches have also served as conduits for dialogue and participation (Acevedo Rojas, 2010; Herrera Huerfano, 2011). Digital platforms such as blogs and "social media" have been used to provide updates and organize meetings and protests. These media have provided opportunities for cultivating notions of development that foreground a conception of the uses of land and water that contradicts the industrial, modernist, extractivist view of companies and goverments (Walter and Martinez-Alier, 2010). Also, the uses of "movement media" follow the pioneering tradition of miners' radio in Bolivia which, starting over half a century ago, served as a channel for the articulation of rights-based discourses around labor conditions, social benefits, and commodities.

Citizens' actions have sporadically received attention from the mainstream press. News coverage followed expected patterns, amply documented in the literature on media and social movements. When citizens' actions offered "newsworthy" elements such as massive protests, conflict, and the occupation of public spaces, they were covered in biased, stigmatizing ways. Coverage focused on violence and paid scant attention to demands and alternative narratives of development. It also reproduced standard inequalities in news-making, namely, the unmatched ability of official sources to determine news (Hopke, 2012).

PULP FRICTION

The second pattern of news coverage of enviromental activism is demonstrated by how the mainstream media covered political mobilization against the establishment of two paper mill plants in the Rio Uruguay in Argentina over the past decade. The movement was a turning point in the historical trajectory of environmental movements in Argentina, a country with weak environmental policies, a weak tradition of environmental reporting, and weak environmental mobilization (Brebbia, Conti, and Tiezzi, 2007; López Echagüe, 2006).[1] The establishment of the citizens' assembly of Gualeguaychú, the city that has been the center of continous protests against the pulp mills, represents an unusual instance of "reflexive modernization" (Beck, 1992) in a domestic landscape in which environmental issues have been largely absent in the reporting of the mainstream media and the discourses of the domestic political elite.

The rise of the *asambleismo* movement was triggered by the decision of two European firms to build two pulp mills on the Uruguayan coast of the Uruguay River, which provides a natural border limit with Argentina.[2] The news generated resistance in Argentine towns opposite the Uruguayan coastline where the mills were to be located. Previously, Uruguayan environmental groups had also protested the decision of their government to authorize the establishment of the pulp mills, and exchanged information on the whereabouts of the planned construction with Argentine groups.[3] Yet, the dynamics that the conflict assumed prevented the formation of a transnational environmental coalition. While the protest initially appeared to be framed as a purely environmental concern, the framing of the issue gradually shifted from an environmental to a nationalistic issue. Such a shift can be attributed in part to the specific repertoire of collective action of the *asambleismo*. It prioritized roadblocks on the international bridges that connect Argentina with Uruguay to gain media and political attention and to exert a major cost on the Uruguayan economy, particularly the tourist industry. The strategy, however, generated widespread resentment among Uruguayans against what was perceived as an unjustifiable and illegal form of protest. This context prevented the formation of a coordinated cross-national movement against the installation of the pulp mills. In the absence of a strong bi-national coalition, the city of Gualeguaychú became the epicenter of the protests. Gualeguaychú, a coastal town of approximately 80,000 in the province of Entre Ríos, witnessed the formation of a massive local movement that actively opposed the installation of paper mills on the other side of the river.[4]

By 2003 self-mobilized neighbors had created the *Asamblea Ciudadana Ambiental Gualeguaychú* (the "Asamblea"). In September 2003 they drafted the "Gualeguaychú Proclamation," which was submitted to the authorities of Río Negro, the Uruguayan state where the mills would be located, expressing their opposition to the installation of the pulp mills. In the same month they also conducted the first blockade of one of the bridges that links the two countries. While the protesters would also resort to street theater, local mobilizations, and caravans, the blockades gave the movement the highest visibility and media attention, and subsequently became the focus of diplomatic debates between Argentina and Uruguay. The movement drew massive support from Gualeguaychú. The local population rallied behind the demands of the Asamblea. The Asamblea's slogan "No to the paper mills, yes to life" quickly became the majoritarian expression of the people of Gualeguaychú. Bumper stickers with the slogan became common in most homes, stores, and vehicles. Local journalists and politicians joined the movement and mirrored its intransigent position in articles and speeches. The town's mayor participated in many of the mobilizations, such as the caravan that took place in October 2003. Governor Jorge Busti was another strong supporter of the movement's claims, and his discourse adopted the radical language of the *asambleístas*. Busti sent complaints to the Inter-American Human Rights Commission, arguing that the Uruguayan government was violating "the right to life" of the people of Gualeguaychú, and also complained to the International Finance Corporation and the office of the World Bank, which was assessing financial support for the projects.[5]

The local environmental movement soon assumed national notoriety, turning what initially was a localized conflict into a major issue of the national political agenda (and a diplomatic conflict as well). In February 2004 the Asamblea brought its claims to the attention of the Kirchner administration. The initial response of the administration was to establish a technical body to analyze the situation. In May 2005 Presidents Néstor Kirchner (Argentina) and Tabaré Vázquez (Uruguay) established a binational commission to evaluate the environmental impact of the paper mill projects. The commission ran until January 2006, but it failed to produce an agreement. The environmental assembly of Gualeguaychú responded by staging a blockade that lasted throughout the summer season and into mid-May of that year.

As the conflict escalated and became nationalized, the Kirchner administration openly sided with the assembly. The intensity of the conflict and the activism of

Gualeguaychú's Asamblea represented a major challenge to the government, which was extremely sensitive to any form of popular protest. The environmental movement brought to the fore a popular assembly, a form of social organization that had acquired visibility and presence in the aftermath of the 2001–02 protests that ended the De la Rúa presidency. In the summer of 2001–02, the major urban centers of Argentina witnessed the formation of spontaneous mobilizations against the government. The movement, which became known as the *cacerolazo* ("pot-banging"), forced the resignation of President De la Rúa, after which neighbors began to organize in "popular assemblies." The assemblies were characterized by their horizontal structure and strong anti-political discourse that demanded the resignation of all elected officials. The presence of the assemblies added a troublesome presence to an already convoluted social landscape where numerous groups, most visibly the organizations of the unemployed, engaged in regular street demonstrations. The eventual threat of a potentially destabilizing protest was a major concern for the administration of Eduardo Duhalde, who was forced to bring forward the calling of elections due to the death of two protesters in 2002, and for his successor, Néstor Kirchner, who was elected in 2003 (Peruzzotti, 2005).

From the beginning, Kirchner showed special concern for the political costs of uncontrolled protest movements. Upon taking office, Kirchner engaged in a series of maneuvers aimed at preventing any challenge from civic actors. His strategy was twofold. First, the administration acted to weaken the network of watchdog journalism and civic organizations that in the 1990s had played an active role in exposing governmental wrongdoing (Peruzzotti and Smulovitz, 2006; Waisbord, 2000). A crucial piece of this strategy was to co-opt and divide the network of watchdog NGOs, particularly human rights organizations, through his active endorsement of retributive politics toward past human rights abuses. The government also devoted substantial energies to controlling the media agenda to prevent the string of media exposés that had shaken Argentine politics in the 1990s. Yet, it was actors such as the environmental assembly that posed the major concern to the administration, given their unpredictable nature and intransigent positions. During its early years, the administration was very successful at controlling the public agenda and defusing social protest, particularly the movement of the unemployed. In fact, in its first years the administration enjoyed a period of relative social calm. Aside from the movement of the unemployed, which has became a constant presence in the Argentine public space of the last decade, the administration has faced only a handful of protest movements. The environmental conflict, therefore, emerged as an undesirable political challenge for the administration, one that rapidly reached national proportions and put the Kirchner presidency on the defensive. In this sense, what was at first a local environmental claim came to be perceived both by Kirchner and his opponents as one of the administration's most difficult political challenges coming from civil society. As it had done previously with similar protest movements, the Kirchner administration avoided confrontation with the Asamblea and openly supported its demands. In so doing, it encouraged the nationalistic discourse to direct civic anger and demands from domestic politics to the Uruguayan government. Foreign Affairs Secretary Rafael Bielsa met with the Assembly and praised the people of Gualeguaychú for their civic organization and courage. Bielsa proclaimed that the issue was now a "national cause" (Aboud and Museri, 2007). After a failed accord between Presidents Kirchner and Vazquez to suspend the construction for ninety days, the Argentine government brought the issue to the International Court of Justice at The Hague in May 2006. The presentation was preceded by a massive action in Gualeguaychú presided over by Kirchner and attended by nineteen governors. Romina Picolotti, a legal counselor for the Asamblea who had intervened on behalf of the Argentina government before the International Court of Justice, was appointed Undersecretary of Environment. The Court's decision favored Uruguay. The escalation of the blockades (which occurred even in Buenos Aires, preventing tourists from boarding ferries to Uruguay) led to the Uruguayan government submitting a complaint to the International Court of Justice, which this time ruled in favor of Argentina. The Court argued that the blockades did not irreparably damage the Uruguayan economy, nor did they prevent the construction of the paper mills. Encouraged by the Court's decision, the protesters continued with the blockades throughout the summer.

The internationalization of the conflict and the polarization of the discourse along nationalistic lines contributed to neutralizing the domestic costs that the conflict could have inflicted on the administration. By redirecting the claims to the Uruguayan administration, Kirchner avoided being the target of the protest. Instead, he positioned himself on the side of the protesters, defining the conflict as a "national cause." This framing of the conflict went hand in hand with the nationalist image that the president had carefully crafted when dealing with the debt crisis. During negotiations with the International Monetary Fund and other financial institutions, Kirchner presented himself as a staunch defender of national interests against the questionable demands that financial institutions were trying to impose on Argentina. For Gualeguaychú's Asamblea, however, the discursive shift had ambiguous consequences.

GUALEGUAYCHÚ'S ASAMBLEA AND MEDIA ADVOCACY

The *asambleismo* faced the same communication challenges that social movements typically confront: how to get the message out; how to persuade news organizations aligned with dominant interests to define the issues fairly; and how to compete with governments and business, which have stable and well-funded mechanisms to affect news coverage (Hutchins and Lester, 2006). Unlike other protest movements that refrain from engaging with the mainstream media, the *asambleismo* approached the media as an indispensable ally. A *Clarín* journalist who has extensively covered the story states, "the Asamblea was keenly aware of the significance of media coverage, especially from journalists and pundits who were sympathetic to its demands" (Toller, 2007). Not surprisingly, the *asambleismo* resorted to street protest to get media publicity. Cognizant of the media's penchant for conflictive and dramatic events, it has staged a variety of public actions to attract media attention such as demonstrations, rallies, parades, and appearances during Gualeguaychú's popular carnival. Media attention has given the movement almost uninterrupted publicity.

The most controversial component of the repertoire was the blockade of roads and bridges between Argentina and Uruguay. The blockades were not simply symbolic moments of public presence, they were also acts of civil disobedience. The Asamblea has organized short-term blockades (particularly during summer vacations and long weekends when large numbers of Argentine tourists cross the border) as well as longer blockades lasting several months at a time. Blockades have a significant economic impact because they prevent the normal flow of commercial traffic in main routes of the Mercosur. Although such forms of occupation of public spaces tap into a long tradition of political street theater in Argentina, they also resemble the recent actions of the *piqueteros* ("barricaders"), an umbrella term applied to groups of unemployed and union activists who, since the 2001 economic crisis, have been regularly staging protests by blocking streets and bridges.

Like other social movements, the *asambleismo* has confronted a typical "catch-22" situation: blockades attract media attention, but they may alienate portions of the public who are annoyed because they feel that street actions infringe upon their rights to use public spaces. While the blockades enjoyed support at the beginning, they eventually became a point of contention within the movement, as well as the target of criticism from public officials who had previously supported them (including President Kirchner and Governor Busti). Despite the criticisms, the blockades retained enormous news value. Even several years after the movement emerged, they are still a favorite "news peg" for news organizations in the coverage of the dispute around the establishment of the paper mill plants.

Also, the *asambleismo* has embraced the increased sophistication of the traditional theater of protest that in environmental politics is commonly associated with organizations such as Greenpeace and Earth First! (DeLuca, 1999). The influence of this approach is visible in the "watercades" in the Rio Uruguay to interrupt water traffic and demonstrate in front of logging ships and the pulp mills, the "chainings" of people to bridges, and the stopping of trucks transporting materials to the Botnia plant. Starting in 2006, the Asamblea has collaborated with Greenpeace Argentina in the planning and staging of publicity stunts such as, famously, the surprise appearance of Gualeguaychú's carnival queen during the 2006 Vienna Summit of Europe and the Americas. In front of fifty-eight heads of state posing for the official picture, she dashed across the podium while holding up a sign that read "No pulp mill pollution. No a las papeleras contaminantes." Images of the bikini-clad carnival queen, surprised officials, and general levity at the otherwise formal event went global via the Internet and news media networks.[6]

The media advocacy strategies of the *asambleismo* have not been limited to public demonstrations. In fact, the movement has developed a multipronged strategy, setting up its own media and actively reaching out to mainstream print and broadcast media. It has actively contacted both local and national media, particularly reporters and columnists who have been sympathetic to their cause. Although the movement has been sensitive to the importance of the media from the beginning, intuition rather than professional expertise informed its news management strategy. The horizontality of the *asambleismo*, one of its most distinctive political features and constantly defended by its members, has also affected news coverage. The absence of clearly recognizable political leaders in the movement prevented the creation of "media icons" who might have become the focus of news coverage.[7] Nor does the movement have permanent spokepersons who might have become go-to, identifiable media sources. In fact, the decision to rotate several people in the position of "press officer" clearly aimed to avoid media focus on individuals at the expense of collective actors. The *asambleismo* explicitly set out to turn all members into potential spokepersons.

OFFICIAL SOURCES AND NEWS COVERAGE

What was the result of the *asambleismo*'s media advocacy efforts? Here it is important to keep in mind that, unlike cases of environmental reporting triggered by disasters and accidents, news coverage lacked the standard focus on crisis. There were neither spectacular spills nor ordinary victims that could have offered the news pegs that are common in coverage of environmental crises (Gaber, 2000). Instead, the conflict was centered around the potential environmental damage that the pulp mills would bring to Gualeguaychú once they began their operations. The local and national media played different roles vis-à-vis the Asamblea's news management strateties. The local media (Gualeguaychú's newspapers, radio, and television stations) have been closely identified with the movement; some stations even decidedly cultivated a xenophobic, anti-foreign line against Uruguay and the European companies (Toller, 2007). Their position has largely reflected the wide support for the Asamblea across socio-economic groups and political parties in the city. They have devoted wall-to-wall coverage to standard news events (e.g., official pronouncements, street theater) as well as to the Asamblea's deliberations. They have often acted as the movement's cheerleaders, condemning the pulp mill companies and the Uruguayan government.

The local media did not cast itself in the role of mediator among involved parties, actors, and audiences; instead, they explicitly took sides. There has been no distinction between professional journalists and activists, a distinction the media purposefully obliterated. Rather than searching for distance and autonomy, the media uncritically embraced the movement and assumed the role of the "voice of the people." They frequently portrayed "an entire city" mobilized behind a common cause. Not surprisingly, the local media rarely covered dissenting voices or "the other side" in the conflict. Print articles seldom quoted Uruguayan officials or spokespersons for the Spanish and Finnish companies. Only when governor Busti began criticizing the blockades did the local media report dissent.

The position of the national news media has been more complex. National coverage started in October 2003, soon after the first blockade. Although they have covered a variety of news events linked to the conflict (e.g., pronouncements by Argentine and Uruguayan officials, deliberations and reactions to the International Court of Justice's deliberations and decisions), they have largely focused on public protest. In fact, the Buenos Aires–based media only started covering the process when the blockades started in late 2003. Although vocal opposition by the Asamblea to the establishment of the pulp mills had started years earlier, it took protest activities to catch media attention. Nor did the media pay close attention to the long-term process of environment redevelopment on both sides of the Rio Uruguay. Since the mid-1990s, the Uruguayan government has embarked on the reforestation of vast areas to provide raw material for the production of paper pulp, a decision with obvious economic and environmental implications for Argentina. Such decisions eventually triggered a burgeoning environmental movement in Uruguay that sought to raise awareness and drum up interest among Argentines. None of these developments were covered by the Argentine national newspapers, which stumbled onto the news story when information about the Asamblea's mobilization circulated in the newsroom grapevine in 2003 .

While the blockades and other demonstrations remained the central "news event," the position of the national media changed over time. Early on, news coverage of the *asambleismo* was largely positive. While the national media did not explicitly adopt the role of community cheerleader, it offered "kid-glove" treatment of the movement. The picture of the *asambleistas* presented by the leading Buenos Aires newspapers differs from the typical coverage that presents a warped, negative image of social movements. Asamblea members and blockade participants were not portrayed as "deviants" or "outsiders" who threatened the status quo or powerful interests; instead, they were presented as everyman, romantic warriors concerned about environmental threats. *Clarín*'s report on the demonstration organized by the Asamblea on May 2005 gives a flavor of the tone of the coverage:

> In a powerful popular assembly, more than 30,000 people demonstrated on the international Gualeguaychú—Fray Bentos.... Businessmen, teachers, clerks walked to unite both margins [of the river] wearing alpargatas [cheap shoes originally worn by rural workers], high heels, sneakers, swollen feet or barefoot, and wearing green vests and orange headbands that read "No to the paper mills." (Toller, 2005)

A *La Nación* story similarly conveyed a sense of widespread popular support for the Asamblea:

> From its beaten-up truck, an older couple offered a mate to a man in a foreign car. A few meters behind them, a woman sitting in her 4x4 truck called her children who were waving an Argentine flag by the side of the road. A young man on a motorcycle, wearing a thick jacket to fend off the cold wind, celebrated. Everyone prayed a slogan as a religion: no to the pulp mills. (Tosi, 2006)

Friendly media coverage of the Asamblea coincided with the support that both state and national governments gave to the *asambleismo* during 2005 and most of 2006. At that time, Governor Busti frequently expressed his public support for the Asamblea. Also, he condemned Botnia for being "the most polluting [company]" and the World Bank for authorizing a $170 million loan to Botnia (*La Nación*, 2006a). He ordered that state employees who attended a massive rally in Buenos Aires in support of the Asamblea could not be penalized with salary reductions. Likewise, the Kirchner administration offered continuous rhetorical support. Foreign Affairs Secretary Bielsa congratulated the "patriotic fervor" of Gualeguaychú. In July 2006 his successor, Jorge Taiana, stated that the government would never accept the installation of the pulp mills, one day after the government justified the road blockades (*La Nación*, 2006b).[8]

The tone of the coverage shifted in late 2006. This shift coincided with the decisions of prominent officials and politicians to distance themselves from the *asambleismo*. News stories reported that the Kirchner administration was searching for a way to prevent blockades, and that the president and Governor Busti were trying to persuade the Asamblea to abandon such actions.[9] Their position was supported by members of the opposition. The dominant sentiment was best illustrated by former President Eduardo Duhalde, who remains an influential figure in national politics:

"Enviromentalists call the attention of the government, but now they make dialogue [with Uruguay] difficult" (*Perfil*, 2006). The shift towards more negative coverage was, arguably, the result of the fact that powerful sources had become more critical of the Asamblea's methods. As many studies have concluded, when official sources shift views on a given issue, the media take cues and echo criticisms (Bennett, 1990). The weight given to official sources in daily reporting was decisive in changing the tone of the coverage.

Initially, this change was evident in *La Nación*, a traditional newspaper that has been critical of the Kirchner administration. A few months later, *Clarín*, the country's best-selling daily, also started to present a more critical perspective. In editorials and opinion pieces, both dailies criticized the Asamblea's decision to continue with the blockades. The news reporting, however, was rarely as critical as the opinion columns; the newspapers continued to report the perspective of the *asambleistas*, and published only a paltry number of stories featuring dissenting voices. The point is neither to make a normative argument about which journalistic norms should be desirable nor to castigate the media for rarely quoting the interests that the Asamblea opposed. Rather, the absence of a wide spectrum of voices and interests deprived the public of fuller access to the issues embedded in the establishment of the paper mill plants.

The conflict also magnified a larger problem: the decision-making process over the multiple uses of a natural resource and its impact on millions of people along the margins of the Rio Uruguay. Just as a strident atmosphere prevented the local media from widening its coverage, the focus on dramatic news events elbowed out a broader perspective about long-term issues that affect environment policies. The focus on immediate news events served the immediate purpose of the Asamblea to get media coverage, and offered an obvious "news peg" for the media to cover the story. It was detrimental, however, for producing more complex reportage.

CONCLUSION

The experience of the *asambleismo* movement against the establishment of pulp mills on the Uruguay River in Argentina sheds light on important issues regarding the relationship between environmental citizenship and media.

First, the *asambleismo* chose a strategy that was effective in getting substantial media attention, yet they did so at the cost of shifting the focus of the conflict from environmental to political demands. Street blockades are a tested tactic in Argentina; they were used by the *piquetero* movement in the last decade to bring an issue into the national spotlight. Subsequently, many different actors adopted this method as a quick and effective way to get the attention of both government officials and journalists. In the case of the paper mills dispute, however, the strategy fueled an international conflict with neighboring Uruguay, which introduced a nationalist tone to what had been a solely environmental issue. In the end, the movement got extensive media attention, but the environmentalist narrative was relegated to the background. The opportunity for generating an extensive debate on environmental policies relevant to the rest of the country, particularly given the expansion of extractive industries, was lost. The movement's environmental activism ended up buried by conventional journalistic ways of news covering and government spin control.

Second, the prevalence of political news frames undermined the potential of the conflict to promote a broad debate about coverage of environmental issues and policies in the press. Unprepared to cover environmental issues, Argentine journalism resorted to its professional default: the adoption of a political, episodic narrative. Rather than exploring multiple aspects of the story, it stayed in the comfort zone of covering the political game. Issues such as sustainable development in the Rio Uruguay, citizens' participation in environmental policy making, the challenge of balancing economic and environmental policies, and collective action around cross-border environmental risks were largely ignored. It ignored the obvious linkages between this case and other examples of environmental citizenship in the rest of the country. Certainly, it would be unfair to suggest that journalism was solely responsible for downplaying environmental issues in favor of old-fashioned politics. Argentine governments have long ignored environmental issues. They have either passively or actively supported the destruction of the environment in large swaths of the country. Yet journalism failed to connect the dots between Gualeguaychú's mobilization and similar citizens' assemblies concerned with environmental issues. It remained focused on the story without providing a wider context for the expansion of extractive industries and citizens' demands for environmental justice.

Third, this case suggests that when government officials side with mobilized citizens, albeit temporarily, media coverage becomes more positive. Because the news media is largely parasitic on prominent official sources, Argentine officials' rhetorical support for citizens' concerns about the establishment of the pulp mills catapulted the movement to national attention and tilted the news coverage in a positive direction. In contrast, when officials become the targets of protests, as in the majority of the conflicts involving extractive industries mentioned earlier, the coverage is decidedly different: the mainstream media, aligned with the "developmentalist" vision promoted by governments and corporations, tend to distort demands and show mobilized citizens in a negative light.

The present case shows that although securing official support may benefit the media tactics of citizens' movements, it also has several limitations. As officials are mostly concerned with the political implications of the mobilization, environmental issues become secondary. For them, the political risks of this case (e.g., electoral politics, diplomatic relations with Uruguay) were paramount, and displaced the environmental concerns of the Asamblea. As the press refused to widen the frame and continued to closely follow the concerns of political sources, the Asamblea's media advocacy became increasingly focused on staging street protests and environmental reporting became virtually absent.

For media scholarship, the theoretical lessons of the experience of Gualeguaychú's Asamblea and other forms of environmental citizenship in Argentina and Latin America are important. The communication backbone of environmental mobilization remains their own old and new media—community radio, local assemblies, and digital sites. In some cases, mainstream news organizations that are not aligned with political and business interests promoting extractive projects are sympathetic to citizens' demands and offer positive news. Eventually, national media direct attention to issues when citizens' actions have elements (e.g., conflict, violence, well-known newsmakers) that fit conventional definitions of news.

As communities continue to mobilize against extractive projects, options for media actions to support environmental citizenship remain clear. Citizens resort to traditional and new communicative traditions for internal communication, and use mainstream media to reach out to key actors—government officials, companies, mainstream media, and global institutions—with the hope of raising the visibility and urgency of environmental issues. Traditional forms such as assemblies and digital platforms offer communicative mechanisms to raise concerns, exchange information, and define courses of actions. Protests continue to be preferred actions to attract mainstream coverage and turn local demands into broad expressions of environmental citizenship.

It would be foolish to suggest that these actions are mutually exclusive, or to focus the analysis only on one set of media tactics. For different reasons, the strategies of environmental activism employ both alternative and mainstream media. Just as environmental movements need their "own" media to gain self-recognition and discuss needs and demands, they also need "other" media to turn their demands into social problems in the public sphere. The mainstream media offer platforms to voice demands and reach out to key decision makers (government officials and corporations). Gaining attention in the mainstream media is filled with well-known challenges: what news organizations consider newsworthy is not exactly what citizens believe is important and necessary. Movements may gain short-term visibility at the expense of having their demands become distorted, stereotyped, and/or stigmatized. Sensationalistic, entertaining dimensions get emphasized over the real issues. Yet, like other forms of citizens' media activism, environmental citizenship is stuck with a news logic anchored in professional values that are disconnected from questions of rights and participation (Waisbord, 2011). Environmental citizenship has no choice but to reckon with the editorial and professional biases of the news media if it wants to change policies and question official conceptions of development. Interacting with mainstream news puts citizens' movements in a slippery communicative terrain. Unlike with their own, self-contained media, they don't control the terms of engagement, and they permanently have to struggle to shape the message. They may have no option, however, if they are interested in influencing policy making. They need to engage with actors, including the media, that have a different vision of environmental and development issues. Effective social change requires various forms of collective communication action.

ENDNOTES

1 This section draws from my co-authored article: Waisbord, S., & Peruzzotti, E. (2009). The environmental story that wasn't: Advocacy, journalism, and the *asambleismo* movement in Argentina. *Media, Culture & Society*, 31(4), 691–709. Used with permission from SAGE.

2 The decision in 2002 to allow two paper mill plants to be built by Empresa Nacional de Celulosa (Spain) and Metsa-Botnia (Finland) is the result of a long-term policy of the Uruguayan state to promote investments in the forestry and paper mill industry. In the 1990s both companies had acquired large parcels of Uruguayan land for forestation. The amount to be invested in the construction of the paper mills ($1.8 billion) represented the largest-ever sum of foreign investment in Uruguay.

3 In fact, Uruguayan environmental groups were the first to warn the people of Gualeguaychú about the risks involved in the paper mill project, in the hope of organizing a coordinated binational strategy against it.

4 The protest also extended to other Argentine coastal towns such as Concordia and Colón. In fact, local protesters in those locations, in coordination with the Gualeguaychú assembly, participated in the simultaneous blockade of the three bridges that link Entre Ríos with Uruguay.

5 The IFC conditioned the financing on the findings of a report. Released in December 2005, the report concluded that the projects comply with environmental regulations and standards.

6 The collaboration between the Asamblea and Greenpeace eventually ended as they pursued different goals regarding the establishment of the pulp mills. While Greenpeace pushed for "clean companies" with restrictions on the volume of production, an enclosed disposal system, and chlorine-free manufacture, the Asamblea adopted a tougher position that opposed pulp production altogether on the Uruguay River. Also, the organizations disagreed on protest strategies, as Greenpeace decided to withdraw support from the blockades.

7 Daniel Perez Molenberg, one of the Asamblea members who has been often quoted in the press, confirmed the rationale behind the media strategy in an interview with Enrique Peruzotti (2007).

8 As previously argued, electoral reasons weighed heavily in the decisions of both the national and state governments to support the Asamblea. The Kirchner administration faced mid-term elections in 2005, and Governor Busti ran for re-election in March 2007. It is also important to note that the Asamblea wrapped its cause in the Argentine flag, protesting against "foreign" interests (both the foreign-owned mill companies and the Uruguayan government). By adopting a nationalistic discourse with clear anti-imperialist overtones, the Asamblea clearly demarcated a patriotic "us vs. them" struggle. Also, it did not criticize the Kirchner administration, but rather sought its support at the beginning of the process. Later, the Asamblea strongly criticized the administration for "failing to defend national interests."

9 Despite President Kirchner's public gestures of distancing himself from the Asamblea, the position of his administration remained ambiguous; Kirchner appeared wrapped in a flag that read "Botnia out" next to a group of *asambleistas* in June 2007, and the Secretary of Media gave free radio and television air time to the Asamblea in private stations.

▪ CHAPTER EIGHT

Online Media, Flak and Local Environmental Politics

KITTY VAN VUUREN

The Internet and online social media are increasingly a standard component of the tactical inventory of environmental movements and have vastly improved their organizational and advocacy capacities, as well as brought local issues to the attention of national and global audiences. Hutchins and Lester (2011), however, point out that protest actions are often short-lived and are primarily directed at gaining mainstream media coverage in an effort to influence public opinion. Their focus is on activists and NGOs such as Greenpeace who employ mobile media technology in their direct action campaigns. This chapter is more concerned with the section of the environmental movement described by Jonathon Porritt (1984: 6) as the "politically oriented greens", pointing to the close relationship between civil society and an emergent green political sphere. Indeed, out of the protests in 1971 against the flooding of Lake Pedder by Tasmania's Hydro-Electric Commission emerged the forerunners of today's Australian Greens, as well as the Wilderness Society, a national NGO that worked closely together throughout the 1970s on conservation issues (Brown and Singer, 1996). Similarly, Jo Vallentine, a long-time peace activist regarded as Australia's first green federal politician, ran for the Nuclear Disarmament Party in Western Australia and entered the Australian Senate in 1985.

Australian political parties and interest groups first took their campaigns online during the 2007 federal election, but the electorate's use of online media fell short of expectations (Gibson and Cantijoch, 2011: 5; Young, 2011). By 2010 about eight million Australians used Facebook and other social media (Australian Greens, 2010), reigniting speculation that the Internet would play a vital role in the federal election. In 2010, campaign social media usage and the volume of content supplied by federal politicians had more than doubled since 2007 (Macnamara, 2011: 32–33). However, the major parties primarily use social media for transmission of one-way communication. Not so the Australian Greens, who made extensive use of online media during the election campaign, attracting 23,000 followers to their Facebook page (Haraldson, 2010). Greens candidates were also encouraged to use their Facebook and Twitter accounts to develop personal contacts with their electorates (Herrick, 2010); a risky approach, as I will demonstrate below. Although online media have been used successfully by the environmental movement to "shine a spotlight" on environmental issues and have influenced the public and political decision makers (Hutchins and Lester, 2011: 161), online media can as easily be deployed by the environmental movement's opponents to undermine its strategies and ridicule its campaigns.

What I observed in a regional electorate during the 2010 Australian election was online behavior that conformed to Garth Jowett and Victoria O'Donnell's definition of propaganda: "The deliberate, systematic attempt to shape perceptions, manipulate cognitions, and direct behavior to achieve a response that furthers the intent of the propagandist" (Jowett and O'Donnell 2012: 7).

Referring to the example of the Clinton administration in the United States, which throughout its term was subjected to a continuous campaign of allegations, often unsubstantiated, Jowett and O'Donnell regard the Internet as a highly effective means for spreading rumor and false information (2012: 160): "The mere appearance of information on the Internet..., no matter how inaccurate, has the potential for giving information a degree of veracity and legitimacy, which then has to be ignored, countered, or challenged". This case study shows that using the Internet to spread rumor and misinformation is by no means the prerogative of elite politics; it is used as effectively at a grassroots level.

The contribution of this case study in understanding the relationship between political communication and environmental campaigning is not immediately apparent. The federal seat of Blair, located 30 km west of Brisbane, the Queensland state capital, was not a key electorate for the major parties or the Greens. The media I discuss in this chapter are also not especially significant. Although the Greens candidate attracted a large number of "friends" to her Facebook profile, the other news websites and blogs discussed here were unlikely to have attracted much attention from the electorate's voters. Nevertheless, most political communication research focuses on national trends, voter behavior, mainstream media and the elite political sphere (see for example, Chen, 2010; Young, 2011; and the contributions to the special election issue of *Communication, Politics & Culture*, 44(2)). There is little published material about the role of the media in the election process in regional electorates, where much of Australia's expanding resource sector is located. Even less is known about non-incumbent candidates, their campaign strategies, and their experiences with online and traditional media (Macnamara, 2011: 22). Furthermore, a focus on non-incumbent, non-elite, grassroots political candidates directs attention to an activity where the boundaries between civil society and the political sphere dissolve.

DOING CITIZENSHIP

Dahlgren notes that low levels of political participation are a chronic feature of modern democracy (2009: 13), and Australia is no exception. Elections are highly mediated, with few voters having direct contact with their political representatives (Young, 2011: 3). Just 5 percent of Australians are very interested in politics, while 12 percent vote only because it is compulsory (Young, 2011: 26). About 25 percent of Australian voters claim politics has little to do with their lives, echoing similar trends overseas (Dahlgren, 2009: 23; Young, 2011: 26). While there are serious challenges facing the democratic project—the lack of citizen engagement, the converging values of the major parties, and the changing role of journalism—a focus on non-incumbents in the Australian political process reveals that many ordinary individuals are prepared to put themselves forward as parliamentary candidates in what are deliberate, public and perhaps courageous acts.

In 2010, 1198 candidates were nominated to run in the federal election for the Senate and the House of Representatives (Australian Electoral Commission, 2010). Many candidates represented the major political parties—the Australian Labor Party, the Liberal Party, the National Party, and in Queensland the Liberal National Party—but even more did not. Some groups—for example, the Climate Sceptics, the Shooters and Fishers, and the Carers Alliance—use elections to put specific issues on the public agenda, especially in the Senate. This exemplifies the heightened political activity undertaken during any election by a broad range of pressure groups, non-government organizations and activists, many of whom make extensive use of online services (see Vromen and Coleman, 2011). Candidates for the House of Representatives focus more on issues affecting their geographical electorates. All use the traditional tools of civic engagement to influence the political sphere: mobilizing around specific issues, working with and against parties and elected officials, supporting legislative changes, letter-writing, canvassing, etc. (Dahlgren, 2009: 15). A focus on non-incumbents during an election campaign can therefore illuminate the blurred boundary between the "domain of the civic" and the political (Dahlgren,

2009: 69; Vromen et al., 2009: 231–286). Successful independent and minority party politicians can have significant impacts upon the political landscape, and this has been a particularly successful approach for the Australian Greens (Brown and Singer, 1996). Indeed, the 2010 Australian federal election resulted in a hung parliament, with the Australian Labor Party seeking the support of three independents and the newly elected Greens MP to form a government. Unsuccessful candidates can also determine the election outcome. Members of the Australian lower house are elected by a preferential system of voting. Each electorate is represented by a single member who must win at least 51 percent of the vote. Voters choose candidates in order of preference. The candidate with the least number of votes is eliminated, and his or her preferences are distributed to the remaining candidates. Candidates and parties negotiate preference "deals" prior to a poll. Candidates then recommend to their voters how best to allocate their preferences, although voters can choose to ignore these recommendations. In marginal seats a candidate may win the seat based on preferences from rival candidates.

THE GREENS CAMPAIGN IN BLAIR

In mid-May 2010 the Ipswich and Lockyer Greens (ILG), a branch of the Queensland Greens, selected Patricia Petersen to run as their candidate for the seat of Blair. She was unknown to the branch's six active members (two had joined the party only a few months earlier), but came recommended by the Queensland Greens executive, even though during the previous month she had been in discussion with the Liberal National Party to run on their ticket (*Queensland Times*, 2010d). Although born and bred in Ipswich, she had run previously as an independent against Tony Abbott, the Liberal Party's federal leader in Sydney. She accepted nomination on the grounds that she would run an active campaign. One branch member had previously stood as a "paper" candidate, referring to a low-key, no-budget, passive campaign strategy. This approach lets cash-strapped minority parties direct their limited campaign resources toward key electorates, while at the same time ensuring that voters can cast a vote for the Greens in less important electorates. The ILG embraced the opportunity to run an active "real" campaign.

The federal seat of Blair includes the City of Ipswich (population 145,000), located about 35 km west of Brisbane, Queensland's state capital. The electorate has about 90,000 enrolled voters. The local daily newspaper, the *Queensland Times* (*QT*), reaches 21 percent of the electorate (APN Australian Regional Media, 2012). Ipswich has a local commercial radio station that broadcasts to the greater south-east Queensland region. There is no local television station. Ipswich does not host a community radio station or alternative newsletters, where grassroots minority political actors are more likely to gain media access. The local commercial press is therefore an important forum for local politics, although it favors sitting members who represent Ipswich's working-class conservative community. The mayor supported the state Labor government's grand visions for the region, which include doubling the population by 2050, massive housing developments and industrial expansion, the construction of a second power station fired by gas from nearby coal seam gas wells, and the re-opening of local open-cut coal mines. Not everyone is happy about these plans. Areas set aside for "hard-to-locate" industries are close to a local koala habitat and the rare and endangered *Melaleuca irbyana*, or swamp tea-tree forest, which is listed under the federal Environment Protection and Biodiversity Conservation Act 1999. Farmers are concerned about the impact of the proposed developments on the nearby Lockyer Valley food bowl, and residents fear a drop in the value of their properties with the re-opening of the mines. The area borders on the Scenic Rim where eight World Heritage–listed national parks are located. For the local Greens there is no shortage of environmental issues, many of national and global significance, that affect the region. However, the electorate and the local media primarily focused on local issues, which were the responsibility of the state government. Nevertheless, state issues influenced voter behavior during the 2010 federal campaign.

The ILG had a campaign budget of AU$1,200, and the membership all had full-time work commitments. With no previous experience, I accepted the volunteer job of campaign manager at what was my second branch meeting. I organized campaign material, monitored local media, and assisted with letter and media release writing. My approach was *ad hoc*, but I kept a journal of the campaign over the four months from when I first became involved until the election on 21 August 2010.

The ILG campaign made use of freely available channels of communication to reach the electorate and promote Greens issues. Activities included writing letters and contributing online comments to the *QT*, radio interviews, writing media releases, organizing public meetings, tending to campaign stalls at local events, and handing out Greens pamphlets at railway stations. Most of the branch's budget was spent on two advertisements in the local press; additional expenses were met by individual members. Letter writing proved successful. Between 15 May and 26 July 2010 the *QT* published 19 letters and text messages promoting Greens policies. Of these, 14 were written by the candidate. As expected, the newspaper also published 12 letters critical of Patricia Petersen that suggested that she lacked integrity and accountability, themes that were repeated in various online forums. After 15 July the *QT* stopped publishing Petersen's letters, but by then her news coverage had increased.

Dr Petersen needed no encouragement to use her Facebook profile in her campaign. She already had an established profile, "votepatriciapetersen", which she managed herself. It had more than 3,200 "friends" with whom she interacted on a daily basis. Although Facebook novices, the ILG assumed that Facebook offered an effective way to reach the electorate: online media would reach younger voters (Macnamara, 2008), and younger voters were more likely to vote for the Greens. This approach was not without problems. It contravened the Australian Greens communication protocol, which demanded some control over candidate websites, blogs and social media. Petersen was asked to migrate her "friends" across to a Greens-endorsed Facebook profile, but no one knew how to do this. Furthermore, the ILG expected to recruit volunteers from her Facebook profile to help with the campaign, so they did not want to risk losing her "friends" as a result of technical incapacity. Petersen had also put considerable effort into expanding the number of friends subscribing to her Facebook profile. Before joining the Greens, she had handed out leaflets to shoppers at local shopping centers, offering them the chance to win a cash prize of AU$1,000 if they subscribed to her Facebook profile (Garry, 2010c). This tactic attracted criticism from the Labor Party incumbent, who claimed the strategy could be in breach of Australian electoral law. Of greater concern to the ILG, however, was the disappointing volunteer recruitment that Petersen's Facebook profile generated: a handful of "friends" agreed to assist with putting up election signs and helping at the polling booths. Her Facebook page also became the target of two hacking incidents, and her Facebook posts generated negative stories in the local press and online.

Between 15 May and 23 August, the *QT* published 21 articles that either mentioned or focused on Petersen. She first made front-page news on 5 July (Korner, 2010). The story covered a public meeting organized by Petersen and the ILG two days earlier to raise awareness about the state government's plan to develop the "high impact" industrial estate near a koala habitat[1] and the endangered swamp tea-tree forest. The meeting attracted about 300 people, and Brisbane-based television stations broadcast the story later that night. The following day, the state infrastructure and planning minister announced that Ipswich people had nothing to worry about (Foley, 2010a), and two weeks later he called on Petersen to apologize to residents for claiming that the development was a "done deal" (*Queensland Times*, 2010f). It was the first and last time that Dr Petersen and the ILG were able to set the election agenda in the media, although the issue remained salient (Foley, 2010c). Patricia Petersen, however, did make further prominent appearances in the *QT*, with the newspaper clearly more interested in Petersen's perceived errant behavior, the risks associated with using Facebook as a tool in a candidate's election campaign, and the preference issue, which could determine the outcome of the election.

A week after the public meeting, the *QT* ran a page 3 story in which Petersen claimed that someone had hacked into her Facebook profile and uploaded pornographic images onto her news feed. The story included a comment from federal Greens senator Scott Ludlam, who suggested it was likely the work of a political opponent (Jackson, 2010). The story attracted media coverage around Australia and the world (especially in the Indian media). Two weeks later Petersen again made the front page of the *QT* (2010e). This story, with no by-line, drew on a report published the previous week in a suburban Sydney newspaper (Morcombe, 2010). The *QT* story quoted a comment Petersen had put on her Facebook profile, refuting a criminal conviction. The story concerned her appearance in the Manly Court (Sydney) in 2004 on charges of dishonesty. She had been cleared of all charges, and in the article, Petersen claimed she was on the receiving end of a dirty political campaign. A week later the *QT* reported a second incident of hacking Petersen's Facebook profile (*Queensland Times*, 2010a).

Meanwhile, Petersen announced her intention to direct preferences to the Liberal National Party, which she claimed could decide the election result (*Queensland Times*, 2010b). Her announcement disregarded a branch decision to run an "open" ticket and not allocate preferences at all. Subsequently, the *QT* reported that Petersen would not allocate preferences, but would "leave it to Green voters to make up their own minds" (Garry, 2010b). The preference issue must have raised concerns in the Australian Labor Party. In the week before the election, they distributed how-to-vote cards instructing Blair's Greens voters to put the Liberal National Party last on their ballot paper (Foley, 2010b). Greens voters were reminded that the Labor Party had a climate change policy, while the federal opposition was led by climate change skeptic Tony Abbott.

The Facebook hacking incident and the preference issue soured relations between the ILG and the Queensland Greens who were managing the Senate campaign. In addition to the unwanted media coverage, the state and federal Greens received a steady flow of complaints against Dr Petersen, by telephone and on Facebook. The state campaign was managed by volunteers with the assistance of an under-paid campaign manager. For a campaign already under considerable stress, the controversies surrounding Patricia Petersen were an unwelcome distraction. The Queensland Greens directed her to "keep a low profile" and desist from further contact with the media, and as the pressure increased, they proposed her de-endorsement. The ILG strongly resisted such a move. They considered Dr Petersen the strongest Greens candidate who had ever contested the seat of Blair. Moreover, to the ILG, de-endorsement meant caving in to unsubstantiated claims and rumors, and risked the demise of the branch, which had been struggling for some time to attract membership and maintain viability.

The first inkling that the ILG could expect flak from Petersen opponents came within a week of her publicly announcing her electoral intentions (*Queensland Times*, 2010d). The ILG branch secretary received an anonymous phone call from "Steve", who suggested that the Greens reconsider selecting Petersen as their candidate. He claimed she had been arrested for shoplifting, was being sued for defamation, suffered from a mental disorder, and did not have a doctorate. The ILG barely knew Dr Petersen but certainly did not know "Steve". The branch wondered why an anonymous caller would be concerned about the Greens' choice of candidate, and interpreted the call as an attempt to undermine branch cohesion.

Meanwhile, and without the ILG being aware of it, a vigorous online campaign was being waged against the candidate, which had started well before her Greens pre-selection, while she was still negotiating with the Liberal National Party. At the end of May the Queensland Greens alerted the branch to Ben Raue's election blog *The Tally Room* and Andrew Landeryou's *Vexnews*.[2] On June 11 Raue announced the removal of about 80 comments from the threads for Blair and the adjoining Wright electorate with a warning that "any future comments slagging off either Neumann or Petersen will be subject to deletion" (*The Tally Room*, 2010). On *Vexnews* (2010) a story about Petersen generated comments over a three-month period, until Landeryou intervened and closed it down on 30 August. The online discussions rarely dealt with the environmental or other issues that were part of the Greens campaign, but rather were preoccupied with exposing Petersen as unfit to run for office. Here, I focus specifically on *Vexnews*, to illustrate the propagandistic features of the comment thread.

GRAY PROPAGANDA

The online campaign against Patricia Petersen gave the appearance of being deliberate and systematic. The charge was led by one individual who made no attempt to conceal his identity: Gilbert Burgh. As a citizen, Burgh claimed his right to demand transparency and accountability of a prospective political representative. He claimed no political affiliation with the Australian Labor Party. His strategy included leaving comments on every available website that covered the Blair election, cross-referencing between these and the local press, and repeating the same concerns over and over again. In April Burgh left a comment on incumbent Member of Parliament Shayne Neumann's Facebook page, questioning the outcome of Petersen's Facebook cash prize contest, the mystery surrounding her employment, and her claims to continuous residency in Ipswich. Others raised these same concerns in letters to the *QT*

(Spark, 2010), together with claims that Petersen supporters "bullied" her critics and that Petersen blocked her critics from her Facebook profile (Worsley, 2010). Burgh appeared regularly on the *QT* website and Facebook page, using every opportunity to criticize Petersen and demanding a response from the Greens. In May, when Petersen flagged her inclusion on the Liberal National Party ticket, Burgh urged the *QT's* online readers to undertake a "Google search before taking seriously anything that the three-times unsuccessful wannabe political candidate has to say!" (Garry, 2010a). Tracking his activity proved time-consuming and was a distraction from other campaign activities. The ILG decided the best response was none. Members had no experience with this kind of counter-campaign and saw little benefit in publicly engaging with Petersen's critics, especially over matters about which they knew little.

On June 10 Burgh announced on *The Tally Room* (2010) and the *QT*'s Ipswich Facebook page that Petersen had been exposed in *Vexnews* (2010). He was first to comment on the *Vexnews* thread, thanking it for "exposing" Petersen. The story concerned Petersen's expression of sympathy for Hajnal Ban, the de-endorsed Liberal National Party candidate for the electorate of Wright, which she had posted on her Facebook profile the previous day, as well as other comments concerning gay relationships. The *Vexnews* story further rehashed previously published news stories and blogs.

How Petersen's Facebook page came to the attention of *Vexnews* is open to speculation, but it brought home to the ILG that their candidate was under close scrutiny. The story's comment thread generated 245 posts. Of these, Burgh contributed 20, which made up nearly 30 percent of the thread's total word count. He reported the results from his own Google search to draw attention to Petersen's questionable behavior; the "mounting evidence" of more people becoming disgruntled with her; and a call on the Greens to more closely scrutinize their candidate. He directed readers to *The Tally Room*'s comments on the Blair thread (removed the next day), and to the blogs *Larvatus Prodeo* (Kim, 2009) and *Crikey* (2004); the postings of SockPuppet (2010) on *Iain Hall's Sandpit*; *Ipswich-City.com* (2010); and the Facebook page "Patricia Petersen should get off facebook", established to discredit the candidate (since removed); as well as the articles published in the *QT*. On 11 July he took the same approach on *Iain Hall's Sandpit* (SockPuppet, 2010). On 29 July Burgh indicated that he had knowledge of the *QT* front-page story—where Petersen denies her criminality—to be published the next day: he concluded a lengthy post with the suggestion that supporters "check tomorrow's news ... I think you'll be pleasantly surprised!" (*Vexnews*, 2010). It is the only occasion where the reader might suspect that Burgh's posts are anything more than a personal crusade. Nevertheless, Burgh was not without supporters, and was joined by "Anonymouse" (41 posts) and "Robbie"

(30 posts) in a sustained critique of Petersen. Petersen, too, did not lack supporters, with "Jennifer Holmes" her main champion with 13 posts.

The *Vexnews* thread attracted comments from 55 identities. At least two were personally known to Petersen: Hajnal Ban (who posted a single comment) and Burgh, whom she knew as a university student. Most contributors hid behind their anonymity, making it impossible to attribute claims and counterclaims to specific political rivals, or indeed to known supporters. Where sources are difficult to identify and the accuracy of information is uncertain, Garth Jowett and Victoria O'Donnell refer to this as "gray" propaganda (2012: 20); this applied to Petersen's supporters as well as her opponents. For example, at *The Tally Room*, Petersen supporter "Wotif" made the unsubstantiated claim that the "ALP have infiltrated all local social networking sites and have been hammering her 24/7" and that Petersen would "chomp into the ALP vote" (*The Tally Room*, 2010). Petersen claimed to know the local ALP identities, based on past membership of that party, but without hard evidence it was impossible for others to determine the political affiliations of anonymous online identities. Furthermore, the use of *noms de plume* gave rise to the accusation of "sock puppetry", where an online identity is used to deceive. The issue attracted 28 comments on the *Vexnews* thread, with both sides accusing each other of sock puppetry. Petersen's opponents were convinced she would resort to such deception, based on an exchange between Mark Bahnisch from *Larvatus Prodeo* (Kim, 2009) and anonymous readers responding to a story about Petersen during the Queensland state election. To an observer it is impossible to determine the veracity of such claims. Nevertheless, a number of Petersen's supporters, including "Jennifer Holmes" on *Vexnews*, seemed to possess very detailed information about the candidate's past, further adding fuel to the charge of sock puppetry and thus the body of "evidence" that demonstrated that Petersen was unsuitable to represent the electorate.

Burgh's appeals to rationality were convincing. In a lengthy response to "Jennifer Holmes" on 20 June, he asserts his "right and obligation as an Australian citizen" to ask a political candidate to respond to questions about policy, as well as establish "the correlation between what she does, the person she is, etc. [that] is relevant to her credibility" (*Vexnews*, 2010). Had he limited his critique to the facts, such as the questions surrounding the AU$1000 Facebook prize, the 2004 Manly court appearance, complaints from those blocked from Petersen's Facebook profile, and even the hyperbole surrounding the "high impact" industrial estate, where she had claimed the development was a "done deal", his criticisms could have served to weaken support for Petersen's campaign. However, the tone of his criticism probably worked against his intention to influence readers' perceptions of Patricia Petersen.

When "Jennifer Holmes" questions Burgh's motivations (17 June), he responds by portraying himself as a victim of a Petersen counter-attack, including threats of legal action. He suggests she is "underhanded, dishonest and untrustworthy" (*Vexnews*, 29 July 2010). He further frames Petersen's political aspirations as deviant behavior. In response to the story about Petersen's 2004 court appearance, Burgh writes "news follows Pat Petersen wherever she goes ... she [sic] prone to recidivism ... a grifter for votes she will rise again to contest another election at an electorate near you!" (Ipswich, 30 July 2010). Burgh does not stop there, but extends his criticism to Petersen's supporters, suggesting that "vocal Petersen supporters are unable to think critically, think for themselves, and need to be abusive" (*Vexnews*, 20 June 2010), while those opposing Petersen are the "reasonable people on this site" (*Vexnews*, 8 August 2010).

In support of Burgh, "Anonymouse" claims "to have seen numerous threats of legal action made by Petersen/supporters [sic]" (*Vexnews*, 26 June 2010), and that "the Petersen supporters [sic] camp respond with the 'she appears to be a nice person why are you picking on her' answers for too long" (*Vexnews*, 28 June 2010). Such responses serve only to polarize the readership, and combined with the tenacity of the online pursuit against Petersen, resulted in the comments backfiring on Burgh and his supporters (see Jowett and O'Donnell, 2012: 159). *Vexnews* readers began to question the motivation behind Burgh's prolonged attack. Some accused him of stalking ("Jennifer Holmes", 20 June); others considered him "unhinged" ("whatever", 3 August). Another claimed Burgh was encouraging bullying, pointing out that the "Patricia Petersen should get off facebook" page was heavily populated with children who were too young to vote: "A quick look reveals that they don't even know who she is! So they're hardly upset with Patricia Petersen. It's a low act that you've engaged in. You're teaching kids to bully" ("rev", 11 August). After the election, the comment thread turned to criticizing ALP politicians in Ipswich, and it was closed at the end of August.

It is not clear what Burgh intended to achieve with his anti-Petersen campaign. In his last post, on 12 August, he states that his sole purpose was "to gather evidence..., and to direct reasonable people to this site so that they can judge for themselves who is abusive" (*Vexnews*, 12 August 2010). There is no doubt that he represented the views of others who were disgruntled with Patricia Petersen, but it is unlikely that his online activity had much impact on the electorate. The Greens vote increased to more than 11 percent in 2010, up from less than 4 percent in 2007, effectively turning a safe Labor Party seat into a marginal electorate (*Queensland Times*, 2010c). Moreover, few people are likely to have taken much notice of the online forums, including the *QT*'s website, since most Blair voters do not read this newspaper. If the intention was to unnerve Patricia Petersen, it did so, with the result that it strengthened her resolve to see the election through to the end, even as an independent, if it came to that. A weaker, first-time candidate might not have withstood the kind of flak that Petersen faced, and might have withdrawn from the campaign and never again run for office. Indeed, a major problem for minority parties is finding candidates who are willing to run more than once, thereby building a long-term relationship with their electorate as well as demonstrating commitment to party policies and principles. This raises the possibility that the online attack against Petersen was intended to destabilize the Greens campaign. It almost achieved this objective. The flak directed against Petersen strained relations between the Queensland Greens and the ILG. The Queensland Greens had little capacity to deal with a candidate who was not central to the federal campaign; their aim was to get Larissa Waters elected to the Senate.

In Ipswich, the criticisms directed toward Petersen threatened the viability of the local branch. Some new members withdrew their support the moment they were confronted with the controversy Petersen attracted. For others, the flak surrounding the local campaign proved to be a valuable experience. Although some of Burgh's claims were not without substance, Patricia Petersen gave the ILG the opportunity to run a "real" election campaign and to test their mettle in the face of opposition. It brought home the importance of maintaining cohesion. Petersen's de-endorsement would have reflected badly on the branch by creating the perception that it was weak and ineffective, and ultimately might have led to its demise. By supporting their controversial candidate as best they could, the ILG were able to demonstrate that they were able to stand by their decisions, and were capable of weathering the political storm. Although Patricia Petersen and the ILG have since parted company, the branch did not regret its decision to endorse her as their candidate. The 2010 federal election represented a significant step forward for the Australian Greens. The party gained nearly 12 percent of the national vote, up by 4 percent from the previous election. The state of Queensland elected its first Green senator. A Green was elected to the House of Representatives (the lower house), and the Greens won the balance of power by gaining nine seats in the Senate (the upper house).

CONCLUSION

This case study served to illustrate the benefits and especially the dangers of online media in an election campaign, with a particular focus on a non-incumbent, minority party candidate. Close up, the events described in this case study seem out of proportion to the Greens' impact in Blair. However, anecdotes from other Greens candidates indicated that they were the targets of similar tactics. Indeed, the events presented here suggest individuals intending to run for public office have possibly a greater impact than they realize, and therefore need to carefully consider their public communication strategies, especially with respect to social networks.

This case study shows that there is considerable co-dependency between the local press, blogs and social media. The local newspaper and the blog *Vexnews* used the candidate's Facebook posts to generate stories. The "I am not a criminal" headline (*Queensland Times*, 2010e) was furnished by the candidate herself on her Facebook profile. *Vexnews* used her posts in a beat-up to question the candidate's credibility, and permitted the generation of a comment thread that gave the issue greater currency than it perhaps warranted. Clearly, political candidates, including non-incumbents, must take considerable care with their Facebook profiles in their campaign. This case study shows that it is all too easy for the mainstream media and opponents to seize upon comments made online and use these against a candidate.

The study also suggests that the online reports and comments were given far greater authority than the local news reports. The Queensland Greens became concerned when Petersen became the subject of comment threads on *The Tally Room* and *Vexnews*, blogs with a national audience. They appeared less concerned with the negative press she was receiving locally, despite the fact that local newspapers publish online and thus are available to national and international audiences. In this study, the cross-referencing between local news websites and national blogs ensured that the controversy surrounding Petersen circulated extensively and rapidly, even though much of the content was trivial, contained sexual innuendo, and was unsubstantiated and sometimes simply incorrect. Whereas journalists can be criticized for publishing inaccurate news, online comments from belligerent anonymous readers are perhaps more damaging in what is "an exercise in faux democracy" (Shepard, 2011). In this case, the target was a seasoned political candidate who was well aware of the kind of flak she could expect from her opponents, but first-time political candidates could easily be discouraged from ever again standing for election, and that would be a great loss to the environmental movement as well as the democratic project.

ENDNOTES

1 The koala is a native bear-like marsupial that feeds on the leaves of the Eucalyptus tree; the Australian government has recently added it to the national list of threatened species.

2 Raue is a former Greens candidate in Sydney, and has written for *Crikey* and appeared on ABC radio in Melbourne. Landeryou was active in the ALP and launched *Vexnews* in 2008.

▪ CHAPTER NINE

Celebrity, Environmentalism and Conservation

DAN BROCKINGTON

The presence of celebrity in environmental affairs provokes ridicule, wonder, appreciation, laughter and disgust, amid other emotions. Celebrity intervention in environmental causes has seen a famous actor get a chest wax to protest against tropical deforestation; it has seen a wealthy, jet-setting actress declare her inability to cease intercontinental travel but vow instead to take more showers and fewer baths; it has seen the establishment of founding populations of rare animals in exotic and quite inappropriate locations; and it has seen months if not years of dedicated and low-profile sacrifice and service to contentious causes. It is because it is so prominent, so colorful and prompts such diverse reactions that the work of celebrity in environmentalism and conservation can prove so interesting.

Celebrity matters as a means of understanding the power and influence of the media in environmental affairs for several reasons. First, the sheer quantity and proportion of celebrity-focused, -led or -mediated articles, films and news reports have increased steadily over recent decades. It is important to demonstrate this. The prevalence of celebrity in environmental affairs is usually assumed, and awareness of it is presumed by observers. It is rarely shown.[1] Nonetheless, it is possible with relatively simple newspaper searches to demonstrate that articles mentioning both celebrity and environmental matters have increased considerably in both absolute and relative terms. Figure 1 shows this increase in the *Guardian* newspaper (UK) between 1985

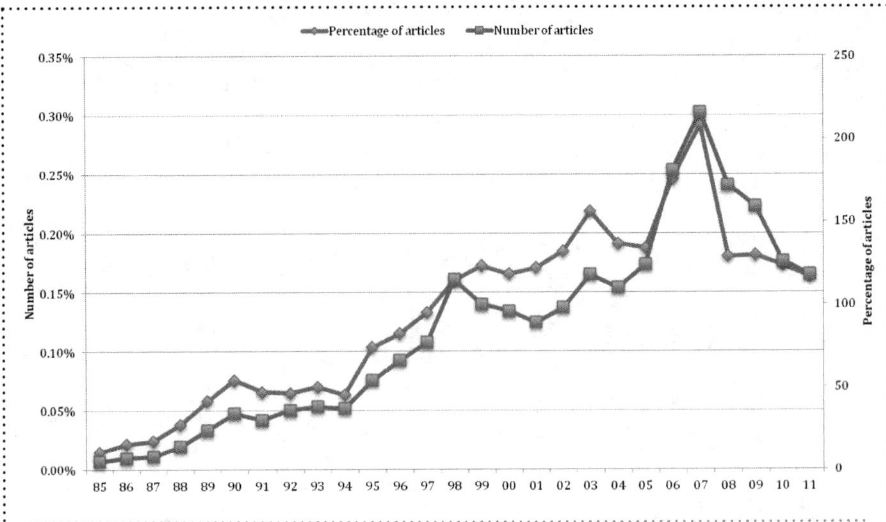

Figure 1: Trends in the mention of celebrity with environment in *The Guardian* newspaper (UK) 1985–2011

and 2005, with articles rising from a mere 5 or so a year to over 200. Note that they appear to decrease after 2005. The increase may not be inexorable.

I chose the *Guardian* not because I believe it to be representative of other newspapers but because its liberal reputation in the UK makes it a useful paper with which to explore trends in reporting about the environment. It is also, as a broadsheet, a paper one would expect to be more resistant to the incursions of celebrity. However, as I have demonstrated elsewhere (Brockington, 2011), it is precisely within UK-based broadsheet newspapers that the rise of celebrity within media culture as a whole is most clearly visible. For where there was so little mention of celebrity in the mid-1980s, there has been a considerable increase up to the present day. This concurs with similar before-and-after snapshots conducted by Turner and colleagues in the Australian press (Turner et al., 2000).

Second, this pattern is important because of the way in which celebrity often tends to work with many audiences—namely, by diminishing the agonistic politics that are normally inherent in many environmental issues and, as Dyer observed, by subsuming the political within the personal (Dyer, 1979). Political choices thus become personal choices, or lifestyle choices, and the bigger issues behind the personalities can be diminished. Dyer went on to argue that in our societies the personal is political. However, if politics are perceived to be not political but merely personal, then the effect will be to depoliticize environmental issues. The other political effect, as we shall see, is the tendency to make things less controversial and promote consensus rather than debate. In other words, celebrity alters the way in which environmental affairs are talked about, thought about and resolved in the public domain.

Finally, celebrity matters because of the sorts of nature and visions for the environment that it can embrace, and, as I will argue below, because of its implications for the relationship between environmental causes and global capitalism. In this reading, celebrity is important not just for the images, messages and ideas conveyed by celebrities themselves. Rather, celebrity in environmental issues and the media is important because of the deeper alliances that it signifies are taking place, and because of the way that these interactions can prove valuable for capitalism even if they appear to be opposed to it.

I present in this chapter some personal reflections on the work of celebrity in environmentalism and conservation, rehearsing some of the findings of my earlier writings. I explain how they came to exist and what shaped them. This chapter constitutes, therefore, a personal perspective on the development of scholarship in this field. There is already a rich literature on celebrity, and articles tracing its rise (Heinich, 2011). There is little literature on celebrity and the environment, in part because the field is still too small and young to merit one. However, there is room for the personal view I offer here. Such a perspective makes it possible to trace the outlines of the epistemic communities being forged as scholars attempt to understand the role of celebrity in environmentalism and conservation.

The central argument of this chapter is that much of the interaction of celebrity, conservation and environmentalism has to be understood as part of the interactions and intertwining of capitalism and conservation. Celebrity is one of the means by which capitalism can become useful to conservation, and vice versa. However, as I make clear, this is by no means the only possible framework for understanding the interactions between celebrity and environmentalism. The field is too diverse to allow a singular interpretation. Nevertheless, framing the interactions between celebrity and environmentalism in that way can make sense of a fair proportion of what we see.

I begin by summarizing the impetus behind my initial interest in this topic and the arguments I first published in *Media, Culture & Society* (Brockington, 2008). I then summarize the main additional arguments that appeared in *Celebrity and the Environment* (Brockington, 2009). These sections are boosted by the findings of more recent research on the interactions between the celebrity industries and the NGO sector (the interviews from this research appear as anonymized numbered sources in the text).[2] I then consider some of the more exciting developments in this field that I believe deserve attention, as well as their consequences, which I look forward to learning about.

CELEBRITY, CAPITALISM AND THE ENVIRONMENT

My own work exploring the role of celebrity in development began with a simple question. I was intrigued by the number of charismatic conservationists operating in East and Southern Africa. All of them were remarkable individuals. All of them were remarkably similar in their public personas, and, above all, in their whiteness. What forces were producing such sameness? Trying to answer that question required that I first explore what was written about charisma, and since many of these figures were also famous beyond their immediate social circles, writings about celebrity. It also meant trying to find out about other famous environmentalists and supporters of environmental causes. Both processes were thoroughly enjoyable. I encountered Weber's writing on charisma (1968), and Boorstin (1992), Marshall (1997), Turner (2004), Gamson (1994), Rojek (2001) and many others on celebrity. I also realized that I had found a topic that proved unusually interesting for colleagues. I was inundated with names when I appealed on the E-Anth Listserv (which links environmental anthropologists and geographers, largely in North America) for suggestions of who I could include in my studies of famous environmentalists.

At the same time I was undertaking this research, my colleagues and I (Jim Igoe, Katja Neves, Sian Sullivan, Rosaleen Duffy, and later, Bram Buscher and Rob Fletcher) were also exploring the interactions between conservation capitalism and the striking ways in which capitalism and conservation were remaking the world together, not, as is often assumed, in opposition to each other. This broader thesis considered the notion that conservation could be remarkably useful to capitalism on occasion and that, conversely, capitalism could serve conservation's purposes (Brockington et al., 2008; Brockington and Duffy, 2010, 2011; Brockington and Schofield, 2010; Igoe et al., 2010). Our ideas were part of the thinking about the neoliberalization of nature that was sweeping radical academia (Castree, 2007a, 2007b; Heynen et al., 2007). This thinking, and the challenges it poses to conservation, has recently been consolidated into a more substantial critique in a recent collaborative paper (Büscher et al., 2012).

The result was what we called "hegemonic conservation" or "big conservation", in which both conservation groups allied with powerful companies and conservation thinkers formed a recognized part of the global elite. George Holmes, who completed his PhD in Manchester in 2009, combined this notion with Sklair's writings on the transnational capitalist class (Sklair, 2001) to develop the idea of a transnational conservationist class (Holmes, 2011). Celebrity was simply part of this bigger picture (as well as integral to the transnational conservationist class). It lubricated the machinery, making it easier for companies and NGOs to work together, and making the parties, meetings and other occasions where they mixed more colorful and enjoyable.

This alliance between conservation and capitalism provided a useful contrast to the then current writings about conservation, environmentalism and celebrity. A significant recent contribution was Brett Hutchins and Libby Lester's examination of media and environmental affairs in Tasmania (2006). They had portrayed environmentalism as relatively weak and vulnerable. They found that environmental causes were treated skeptically by the media, for whenever environmentalists produced media events these staged protests (sometimes involving celebrity) were given unfavorable treatment. This they attributed to the vulnerable and marginal position of environmental affairs within the broader economy. As environmentalism was marginal to business and government, so also environmental protest and environmental reporting were marginalized by the media.

A similar vulnerability formed the basis of the work of Meyer and Gamson in their analysis of the role of celebrity in supporting environmental causes in the USA (1995). These authors found that celebrities tended to shy away from more controversial environmental affairs, and indeed tended to render the more controversial issues in whichever causes they were engaged with less controversial and therefore more palatable for mainstream audiences. The goal of celebrity interventions was clearly to avoid criticism in the media. This desire and behavior was attributed by Meyer and Gamson to the more general vulnerability of celebrity to critiques of inauthenticity. Celebrity roles are performed in the public eye and represented to the media. Therefore, no matter how heartfelt the speeches are or how well qualified the celebrities are to speak, there is a perennial skepticism among the audiences that this is just another act, that it does not spring from the heart, and that these passions are merely lines that have been scripted. It should not be surprising, then, especially if celebrities do lack formal qualifications or deep personal roots in a cause, that they will shy away from anything that may invoke a hostile response in the media.

My interpretation of the role of celebrity in environmental affairs differed from those of Hutchins and Lester and Meyer and Gamson because I was struck by the vitality of mainstream conservation under contemporary capitalism. There were numerous signs of this vitality. For example, protected areas have grown rapidly in the last two decades and thrived particularly well during the heights of neoliberal dominance from the mid-1980s to the mid-1990s. Protected areas could be integral to capitalism. They were useful to the ecotourism business, and they were useful for the environmental offset industry whenever damage caused by mining or oil pipelines needed to be mitigated with concrete conservation activity. It was also clear that conservation NGOs have become much more effective at forming strong and profitable alliances with large multinational companies (MacDonald, 2010).[3] The value of conservation groups to capitalism is also visible in their embrace by powerful political interests. Corson's fascinating research into the International Conservation Caucus in Washington, for example, describes a litany of comfortable relationships between powerful politicians, leading conservation NGOs and corporate power (Corson, 2010).

This state of affairs produces quite a different interpretation of the relative weakness of environmental causes. Where the earlier authors had seen weaknesses, this reading finds strength; where they saw vulnerability and isolation, here is evidence of a far-reaching and powerful hegemony. These are business interests that are aligned to environmental causes, green products providing new forms of consumption that are profitable to capitalism, and credibility and authority imparted to environmental spokespeople by corporate power. Celebrity added to this mix, made it stronger and was itself strengthened by it. There was little weakness here, and much more power and strength.

There is marked contrast, therefore, between that state of affairs and the writings on environmental causes and environmental protest that emphasize their weakness. Such causes and protest have often been characterized as rooted in a love of local places and driven by residents. Yet, in conjunction with these parochial concerns, it is also possible to observe a global environmental movement, and particularly a global wildlife conservation movement, that is concerned with preserving species and habitats that often are far removed from the homes of the movements' wealthy patrons (Prendergast and Adams, 2003). The conservation urges of wealthy hunters, or rich people who want to preserve the beauties of African scenery for their safaris and Indian forests for their elephant rides, are certainly rooted in a love of place. But these are not locally driven movements. In fact, they are often in conflict with very different local understandings of what the environment should be like and how and by whom it should be used (Guha, 1997, 2003). Simply put, the presence of a national park will impede peasants' access to fuel wood, grazing, roofing grass and wild meat. This means that there are in fact many different "varieties of environmentalism", to use Guha and Martinez-Alier's phrase (1997). In many sites around the world there is conflict between two or more such varieties. As I have documented for the Mkomazi National Park in Tanzania (Brockington, 2002), it is quite possible for the locally rooted environmentalism to be displaced and usurped by a global international environmentalism, which is thoroughly well connected to the heights of economic and political power.

The presence of such strength does not of course refute the theses advanced by Hutchins and Lester and Meyer and Gamson. These writers' explanations were clearly correct in the particular circumstances described, and would be the best explanations of celebrity interactions with environmental causes in other situations as well. But we need to consider another dimension in which the interactions between celebrity and environmentalism could be explored.

These powerful interactions of celebrity might be particularly important where environmental causes are not local, but part of global movements. Then they are not so much driven by physical interactions with the environment as they are dependent upon representations of those natures and environments. One does not have to study celebrity to understand these representations, yet celebrity is often part of the representations. Understanding these representations is enriched by understanding the variety of roles of celebrity within them.

There is a broad gamut of celebrity active here. There are famous people who have lent their fame to environmental causes, people who became famous because of their involvement in conservation and environment, and people who were involved primarily in producing film and television productions about the environment (Brockington, 2008).

Specific examples from each of these domains may help to demonstrate some of the arguments developed here, which have been reinforced by ongoing research. There are two general types of celebrities who lend their names and reputations to environmental causes. There are those who volunteer their services free of charge as part of their commitment to a cause. All associations negotiated by NGOs in the first instance, with almost no exceptions, arise from such a basis. NGOs will pay travel expenses, but they will not pay celebrities or their managers and agents any fees.[4] This policy can restrict the amount of time available for the work, unless the stars are fading, in which case their agents may contact the NGO asking for their client to be asked to do something.[5] Celebrities are paid for their activities when they are also endorsing a company or a corporate product. In such circumstances, while the benefit to the charity may be valuable, the prime mover is the company, and normal advertising charges apply. Corporate supporters may well get involved in charitable activities which have associated celebrity support, and may enjoy effectively free publicity as a result.[6] However, where there is a risk that celebrity support for a charity will become free corporate endorsement, celebrity liaison officers will usually ask the company involved to negotiate directly with the agents and interests of the celebrities concerned.[7]

An example of a celebrity who has forged long-term links with an organization for no material gain is Harrison Ford, whose links with Conservation International demonstrate particular dedication. He serves on their board as a vice president and has released a series of public service broadcasts with them. These have included a widely disseminated video of Ford having his chest waxed to raise awareness about the problems of tropical deforestation,[8] and a more recent announcement heralding the sustainable development summit in Botswana which proclaimed the success of development in Africa (which is unfortunately described as a continent where "Nature and People are one").[9]

An example of paid support is Lily Allen's tour of rainforest conservation projects run by the World Wide Fund for Nature (WWF), which was undertaken as part of BSkyB's Corporate Social Responsibility work.[10] The trip was covered by journalists who were also part of the Murdoch media empire, which owns Sky.[11] This sort of engagement may well lead to voluntary service, but its role here was commercial in the first instance.

When NGOs seek celebrity supporters they are primarily looking for people who are likely to have an authentic connection with their cause in the public mind.[12] In this respect, presenters of natural history programs are particularly valued spokespeople and ambassadors of environmental and conservation organizations. These figures, having presented nature on television, can easily become nature's spokespeople beyond the screen.

Presenters are people with whom, or through whom, it is good to experience nature. Hence it should be no surprise that Jonathon and Angie Scott, both brilliant photographers (and in the case of Jonathon, also a wildlife presenter), offered guests the opportunity to accompany them on safari to wonderful locations in Africa, Antarctica and India. However, the most accomplished combination of business with conservation nous was, and still is, visible in the work of Steve and Bindi Irwin. The late Steve Irwin's conservation work (and now his daughter Bindi's) was combined with a large variety of different products that could be purchased on the Australia Zoo website, including Bendy Bindi dolls and the "Steve Lives" surfwear range.[13] The latter, released when Irwin was still alive to counteract rumours of his death, were still advertised years after his death. Purchasing these goods rewarded both the green intent and the material satisfaction of the consumer (Brockington 2009).

Famous environmentalists and conservationists are remarkable for the persistence with which they have flourished over time. There have been prominent public figures in the work of environment and conservation affairs in just about every decade in the last 100 years (except the 1940s). The roll call of great names over this period includes Swampy, Jane Goodall, Ian Douglas Hamilton, Saba Douglas Hamilton, David Bellamy, Jonathon Porritt, Diane Fossey, Jacques Cousteau, David Brower, the Adamsons, Peter Scott, Rachel Carson, William Beebe, Grey Owl, John Muir and, posthumously, Henry Thoreau. All these people have been among the numerous environmental spokespeople who have been accorded space in the media and who excite the public imagination as a result of their campaigning and advocacy.

Environmentalism and conservation are causes that have to be understood as much through the work of their prophets as through the hands-on, down-to-earth work of their publics.

The relationship of these figures to capitalism is interesting. Muir virulently, if unsuccessfully, opposed the construction of a large dam at Hetch Hetchy in California. His role almost defines the anti-capitalist stance that once characterized many forms of environmentalism. However, celebrity conservationists also capture a number of ways in which environmentalism and capitalism can be quite compatible. First, there is the rather simple point that defense of wilderness is no threat to capitalism if those wildernesses are useless to capitalism. The worthless lands hypothesis suggests that the interests of capitalist development have been quite content to allow conservationists protected areas so long as they are high up, rugged and/or cold. It is quite possible for conservation interests and capitalist progress to proceed in such instances. Some authors have observed that fixing on wilderness areas (e.g., unspoiled cedar habitats in the western USA) has resulted in an inappropriate disengagement from tainted lands (e.g., logging concessions) that could be ecologically useful and valuable to conservation interests (especially spotted owl conservation; see Proctor and Pincetl, 1996). Second, conservation activities can provide an outlet for capitalist philanthropy that can neatly ignore contradictions and activities incompatible with other conservation goals that have produced the wealth in the first place. Here, celebrity conservationists can be useful conduits for fundraising. A particularly useful example is the Wildlife Conservation Network, which puts wealthy Californians in touch with specially selected conservationists promoting the survival of charismatic wildlife. Celebrity won through conservation activity can be a most useful means of generating funds from wealthy patrons in such networks.

CELEBRITY AND THE ENVIRONMENT

The world of celebrity and the environment can be troubling. As I stated in 2008:

> Conservationists insist that conservation can only succeed if it builds on people's close relationship with the wild (Milton, 2002; Adams, 2004). But it appears that these relationships need not be built on physical interaction. Indeed, in a world of para-social relations they cannot be. Rather, when people consume celebrities' support for conservation, when they watch Sting on television, buy an acre of Africa online, send an email that Robert Redford wrote to their senator, watch Leonardo DiCaprio's website video on global warming and enthuse about the patronage of wild Africa by diverse members of European royalty, they are restoring their relationship with the wild. (Brockington, 2008: 563)

The implication is that as celebrity interactions with environmental affairs come to proliferate within environmental activism, so people's relationship with the wild will diminish. Exploring the work of celebrity would help us to understand the changing character of environmental movements as they became less rooted in particular places, more global and more dependent upon representations.

However, the emphasis of that argument changed in *Celebrity and the Environment* (2009). Instead of seeing celebrity interactions with environmentalism as distorting an environmental movement, I noted that quite possibly they had been *constitutive* of those movements for many decades. This position was supported by the role of natural history film in forming and shaping environmental consciousness in much of the developed world, as well as the early role of NGOs such as the WWF and the African Wildlife Foundation (AWF) in environmental campaigning. It became clear that there was a significant element of the environmental movement which had hinged upon vicarious engagement in environmental matters for a considerable time. The hands-on activities of the naturalists that Allen (1994) described in his history of their activities were winding down dramatically after World War II. The environmental movements that have become popular in the West and global North in the last 50 or 60 years have been the movements of publics who, in many aspects of their lives, are largely alienated and separated from the natures that they seek so passionately to conserve. Celebrity conservation is not a new development within the field environment activities, but the continuation of a long tradition.

Celebrity and the Environment offered an answer to the dilemma that had prompted my exploration of this topic in the first place, namely, the proliferation of charismatic white conservationists in East and Southern Africa. Part of the explanation lay in the long histories of racialized wildlife conservation, but the more powerful cause lay in the demands and expectations of the Western audiences who consumed those images. There was a ready market for great white people saving African wildlife, which was why they were popular in newspapers and television films in Europe and North America.[14] In other places, however, wildlife conservation was much more powerfully driven by national governments and national audiences. Thus the Indian conservation scene is dominated by great Indian conservationists—not by white heroes and heroines.

The other contributions of the book explored the concentrations of wealth and power that had given celebrity such influence within environmental circles. It considered changing NGO practices and configurations that made celebrity involvement so useful. It examined the changing funding regimes of conservation organizations, and in particular, the influence of extremely wealthy individuals on conservation funding. The book refrained, however, from insisting that such associations were of themselves problematic. Celebrity is certainly elitist (Kurzman et al., 2007), but the presence of celebrity environmental affairs does not necessarily reinforce the power of elites. In fact, celebrity may be a useful way of bringing elites to heel (Cooper, 2008). Celebrities can voice in elite circles popular challenges that otherwise might not get heard. I also did not agree with the complaint that celebrity involvement is hypocritical and self-promoting. Celebrity activities are by definition undertaken in the public domain, and to observe that the celebrity's image is promoted by their activism is simply banal. This is particularly the case given that "their conspicuous consumption is performed for societies that are already wedded to consumerist ideologies" (Brockington, 2009: 125).

There were some much more important criticisms of the work of celebrity in environmental affairs, developed in the last chapter of the book. Building on the work of Guy Debord (1967/1995) and Jim Igoe (2010), I argued that the problem of celebrity involvement was that it could become part of a larger scheme of conservation iconography in which origins and histories could be obscured or erased. Celebrity is integral to producing a mediagenic world in which the social problems and awkward issues can be smoothed over. However, these representations do not just shape our understanding of these places; they forge our expectations of what they will be like. They act, in other words, as "virtualisms": models that come to define how the world should be and to which the world is expected to conform (Carrier, 1998).

All of this analysis could present a rather depressing picture of sofa-based conservationists struggling to create worlds existing only in their imaginations or on their television screens. Celebrity becomes the handmaiden to the inevitably violent processes initiated by such jarring visions. Yet, as argued at the end of the book, this would be unduly pessimistic. In the first place, even powerful organizations like wealthy NGOs have proved remarkably inefficient in forcing their virtues into existence (Carrier and West, 2009). In the second place, there are some remarkable uses of wildlife film that promote effective policy change. There are also some very interesting experiments promoting audience participation in programs like *Springwatch* in the UK. Such programs have produced not only many more viewers but also, crucially, many more *participants* in observing local natures in our own back gardens. Ultimately, the varieties of celebrity conservation and celebrity environmentalism are too great to allow any singularly pessimistic (or optimistic) conclusion.

FUTURE DIRECTIONS

It is important to recognize that these statements and analyses of the work of celebrity in conservation did not initiate a new body of research into the topic. A number of colleagues have simultaneously engaged in the study of the role of celebrity in environmental affairs. Graham Huggan began his analysis of the influence of leading naturalists who are also authors and filmmakers. Max Boykoff and Mike Goodman analyzed the role of celebrity and the media in climate change debates and the promotion of fairtrade food (Boykoff and Goodman, 2009; Goodman, 2010). Malcolm Draper had already looked at the work of charismatic conservationists in Southern Africa (Draper, 1998; Draper and Maré, 2003). Jim Igoe (2010) has been exploring the role of spectacle (following the writings of Guy Debord) in creating conservation awareness and causes. Libby Lester has set the role of celebrity in environmental affairs within a broader context that includes iconic non-human nature (2010b). William Beinart has been exploring the history and creation of wildlife film, and in particular the black African networks that underlay (and underlie) the whiter, often more European and American networks of natural history film makers. Our research was all aided by an AHRC research network grant (from the UK government) that allowed us and others to compare notes and report findings and new initiatives. In other words, this has been a fertile and dynamic field.

There is a tremendous diversity of and within celebrity. While it makes sense to talk about *celebrity* as a concept, industry and idea, as soon as we begin to talk about the variety of *celebrities*, it can become very difficult to generalize. This problem becomes immediately apparent if we consider the number of celebrities found in different parts of the world. The British "Red Pages" database tracks agent and manager details for more than 25,000 public figures. The US-based "Ask a celebrity.com" has more than 60,000 names on its books. There is a vast and fragmented collection of different people famous for different reasons and to different degrees. Generalizing about the work of "celebrity" in environmental affairs may be possible at the crude level, but effective insights require some sort of typology of involvements and activities.

A second point follows from the first—if there are so many celebrities out there, we have to query the reach and audiences that celebrity (which celebrities?) can command. Early research into the prominence of celebrity coverage in newspapers in the UK suggests that the relentless increase of celebrity reporting in all aspects of the news appears to have stagnated. Indeed, it is possible that it has now begun to decrease (Brockington, 2011). Figure 1 demonstrates a pattern found commonly in research in other newspapers and on other aspects of celebrity culture: there has been a decline in reporting on celebrity in recent years, at least as measured by these data. If these results are confirmed, then it will suggest that the market for celebrity reporting is not completely elastic. What had proved to be an apparently ever-increasing and popular means of getting into the news may have struck some boundaries.

Finally, it is important to remember that any work on celebrity must be as attendant to the genres and products from which celebrity is deliberately absent. In this respect, one of the most exciting and enjoyable things to research has been alternative and potentially radical film making. It was already obvious that bespoke films aimed at small but powerful decision-making audiences could be incredibly effective. Again, there is a lot of variety here worth exploring. As part of the AHRC network, a number of us attended the Wildscreen Film Festival in Bristol in 2010. There we were privileged to see the remarkable award-winning film *Green*, written, produced, filmed and directed by Patrick Rouxel. *Green* portrays a complex set of stories of commodity chains that tie consumers to forest destruction in Indonesia and to the death of orangutans, and yet it has only one word of dialogue in the entire film. It captures the stillness and quiet of forest ecosystems, with no music for the first ten minutes or so, and is, in my opinion, far more exciting and chilling than dramatic footage of animals killing each other or fighting to mate. It makes its audiences grieve. *Green* is also a remarkably successful film. It has won many prizes and is available for free on the Internet, as it was not made for a commercial company.[15] *Green* is also, need I say, devoid of celebrity. What is interesting about *Green* is its potential to excite audiences, and the ways in which a film like that can be made accessible to and more engaging for those who watch it. In collaboration with Patrick Rouxel, my colleagues and I have created a website that makes the film more accessible to audiences of university students by providing short essays that will help situate the film within their studies.[16] This initiative may lead to further user-generated content that makes audience interactions with celebrity and other forms of media such a fascinating field to explore.

There are a huge variety of celebrity engagements with environmentalism that will excite great variation in responses from audiences and have diverse impacts on environmental and conservation causes. Exploring that variety will require several theoretical frames in which to explain the nature of celebrity interactions with environmentalism. It is an expanding and welcoming field, and one that continually rewards further investigation.

ENDNOTES

1 Boykoff and Goodman (2009) is a useful exception to this trend.

2 The research was sponsored by ESRC Fellowship RES-070-27-0035, and I am most grateful for this support.

3 A particularly polemical example of the interactions that can take place between NGOs and international companies is provided by a video produced by environmental activists who tricked Conservation International (CI) into thinking that they were negotiating with Lockheed

Martin, a major arms manufacturer. The recordings of these interactions included activists posing as Lockheed Martin representatives asking to discuss the recycling of their cluster bombs and the adoption of a suitable raptor that could become a useful symbol promoting the sale of their warplanes. CI was quick to insist that the video had been too selectively edited to be reliable (http://www.dontpaniconline.com/DPTV/undercover-with-conservation-international, accessed 4 July 2012).

4 Source 35.

5 Source 14.

6 Source 51.

7 Source 61.

8 http://www.conservation.org/fmg/pages/videoplayer.aspx?videoid=30 (accessed 4th July 2012).

9 http://www.conservation.org/fmg/pages/videoplayer.aspx?videoid=128 (accessed 4th July 2012).

10 Source 33.

11 http://www.thesun.co.uk/sol/homepage/news/Green/2970915/Lily-Allen-backs-bid-to-save-rainforest.html (accessed 4th July 2012).

12 Sources 35, 37.

13 https://shop.australiazoo.com.au/bindi/dolls-plush-and-games (accessed 18 February 2012). https://shop.australiazoo.com.au/steve-lives-surfwear (accessed 18 February 2012).

14 Some time after writing the book I received a wonderful confirmation of the thesis from a former commissioner who described to me a meeting between BBC executives and executives of a major American broadcasting company. The members of the meeting were looking at a film they had commissioned about environmental and conservation issues in Nigeria. Unfortunately, when the film was commissioned, it had not been made clear to the American commissioners that the activists concerned were Nigerian. The program was full of Nigerian voices talking about Nigerian issues. The American commissioners were horrified; this sort of content would be completely inappropriate for their American audiences. "Where are the North Americans?" they complained to the filmmakers, "couldn't you even find some Canadians?"

15 It is available here: http://www.greenthefilm.com/

16 http://studyinggreen.wordpress.com

▪ CHAPTER TEN

Dodgy Science or Global Necessity?

Local Media Reporting of Marine Parks

MICHELLE VOYER, TANJA DREHER,

WILLIAM GLADSTONE AND HEATHER GOODALL

The digital age and globalization has brought international issues to our doorstep and placed the local in the context of the global. News media have played a crucial role in allowing recognition and exploration of the global origins and outcomes of many environmental crises such as climate change, deforestation, threatened species management and biodiversity loss (Cottle, 2011c). The modern environmental movement has responded to the global scale of these crises with campaigns for global solutions. Many of these campaigns rely heavily on coordinated, collective action across a multitude of jurisdictions around the world, with the success of global campaigns dependent on the success of multiple local-scale actions. The slogan "think global, act local" has become the rallying cry of the modern environmental movement. Yet the individual success of these actions depends significantly on local conditions, particularly community and political support. The media, including local news and community-based media, play a crucial role in influencing both these factors.

Marine Protected Areas (MPAs)[1] are one example of a highly contested conservation goal that is being vigorously pursued on a global scale but meeting significant resistance at the local level (e.g., Banks and Skilleter, 2010; Carneiro, 2011; Weible, 2008; Wescott, 2006). In response to large-scale loss of marine biodiversity and the collapse of a number of fisheries, a range of international agreements have been developed which commit signatories to developing a system of MPAs covering between 10 and 30 percent of their marine habitats by 2012 (Spalding et al., 2010). Australia has responded to its commitments under these agreements by progressively implementing a network of MPAs in both Commonwealth and State waters. Throughout the country, however, the establishment of new MPAs has consistently led to intense local resistance.

The success of the transnational response to a global marine environmental crisis relies heavily on its implementation at local level. Resistance from local communities and key stakeholders has led to the failure of many attempts to establish MPAs throughout the world (Agardy et al., 2003; Fiske, 1992; Weible, 2008; Wolfenden et al., 1994). Once an MPA is established, its success depends enormously on the support and goodwill of key stakeholders, especially fishers (Agardy et al., 2003). This chapter will examine the role of local news media in digital and traditional forms in the move to translate the call for a global approach to marine conservation to local-level implementation of MPAs. In particular, it will examine how key stakeholders at a local level can influence wider debates over marine conservation through local news media. It will specifically examine two marine parks established on the north and south coasts of the state of New South Wales (NSW), on the east coast of Australia. NSW marine parks are large "multiple use" MPAs zoned for different types of use. The highest level of protection within a NSW marine park is the "sanctuary zone", or "no-take" zone, where all forms of fishing, extraction of marine life, and damage to habitat are prohibited. This zone type is the most restrictive, and is therefore often the most controversial aspect of marine park planning.

In December 2005 and April 2006, respectively, the NSW state government established the Port Stephens–Great Lakes Marine Park on the mid-north coast and Batemans Marine Park on the south coast. The process by which these parks were gazetted and zoned was virtually identical. Both parks are roughly the same size, have similar levels of sanctuary zone protection, and had extensive public consultation processes. Despite the similarities, each community responded differently to its local marine park, with the Batemans Marine Park generating significantly more resistance from recreational fishing interests. Our study commences in 2005 with the announcement of the intention to declare the parks, and incorporates the development of zoning plans for each. The investigation concludes in 2010, two years after the implementation of the final zoning plans.

THE NSW MARINE PARK DEBATE— WHO HELD A STAKE, WHO HAD A VOICE?

Australia's oceans are used and valued for a variety of reasons. They are important economic and social resources, and therefore the debate over their use incorporates a wide variety of perspectives. At a local level, marine park planning processes generally involve consultations with commercial users (including fishing, mining, tourism, shipping and ports), non-commercial users (recreational fishers, divers and researchers), Indigenous and community groups. Conservation groups are one form of community group that are active within marine park planning processes and may include state-wide/national non-governmental organizations (NGOs) or "grassroots" organizations working on environmental issues at a local level. All the recognized user groups could argue that they have a direct and immediate stake in how their local marine environment is managed, and particularly in zoning arrangements which may restrict their use. In the cases of the NSW marine parks, however, not all can claim to have had a voice in the public debate.

STAKEHOLDERS IN THE NEWS MEDIA

The concept of "voice" and in particular how voice functions within the media operates at many different levels (Couldry, 2010). For the purposes of this chapter the exploration of "voice" is limited to its political use, namely, the "expression of opinion", or the expression of a "perspective on the world that needs to be acknowledged" (Couldry, 2010: 1). News media is commonly portrayed as a "battleground" on which competing voices fight to gain access and prominence (Cottle, 2000a; Hall et al., 1978; Lester, 2007). Professional norms lead journalists to seek out spokespeople from "credible" sources in order to present their stories as well-grounded and objective. This will often mean that government or other elite groups are able to gain access to the media more readily than other smaller marginal and minority groups, and can use this access to set the terms of the debate in the media, becoming what is known as "primary definers" of news topics. By gaining an authoritative and dominant position within the news, a primary definer can not only influence the way a problem is presented, they can also strongly influence the preferred potential solutions to the problem (Cottle, 2000a; Hall et al., 1978, Lester, 2010b).

More than 500 news articles published between December 2005 and December 2010 in local newspapers in the immediate vicinity of the two NSW marine parks were analysed (Table 1). The primary definer or dominant spokesperson was noted for each article and assigned to a key stakeholder group (Figure 1). Primary definers were classified as those spokespeople who set the agenda or theme of the article. In most cases this was the spokesperson first quoted or referred to in the article. However, in some circumstances it was the spokesperson given the greatest exposure or prominence in the article.

Table 1: Local Papers Within Marine Park Area

Park	Newspaper	Circulation*	Readership*	Frequency
Port Stephens Marine Park	*Newcastle Herald*	48,000+	Mon–Fri: 131,000 Sat: 186,000	Mon–Sat
	Great Lakes Advocate	5,862	18,028	Weekly
	*Port Stephens Examiner***	28,123	28,123	Weekly
Batemans Marine Park	*Bay Post/Moruya Examiner*	3,769	8,589	Bi-weekly
	Narooma News	2,341	6,259	Weekly
	Milton Ulladulla Times	5,050	15,814	Weekly

*Readership figures from http://www.ruralpresssales.com.au/index.asp (accessed March 2011) or http://www.adcentre.com.au/ (accessed March 2011).

** Free paper

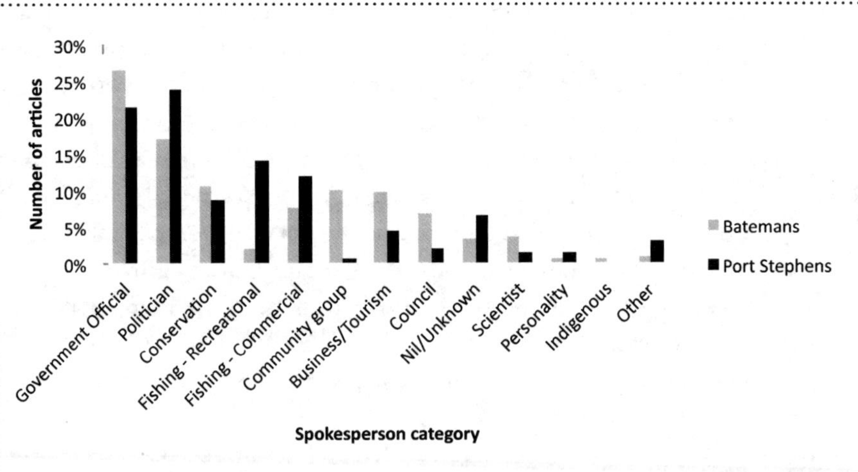

Figure 1: Representation of major stakeholder groups as primary definer in marine park news articles in the areas covered by the Batemans Marine Park and the Port Stephens Great Lakes Marine Park.[2]

Coverage was given to a broad diversity of views, but government sources and politicians dominated the role of primary definer (Figure 1). Primary definers from non-government sources included fishing interests (commercial and recreational), business or tourism and conservation groups. Marine scientists and Indigenous spokespeople were rarely primary definers of news articles relating to the marine parks.

STAKEHOLDERS ONLINE

A Google search of "NSW Marine Parks" was conducted on two separate computers. The first 100 results of each search were analysed and categorized into stakeholder or interest groups and the results were averaged (Figure 2). This search suggested that conservation and recreational fishing groups dominated the non-government sites relating to NSW marine parks. There were also a large number of tourism or "destination" sites promoting the area as a marine park as part of their marketing pitches.

The websites of conservation groups included active campaigns for increased MPA coverage of Australian waters, including in NSW, supported by e-lobby forms and online donations. The websites of recreational fishing organizations or clubs often incorporated fishing forums or chat rooms where MPAs continue to be a frequent topic of discussion and debate.

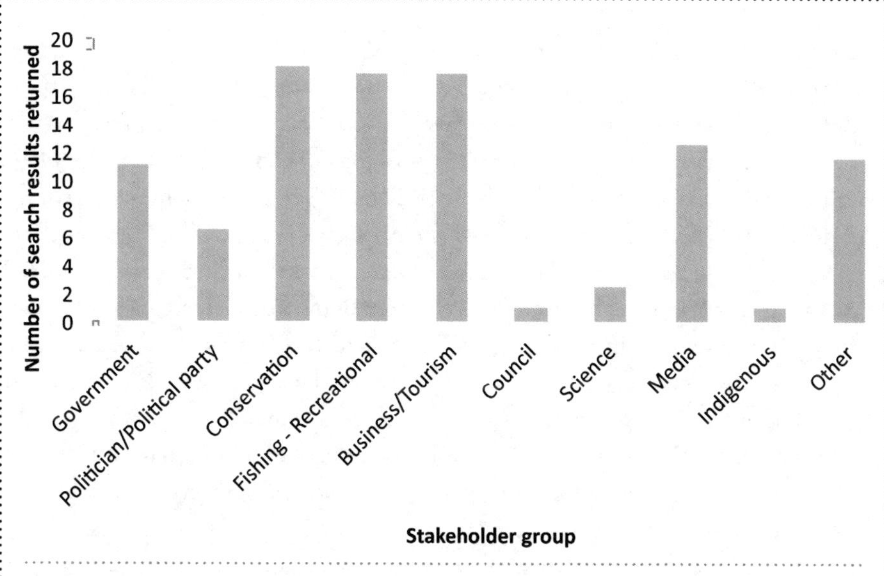

Figure 2: Representation of major stakeholder groups in the first 100 results of "NSW marine parks" (date of search: October 2011).

STAKEHOLDERS IN THE PLANNING PROCESS

Stakeholders had the opportunity to be involved in the marine park planning processes through advisory committees, stakeholder and public meetings, and written submissions (Voyer et al., 2012). A comparison was made of the submissions received in the second period of public consultation in each park, following the release of a draft zoning plan for comment. Despite the Port Stephens Marine Park having a neighbouring population nearly three times the size of the Batemans Marine Park, the Batemans Marine Park received more submissions (Table 2).

Table 2: Population Size of Marine Park Area Compared with the Number of Submissions Received During the Planning Process

	Port Stephens Marine Park	Batemans Marine Park
Population size*	87,972	32,633
Total number of submissions**	4,399	4,988#

* (Powell and Chalmers, 2005, 2006).

** (Marine Parks Authority, 2006a, 2006b).

Statistically significant difference

People making submissions were requested to nominate their interests in the marine park from a range of categories. For the Batemans Marine Park, recreational fishers dominated the submissions (81 percent of respondents), compared with 52 percent for the Port Stephens Marine Park. By way of contrast, the Port Stephens Marine Park received significantly more submissions from people nominating conservation as one of their main interests in the marine park (46 percent), compared with 22 percent for the Batemans Marine Park (Marine Parks Authority, 2006a, 2006b).

A number of submissions were received from individuals referring to alternative zoning plans developed by conservation groups which called for an increase in protection through sanctuary zones. For the Port Stephens Marine Park 27 percent of the total submissions supported this alternative plan, compared with a similar plan in Batemans Marine Park that was supported in 15 percent of submissions. By contrast, the Batemans Marine Park was dominated by submissions received from recreational fishers in the form of two form letters circulated by fishing groups which called for a decrease or complete removal of some or all sanctuary zones; they made up 71 percent of the total submissions received. No comparable form letter was mentioned in the Port Stephens Marine Park report (Marine Parks Authority, 2006a, 2006b).

Analysis of the place of residence of the people who made submissions to the government during the marine park planning processes provides an insight into whether these parks were seen as a purely local issue or had greater resonance on a regional, national or global scale (Table 3). The majority of the submissions received on the Port Stephens Marine Park came from the local area (50 percent), while for the Batemans Marine Park the largest proportion of submissions came from neighbouring regions and other parts of NSW (43 percent), compared with only 28 percent from the local area.

Table 3: Place of Residence of Individuals or Organizations Making a Submission on the Marine Park Zoning Plans

Location of respondents	Port Stephens Marine Park		Batemans Marine Park	
	Actual	% total	Actual	% total
Marine Park area	2,216	50#	1,412	28#
Neighbouring region	199	5#	461	9#
Sydney metropolitan	787	18	877	18
NSW (all other areas)	531	12#	1,713	34#
Other states and international	304	7	188	4
No postcode given (incl. organizations)	362	8	337	7
Total	**4,399**	**100**	**4,988**	**100**

Statistically significant difference

THE ROLE OF SPOKESPEOPLE: CONSERVATION VS. FISHING, GLOBAL VS. LOCAL?

In the planning process for Port Stephens Marine Park a large proportion (83 percent) of the conservation spokespeople came from one of a number of major NGOs that were active in the planning processes of both marine parks, but were also involved in a range of other campaigns at state, national and global scales (Figure 3). In contrast to this, in the Batemans Marine Park greater prominence was given to spokespeople affiliated with local conservation bodies focused on community-based responses to local environmental issues (52 percent).

A similar pattern is clear when examining those articles classed as having fishing interests (recreational or commercial) as primary definer. The category of "community groups" was also included in this analysis, as this category was heavily repre-

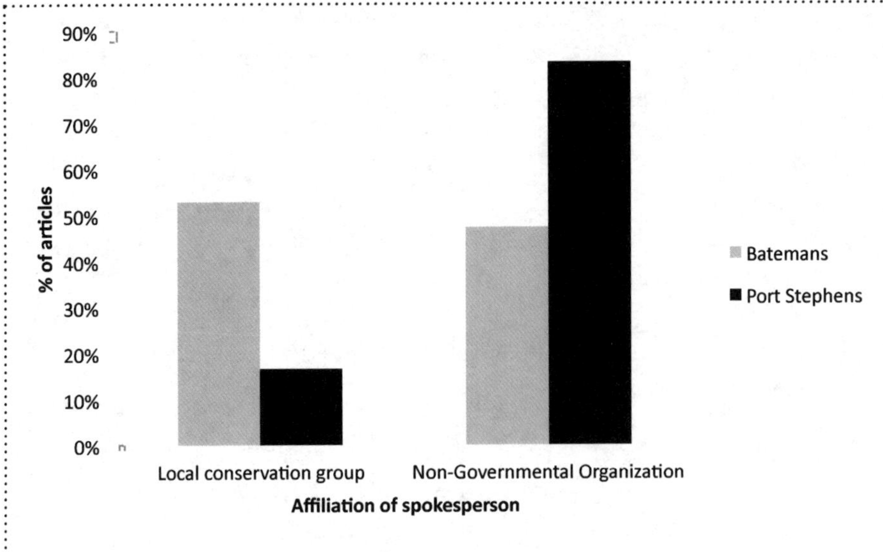

Figure 3: Media sources for conservation spokespeople in Batemans and Port Stephens Marine Parks in the period 2005–2010.[3]

sented in the Batemans media (see Figure 1). For the Batemans Marine Park the category "community group" was comprised of two local groups that were aligned with recreational fishing interests but also claimed to represent a broader constituency of commercial fishers, business owners and the general public. Together they made up 48 percent of the Batemans articles in which the primary definer was a fishing interest (Figure 4). Representatives of local fishing clubs (usually aligned with local pubs or clubs) also had a significant voice in the Batemans media. Recreational fishing spokespeople in the Port Stephens Marine Park process were predominately affiliated with the local branches of the state-wide recreational fishing lobby group Ecofishers. Ecofishers were active in both marine park areas, but were far less prominent in the Batemans articles.

Figures 3 and 4 suggest a difference in the editorial approach to media coverage of the marine parks in each community, with editors in the Batemans newspapers showing an apparent preference for local sources in their coverage of the issue.

COMPETING FRAMES OF SUPPORT AND OPPOSITION

A content analysis was conducted using purposive sampling of the competing frames found in local news articles at critical events within the marine park planning pro-

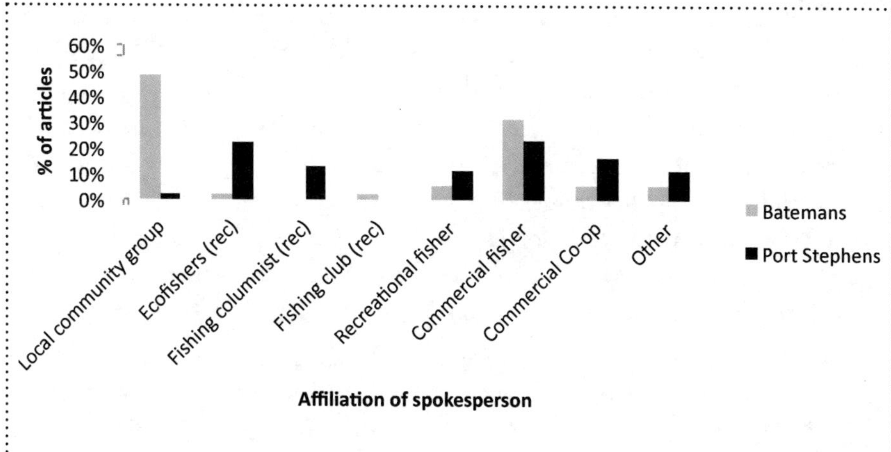

Figure 4: Media sources for fishing spokespeople in Batemans and Port Stephens Marine Parks in the period 2005–2010.[4]

cesses, namely, the announcements and declarations of the parks, the release of the draft and final zoning plans and the commencements of the final plans. This analysis examined the messages associated with the key competing stakeholder groups, outlined above. These messages were examined and coded for what features or aspects of the proposed marine park they supported or criticized, and what these messages excluded. For example, supportive frames might include reference to "fishing benefits" or "improved tourism opportunities", while critical frames might highlight aspects of the proposal to do with its costs to taxpayers, the political motivations behind it or the "socio-economic impacts" of the park. These codes were then compared to the online material of key stakeholders.

WHY DO WE NEED A MARINE PARK? FISHING BENEFITS VS. BIODIVERSITY THREATS

A number of surveys into community attitudes towards MPAs have demonstrated that people are more likely to support an MPA if they believe it is needed (Sutton and Tobin, 2009; Thomassin et al., 2010). Therefore, debates over an MPA proposal often involve arguments for and against its necessity, and these are important aspects of the messages of supporters and opponents of MPAs, given they are likely to have a major influence on community opinion. Analysis revealed two main "need" frames in the messages of supporters and opponents in the local media of the marine parks. They were the "benefit" frame, focusing largely on the benefits the park would (or would not) deliver to local communities, and the "threat" frame, which included identification of threats to the marine environment and/or identification of if/how the marine park would address these threats.

Supporters of the marine parks, including conservation groups and government politicians, tended to be associated with the "benefit" frame in the news articles included in this analysis, rather than a "threat" frame. In part this may be due to the fact that the national policy frameworks for MPAs in Australia define MPAs as primarily a precautionary measure, or a system-wide "insurance policy" against loss of biodiversity rather than a response to site-specific threats. Government policy states that the primary objective of the national system of MPAs is:

> ...to contribute to the long-term ecological viability of marine and estuarine systems, to maintain ecological processes and systems, and to protect Australia's biological diversity at all levels. (Australian and New Zealand Environment and Conservation Council Task Force on Marine Protected Areas, 1998: 5)

MPAs are therefore not intended as a fisheries management tool aimed at improving fishing productivity or fishing experiences. Whilst improvements in fish stocks may be a *possible secondary benefit* of MPAs, this is not an *objective* of the Australian MPA system (Department of Sustainability, Environment, Water, Population and Communities, 2011; Marine Parks Authority, 2001). Despite this policy, the "benefit" frame found in the local news media seldom related to biodiversity protection, but rather to fishing and tourism improvements, as seen in the excerpts below (emphases added):

> ... conservation groups such as The Wilderness Society, the National Parks Association of NSW and the Nature Conservation Council of NSW claim marine sanctuaries and the no take zones within marine parks *enhance fish populations and fishing experience* in surrounding areas.
>
> *Great LBQakes Advocate*, 30 November 2005: 7

> The park could also bring with it *widespread boons for the Bay's human population*. Mr Fleming said it would *ensure the long-term viability of tourism* on the Nature Coast and *boost charter boat business opportunities.*
>
> Spokesperson, Local conservation group (Coastwatchers), *Bay Post*, 2 December 2005: 4

> The final result is that you can fish in 80 percent of the marine park and that *fishing is certain to get better.*
>
> Minister for the Environment, *Bay Post*, 15 December 2006: 8

While references to the biodiversity benefits of marine parks was not entirely excluded from these frames, they were usually included as an aside or postscript to references to the benefits to fish stocks or fishing experiences. The main "need" frames used to support the marine park emphasized a possible secondary benefit of MPAs and de-emphasized their primary objective. Perhaps more importantly, however, a focus on the "benefit" frame over a "threat" frame meant there was very little serious discussion about local threats to the marine environment, and how the marine park would address them. These threats become implied rather than explicitly scrutinized. Frames that focused on improvements in fish stocks implied that fish stocks were in decline, or under threat, and defined fishing (or overfishing) as a "problem" that the marine park would address without providing any specific evidence to the reader to support this claim.

Much more explicit links are made to these threats in the online material of conservation groups, which has a greater emphasis on the "threat" frame. Here they list overfishing as the first in a long list of threats to a unique and vulnerable environment:

> The NSW marine environment is home to thousands of different species of aquatic plants and wildlife. A unique mixing of coastal currents means that about 80 percent of these marine plants and animals are found only in Australian waters.
>
> *But these plants and animals are under threat.* Overfishing, pollution, climate change, introduced marine pests, emerging diseases, inappropriate development and lack of adequate protection are all putting our marine wildlife *at risk.*
>
> Nature Conservation Council of NSW (www.nccnsw.org.au/marine, accessed October 2011, emphasis added)

Not surprisingly, people involved in fishing in the communities in question challenged being framed as the "problem" and sought to redefine the way both the "problem" and fishing were presented in the media frames of marine park supporters. This was done through the use of a "threat" frame that highlighted alternative threats to the marine environment (apart from fishing) which they suggested the marine park would be ineffective at managing, including pollution, urban development and habitat destruction. These threats were contrasted with fishing, which was presented as a benign activity, implying that restrictions on fishing were unnecessary, unfair and inequitable. The media frame of park opponents emphasized alternative threats to the marine environment and de-emphasized the threats posed by fishing. They also de-emphasized the value of MPAs in managing threats apart from fishing.

> Newcastle's Commercial Fishermen's Co-operative has made public 20 years of fish-catch records it says shows *the region's waters are alive and well.* Co-op manager Bill Pearce said the figures showed the State Government's push to lock up rivers and oceans in marine parks was *misguided and unjustified.*
>
> *Newcastle Herald*, 6 December 2005: 6 (emphases added)

> ... "all they're doing is *diverting the attention away from the problems with pollution* and the massive expense it would be to fix it," he said....
>
> ... "Recreational fishing had *a low impact on biodiversity*", he said. "So...*why are we being targeted?*"
>
> Spokesperson, Ecofishers, *Newcastle Herald*, 9 May 2006: 4–5 (emphases added)

Fishing spokespeople also challenged the validity of the "benefit" frame by focusing on possible social and economic costs, particularly as they related to loss of tourism, loss of income to local businesses or jobs to commercial fishers, and the loss of family recreation.

> "They're very poor reasons to put a lot of people out of business, not just charter operators but motels, tackle shops and tourism businesses in general," Mr Stuart said. "Everyone suffers."
>
> Bait and Tackle store owner, *Narooma News*, 19 July 2006: 2

> Dr Creagh said family fishing holidays within the park area would become a thing of the past, while commercial line fishers would be hardest hit by the draft zoning plan.
>
> Spokesperson, Community group (Narooma Ports Committee), *Bay Post*, 19 July 2006: 4

Finally, marine park opponents reinforced their arguments against the necessity of a marine park by questioning the political motivations behind the declarations, suggesting they were purely a political exercise rather than one designed to address any environmental outcomes.

> The State Government has been losing popularity in the polls and is not confident in winning next year's state election without the extreme Green preferences.
>
> Spokesperson, Ecofishers, *Port Stephens Examiner*, 11 May 2006: 2

Many of these themes are repeated in the online material of fishing groups such as Ecofishers. However one message the web material contains that was seldom encountered in the media is the concept of MPAs as an infringement of fishers' "rights".

> Our alignment is for the *rights and responsibilities of recreational fishers*. We are totally committed to you and *preserving your rights*.
>
> Ecofishers homepage (www.ecofishers.com, accessed October 2011, emphasis added).

The introduction of the concept of fishers' "rights" elevates the debate from a complex dispute over access and use to a clash of ideology, and it highlights a perceived shift of the conservation movement from the fringes of power to the empowered. It demonstrates that these resistance movements positioned themselves as the "repressed" in this debate, fighting to protect their way of life and their voice from the government, the conservation movement and those they saw as eroding their rights.

WILL A MARINE PARK WORK? DODGY SCIENCE VS. SCIENTIFIC CONSENSUS

Closely related to questions surrounding the necessity of MPAs are questions around their effectiveness in achieving their stated goals. As seen in the previous section, arguments from supporters and opponents about the need for the marine parks were framed around an area outside the actual objectives of the parks. Frames relating to the likely effectiveness of the two parks followed a similar trend by focusing on the fisheries management credentials of MPAs, one of the most contested areas of MPA science (Barrett et al., 2007; Gladstone, 2007; Kearney, 2007a, 2007b, 2009). This contestation was reflected in the media coverage, and allowed for an inflated sense of uncertainty surrounding MPA science generally. In the majority of cases, however, sources were politicians, conservation groups and fishing groups rather than academics or scientists.

Conservation groups sought to frame the marine parks as being supported by science by emphasizing "big picture" scientific consensus statements about the value of MPAs in general, on a global scale. They cited international examples of the so-called "spillover effect" whereby it is hypothesized that fish numbers within no-take MPAs increase to the point where excess individuals migrate (or "spillover") into surrounding areas, improving fish stocks and the fishing experience in areas where fishing is allowed.

> More than 1600 international scientists and conservationists have backed a call for at least 20 percent of the seas to be *protected from fishing* by the year 2020.
>
> Spokesperson, Conservation NGO (The Wilderness Society), *Great Lakes Advocate*, 30 November 2005: 7 (emphasis added)

> A recent report from New Zealand claims the establishment of marine reserves has led to a boost in tourism, a significant improvement in fish stock and the re-establishment of the natural food chain stimulating the return of the larger predator fish.
>
> Spokesperson, Conservation NGO (National Parks Association), *Narooma News*, 7 December 2005: 9

Fishing groups challenged this aspect of MPA science with competing "dodgy science" frames. They refuted the existence of any evidence of the "spillover effect" and challenged the "big picture" statements by focusing on the scientific process within the local marine park area, suggesting that the body of international MPA science was irrelevant or inapplicable to the local area. Protection targets such as the 20 percent referenced above, which is derived from scientific consensus statements and international agreements on global MPA targets, were redefined as being aligned with a Green agenda or government policy and therefore represented a purely political rather than scientific goal.

> "There is *no proven scientific evidence anywhere in the world* that locking the general public out of sanctuary zones will make any significant improvement to the fish stocks" he said.
>
> Spokesperson, Ecofishers, *Port Stephens Examiner*, 11 May 2006: 2 (emphasis added)

> It is crucially important for the future of the Marine Park that the zoning follow logical and scientific reasoning rather than a desire to meet the *hypothetical figure dictated by Sydney bureaucrats of the NSW Department of Conservation.*
>
> Spokesperson, Community Group (Coastal Rights Association), *Bay Post,* 19 July 2006: 4 (emphasis added)

CONCLUSION

The marine park debate in NSW points to the complexity of modern environmental campaigns, particularly in translating global conservation messages to a local audience. The Port Stephens planning process was characterized by significantly more submissions from locals and "conservationists" and a greater level of support for increased sanctuary zone protection than the Batemans process. However, the media coverage of the issue was dominated by established state- and nation-wide NGOs and lobby groups. It is possible that the prominence of the larger conservation NGOs in the Port Stephens local news media allowed them to gain a new local audience for their more globally focussed online messages, and this may have had a direct impact on the success of their campaigns in that park. This analysis, however, highlights the difficulties in effectively localizing these global messages. Local media outlets tended to use a "benefit" frame to report the messages of conservation groups and government politicians, emphasizing fisheries management and tourism improvements to local communities over the wider goals of biodiversity protection. This may have been a deliberate strategy by conservation groups to make their message relevant to local audiences, or it may have been an editorial decision to concentrate on those aspects of their message most relevant to their readership. Regardless, it effectively shifted the focus of the debate onto an area which is not a primary objective of MPAs, and is one of the most contested areas of MPA science. This may have fed perceptions of mistrust in the messages of government and conservation groups in relation to the marine parks, and confusion as to their role and function in wider marine conservation management. While it is understandably difficult to communicate the somewhat intangible benefit of "biodiversity protection", this example demonstrates how global messages can be compromised by focusing on those components which make it more easily understood and relevant to local readers. In addition, by focusing on benefits at the expense of a rigorous explanation of threats, arguments relating to the "need" for a marine park were weakened and concentrated on an area on the periphery of the fundamental purpose of MPAs.

In contrast to the Port Stephens Marine Park, submissions received in the Batemans planning process were heavily dominated by recreational fishing interests, as well as people outside the local marine park area, yet the media coverage in the area was highly localized. This suggests that fishing interest groups were successful in regionalizing their local messages, most likely through the use of online tools such as Internet forums and websites. Wider relevance was given to local arguments by linking the local battle to a state- and/or nation-wide agenda by the government to attract Green votes and strip fishers of their "rights". This concept surrounding the rights of fishers, coupled with an emphasis on the impacts of the parks on fishers, points to a positioning of fishing groups as representatives of the average fisher, the "underdogs" or "victims" in the marine park debate, fighting against the establishment to protect their way of life.

The growth in power and influence of conservation groups and the Greens political party during the study period was matched by a corresponding growth of environmental "resistance" groups who borrowed the techniques perfected by the larger conservation NGOs to fight what they perceived as an "extreme" conservation agenda. These techniques included harnessing grassroots support, leading protest actions and conducting online and traditional lobbying campaigns. Digital media appears to have been used as a tool to enable conservation NGOs to make a transition into a powerful lobby group and political force, but with this shift in power came the emergence of a new "fringe" group claiming the position in the debate of an anti-establishment protest movement.

It is likely that the shifting dynamics of power in the debate over marine parks in NSW played an important role in the 2011 NSW state election. The continued growth of a grass-roots movement around recreational fishing allowed politicians to capitalize on the conflict by promising a redistribution of power away from the Greens (e.g., Gay, 2009). As a result, since its election the new NSW conservative state government has made some dramatic changes to the management of NSW marine parks, many of which appear to respond directly to the critical "threat" and "dodgy science" media frames seen during the planning of the Batemans and Port Stephens Marine Parks. Digital media is unlikely to have been a direct cause of the ebb and flow of political influence of key interest groups in this debate, but rather a powerful enabling tool to assist their campaigns and capitalize on the political climate of the time. It remains to be seen whether fishing groups will be able to sustain a position of repressed protest movement in the new political climate, although the trends seen throughout this debate provide insight into the constantly changing face of environmental politics.

ENDNOTES

1 For the remainder of this chapter we will use the term *MPA* in a generic sense to describe marine managed areas at a global or regional scale, and the term *marine park* in a specific sense to discuss the form of MPA used at a local scale in the two study areas.

2 These data have been standardized to take into account the differences in publication frequency of the newspapers included in the study.

3 These data have been standardized to take into account the differences in publication frequency of the newspapers included in the study.

4 These data have been standardized to take into account the differences in publication frequency of the newspapers included in the study.

PART 3

Communicating Crises

▪ CHAPTER ELEVEN

Greening Wildlife Documentary

MORGAN RICHARDS

> *The loss of wilderness is a truth so sad, so overwhelming that, to reflect reality, it would need to be the subject of every wildlife film. That, of course, would be neither entertaining nor ultimately dramatic. So it seems that as filmmakers we are doomed either to fail our audience or fail our cause.*
>
> — Stephen Mills (1997)

Five years before the BBC's *Frozen Planet* was first broadcast in 2011, Sir David Attenborough publicly announced his belief in human-induced global warming. "My message is that the world is warming, and that it's our fault," he declared on the BBC's *Ten o'Clock News* in May 2006. This was the first statement, both in the media and in his numerous wildlife series, in which he didn't hedge his opinion, choosing to focus on slowly accruing scientific data rather than ruling definitively on the causes and likely environmental impacts of climate change. *Frozen Planet*, a seven-part landmark documentary series produced by the BBC Natural History Unit and largely co-financed by the Discovery Channel, was heralded by many as Attenborough's definitive take on climate change. It followed a string of big budget, multi-part wildlife documentaries, known in the industry as landmarks,[1] which broke with convention to incorporate narratives on complex environmental issues such as habitat destruction, species extinction and atmospheric pollution. David

Attenborough's *The State of the Planet* (2000), a smaller three-part series, was the first wildlife documentary to deal comprehensively with environmental issues on a global scale. A few years later, BBC series such as *The Truth About Climate Change* (2006), *Saving Planet Earth* (2007) and *Frozen Planet* (2011) finally gave environmental issues the mainstream prominence and high production values they were lacking.

For over fifty years the BBC Natural History Unit has produced some of the most powerful and iconic visions of wildlife and nature. But the blue chip programmes for which it is renowned, named for their ability to sell well in international television marketplaces, have been largely untroubled by the consequences of climate change and other environmental issues. Instead, these issues have been relegated to the margins of the genre, while spectacular or action-packed visions of animal behaviour have taken centre-stage. In 2004 Simon Cottle suggested the wildlife genre's "failure to produce programmes informed by environmental and political issues relates to the shelf-life, and hence longevity, of these programmes as a commodity, as well at their potential international appeal" (2004: 96–97). Whilst acknowledging that occasional series dealing with themes of global environmental threat have been produced, he criticises the genre's "chronic lack of engagement" with ecological politics as "inexcusable" within the context of the rise of environmental social movements and a growing environmental consciousness (2004: 97). It is a central argument of this essay that wildlife documentary has more recently undergone a green transformation. Since the new millennium, wildlife documentaries have incorporated environmental politics and issues in new ways, allowing them to gain a greater level of prominence, thus countering the view that the dynamics of international television have rendered environmental messages incompatible with big-budget documentary series.

The emergence of what I call "green chip" programming represents a key turning point in the wildlife genre's engagement with the science of climate change and environmentalism. But the rise of green chip programming has been accompanied by a shift in how environmental issues are produced and framed. Images of catastrophic landscapes and poignant stories of gorillas and tigers on the verge of extinction are now accompanied by narratives that stress the audience's potential to enact change by donating to individual conservation projects and engaging in waste reduction schemes, renewable technologies and adaption to global warming. This shift in wildlife documentary's engagement with these issues—from the condemnatory "lectures" on environmental degradation and species extinction, which began to appear in the margins of the genre in the 1980s, to the construction of aware consumers in programmes like *Saving Planet Earth* and *The Truth About Climate Change*—has been shaped by wider environmental politics and other media representations of climate change.

The greening of wildlife documentary coincided with the release of a report by the International Panel on Climate Change (IPCC) in early 2007, based on a near consensus of scientific opinion on the causes and probable impacts of anthropogenic global warming. This report, which included grim forecasts about rising sea levels, extreme global weather patterns and soaring temperatures, marked a shift in the wildlife genre's treatment of environmental issues, much as it "proved to be a transformative moment in the news career of climate change" (Cottle, 2009a: 506). The near consensus of the world's climate scientists saw climate change gain recognition as a "global crisis", shifting the news values of balance and impartiality which had allowed a small but media-savvy contingent of climate change skeptics and deniers to cast doubt on the science of global warming. As news media embraced climate change as a global concern and began deploying spectacular images in their reports (Cottle, 2009a; Lester and Cottle, 2009), wildlife documentary's long-standing avoidance of controversial issues began to give way to more nuanced, if upbeat, explorations of climate change and other environmental issues. As I will demonstrate, the exclusion of environmental issues from wildlife documentary stems in part from the wildlife genre's presentation of uncontroversial science.

My argument will proceed in three parts. First I consider the politics of blue chip programming, examining how the difficulties of filming animals in the wild and the expense of obtaining detailed footage of animal behaviour led to the dominance of the blue chip format. While this format was technically and economically expedient, it meant that environmental issues were routinely excluded from the majority of wildlife documentaries. Next I investigate how the absence of environmental issues in BBC landmark wildlife series, which attract the largest international audiences of any wildlife documentary, is implicated in the very narrow scientific paradigm of "natural history" programming and the international co-production deals that underwrite the multi-million-pound budgets needed to produce these series. Finally, I examine the rise of green chip programming and consider some of its problematic aspects, namely, its implications for how environmental issues are constructed on screen and how, and in what form, these issues reach international audiences.

THE POLITICS OF BLUE CHIPS

Wildlife documentary has come to assume a key role in the public understanding of science and environmental issues, generating popular awareness and helping to shape public engagement with environmental politics and conflict. As our contact with the wild has become more remote, wildlife documentary has become the primary frame through which industrialised people view wildlife and nature. To give just one example, 48 percent of the UK population watched at least 15 minutes of *Frozen Planet* (2011), a remarkable figure considering the fragmentation of audiences brought about by the rise of digital broadcasting and online media. But, prior to the recent greening of wildlife documentary, the rise of the blue chip format meant that environmental programmes remained, for the most part, on the margins of the genre.[2]

Derek Bousé outlines seven key characteristics of blue chip programming. Blue chip programmes depict charismatic mega-fauna such as big cats, primates and elephants; they contain spectacular imagery of animals in a "primeval wilderness"; they incorporate dramatic and suspenseful storylines; they generally avoid science, politics and controversial issues such as wildlife conservation; they are timeless, carefully framing out any historical reference points which might date the programme or effect future rerun sales; and they avoid people, including presenters and all artefacts of human habitation (Bousé, 2000:14–15). These elements are not hard and fast, nor have they always co-existed. But the commercial success of this format, which was first realised in Disney's *True-Life Adventure* films (1948–1960), set the precedent for wildlife documentary's persistent marginalisation of environmental issues.

In the late 1940s Disney hit upon a lucrative formula that brought wildlife documentary to mainstream cinema audiences for the first time. The ten short and four feature-length films in its *True-Life Adventure* series were influential and innovative; they were also thoroughly anthropomorphic and sentimental. Despite these drawbacks, Disney's legacy, even to BBC wildlife programming, is undeniable. Bousé argues that Disney effectively codified the genre, bringing its conventions into focus as "a discrete and recognisable cinematic form" (2000: 62). There was nothing inherently new about this approach. It drew upon earlier forms of wildlife filmmaking, synthesising elements of safari films, scientific-educational films and ethology films, and incorporating them with aspects of other, more popular genres such as cartoons, comedy and Hollywood westerns.[3]

Disney's breakthrough lay in its ability to dramatise the natural world and bring wild animals and nature to life using full-colour cinematography and lavish musical scores—the full theatrical works, designed to bring wildlife into the mainstream. It was their glossy finish and sense of drama, more than anything else, which essentially distinguished Disney's films from other wildlife fare and gave them a commercial edge, an edge that was further honed through Disney's monopoly over distribution. Despite their high production values, the *True-Life Adventures* had excellent profit margins.[4] The success of the films, as Cynthia Chris points out, was also directly linked to Disney's "distinctive brand identity," which allowed it to pair a Disney live-action short with a Disney animated feature (2006: 29). Films such as *Seal Island*

(1948) and *The Living Desert* (1943) were entertaining and exciting, but they also represented nature as an infinitely renewable and abundant resource.

In spite of the absence of explicit conservation messages in its wildlife films, Disney won the support of conservation organisations like the Wilderness Society and the Audubon Society in the 1950s. Greg Mitman notes that by bringing beautiful visualisations of nature into people's homes, Disney "established film as an important propaganda tool in the enlisting of public support for environmental causes" (1999: 130). One reviewer in the Wilderness Society's publication, *The Living Wilderness*, praised Disney's portrayal of "the simple beauty of untouched woodlands and their wild inhabitants". The Audubon Society even saw fit to award Walt Disney with the Audubon Medal in 1955, for "distinguished service to the cause of conservation" (quoted in Mitman, 1999: 123). Yet, the success of Disney's blue chip model proved that making nature entertaining and popular was, by and large, incompatible with the depiction of more complex ecological environments that included people.

Audiences were attracted by the *True Life Adventures*' presentation of a sentimental and sanitised vision of nature, which, although not always harmonious, could be understood and rationalised in simple terms. They were entertaining and educational, but not too scientific. Disney instinctively favoured filmmakers with an "experiential" connection to nature, based on the craft of the woodsman or that of the amateur naturalist and acquired through time spent in the field, rather than those with a more purely scientific bent (Mitman, 1999: 118). Scientists were engaged in making a number of the films, but only a few films included scientific advisors in their credits. The preferred narratives of many of the films, with their motifs of young animals struggling to survive and of journeys undertaken in harsh and unforgiving environments, were more theological than scientific. This is best demonstrated by *Nature's Half-Acre* (1951), a two-reel film ostensibly about the origin of species, which manages to make no mention of evolution. Instead, as Mitman observes, the "web of life" is explained in theological terms reminiscent of the nineteenth-century Linnaean notion of the balance of nature, in which species vary and keep one another in check (but never explicitly evolve) under "Nature's" watchful eye (1999: 128). As such, they were designed to keep conservationists, scientists and evangelicals onside. And for a time, they succeeded.

Disney provided a tried and tested format that was endlessly remodelled on television by different practitioners keen to capitalise on its value as both an entertaining and educational resource. It was some time before documentaries with similarly high production values to those of Disney's wildlife films proved viable in Britain's fledgling television industry. With practically non-existent budgets and acres of broadcast schedules to fill, wildlife television developed its own distinctive forms. The BBC

Natural History Unit (NHU) was established in Bristol in 1957, but producers in the BBC West Region Film Unit had been making wildlife television programmes since the early 1950s. The BBC's first wildlife television series, *Look* (1955–1965), was essentially a naturalist's lecture with excerpts of film. The acclaimed naturalist Peter Scott casually chatted with his guests, who included scientists, naturalists and amateur naturalist filmmakers, on a set mocked up to appear like a naturalist's study. The BBC clung doggedly to its naturalist's lecture format throughout the 1960s, and, somewhat predictably, it fell behind its competitors in terms of ratings.

Survival Anglia, a production unit that was part of the British commercial network ITV, had begun making wildlife programmes loosely following a blue chip style in 1961. Programmes in the long-running *Survival* series (1961–2001) on ITV were carefully packaged as family entertainment and aimed at a mainstream audience, in a similar manner to Disney's *True-Life Adventures*. When the BBC finally embraced the blue chip format in the late 1960s, it cultivated a more rigorously scientific approach that served to differentiate its programmes from those of Survival Anglia, but still retained many of the trademark features that underscored the success of this format. The BBC's blue chips combined spectacular cinematography with scientific narratives that incorporated moments of drama and suspense. This scientific but still dramatic formula was built on the NHU's close association with scientists and amateur naturalists, which had been fostered since its outset. In a report on the first five years of the unit's operation, Desmond Hawkins, then a senior producer in the NHU, argued that science should be the driving force behind the BBC's wildlife programming:

> The spirit of scientific enquiry must have pride of place. In handling this subject we expose ourselves to the critical scrutiny of scientists, and their approval is an important endorsement. Moreover, it is their work that throws up the ideas and instances and controversies from which programmes are made. We look to them as contributors, as source material, as consultants and as elite opinion on our efforts. In short, we need their good will. (Hawkins, 1962: 7)

By placing scientific narratives at the centre of its programmes and simultaneously capitalising on the dramatic potential of the blue chip format, the BBC were able to craft a unique niche in the international television market. Blue chip programmes produced or commissioned by the NHU for BBC strands such as *The World About Us* (1967–1983), *The Natural World* (1984–present), and *Wildlife on One* (1977–2005) have been sold internationally since the late 1960s, and more recently broadcast on the Discovery Channel and Animal Planet.

David Attenborough's landmark series, beginning with *Life on Earth* (1979) and continuing through *The Living Planet* (1984) and *The Trials of Life* (1990) to his latest series *Frozen Planet* (2011), constitute the BBC's quintessential variant of the blue chip format. They differed from conventional blue chip programmes in their overarching scientific narratives and in their use of Attenborough as a trusted guide and presenter. Rather than focusing on a particular species or exploring the ecology of a particular environment, as many blue chip programmes had done before, landmarks had the space to develop and dramatise complex scientific ideas, weaving together footage and narratives from around the globe. Yet, in spite of the BBC's "scientific" approach to wildlife programming, the rise of the blue chip format meant that the vision of the natural world that became standardised on television sets worldwide was largely apolitical. In the late 1960s the blue chip format, with its studious avoidance of all aspects of human culture, became the industry standard primarily because it was a format that could be adapted for sale in different television marketplaces. The dominance of blue chip programming in its various forms ensured that the more complex "realities" of environmental politics rarely encroached on the constructed reality of the wildlife genre. In many respects, the BBC's dual focus on science and ever more spectacular visualisations of nature proved just as blind to environmental issues as Disney's sugar-coated and sensationalised *True-Life Adventures*.

KEEPING IT BLUE: FRAMING SCIENCE AND ENVIRONMENTAL ISSUES ON SCREEN

In 2002 environmental campaigner George Monbiot published an article in *The Guardian* in which he criticised the exclusion of environmental issues in wildlife documentaries. "There are two planet earths," he wrote. "One of them is the complex, morally challenging world in which we live, threatened by ecological collapse. The other is the one we see in the wildlife programmes". He singled out David Attenborough for his harshest criticism:

> He shows us long loving sequences of animals whose populations are collapsing, without a word about what is happening to them. Indeed, by seeking out those places, tiny as they may be, where the habitat is intact and the population dense, the camera deliberately creates an impression of security and abundance. (Monbiot, 2002)

In response, Attenborough defended his programmes by citing *The State of the Planet* (2001), his recent assessment of the "present ecological crisis", and arguing that the main focus of his other series was "zoology", an academic discipline which he clearly viewed as separate from environmental politics and conservation (Attenborough, 2002).

This episode sheds light on one of the central paradoxes of the wildlife genre. BBC wildlife documentaries, particularly those narrated or presented by David Attenborough, are invested with scientific authority. But following the narrow paradigm of zoology or natural history, they represent a very particular brand of science: that which is already proven and beyond doubt. Safe science. By focusing on scientific theories from within the branch of biology that relates to the anatomy and classification of animals and plants, wildlife documentaries have remained fixated on scientific theories that are supported by a majority in the scientific community and are subject to uncontroversial media treatment. Attenborough's defence that his programmes cater to an interest in "zoology" serves to emphasise the point that the exclusion of environmental issues from the majority of wildlife documentaries arises in part from the wildlife genre's focus on uncontroversial science.

This view is in keeping with Michael Jeffries' assessment that "the science of natural history not only occupies its own broadcasting niche; it works to a different paradigm" (Jeffries, 2003: 527). According to Jeffries, the wildlife genre, as epitomised by Attenborough's programmes, is stuck in an "old ecology of equilibrium and adaptation combined with romantic awe and wonder", while science documentaries, particularly those in the BBC's flagship science series *Horizon* (1964–present), "represent the world (and the rest of the universe) as changeable, challenging, contingent" (2003: 543).

Horizon has received praise for its willingness to scrutinise science as a dynamic and contested field shaped by wider social and political processes (Jeffries, 2003; Secord, 1996; Darley, 2004; Silverstone, 1984). Programmes in the *Horizon* series make use of "talking head" interviews from scientists and other experts to construct the contours of a particular scientific debate, in contrast with the uniform parade of spectacular imagery in blue chip wildlife documentaries. However, as Darley and many others acknowledge, despite its willingness to represent scientific disputes, *Horizon* still presents viewers "with assured and univocal stories of discovery and progress" (Darley, 2004: 232). In other words, the practice of science is still portrayed as a heroic, if contested, struggle that ends in certainty. The sheer breadth of topics covered by *Horizon*—theoretical physics, biomedical science, palaeontology and archaeology, to name just a few—contrast with the narrow paradigm of natural history, which seems almost risibly Victorian. Jim Secord points out that "within the realm of scientific practice, the term 'natural history' is now itself something of a museum specimen"; he even suggests, "to call someone a 'natural historian' sounds quaintly old-fashioned or even abusive" (1996: 449).

One of the results of the generic and industry-based separation of science and natural history programming is that *Horizon* programmes have been much better at tackling controversial environmental issues.[5] David Attenborough's landmark series, by contrast, with their focus on anatomical adaptations and concise explanations of animal behaviour, evoke nature as "balanced and ordered", and deliberately avoid controversy (Jeffries, 2003: 529). This is best illustrated by the first landmark wildlife series, *Life on Earth* (1979), which outlined the story of evolution in thirteen parts. Rather than focusing on new discoveries in the emerging field of ethology (the study of animal behaviour), for which Konrad Lorenz, Niko Tinbergen and Karl von Frish had won the Nobel Prize in 1973, this series used living animals to chronologically chart the evolution of life on earth. At the time, producers in the BBC Natural History Unit criticised *Life on Earth* for presenting nineteenth-century science. They were critical of the fact that the most expensive wildlife series to date, with a budget of £1 million, disregarded the latest scientific discoveries (Parsons, 1982). The landmark format proved to be a hugely popular format. *Life on Earth* attracted average UK audiences of 15 million—an exceptionally high figure for a documentary at that time on BBC2—and an even larger global audience. Subsequent landmark series, which retained the sense of awe and wonder at the beauty of the natural world, were routinely broadcast in more than 100 territories.

The stability of the blue chip format, with its reliable economic returns, meant that BBC landmark series shied away from controversial topics in science and environmental politics. In any case, Attenborough regarded the narrow focus on zoology in his landmark series as entirely justified. When asked in an interview in 1984 about his responsibility to the environment as a filmmaker, he argued:

> As a conservationist, I think I would be doing the world a great disservice if I tacked onto the end of every single programme that I did, a little homily to explain yet again that mankind is wrecking the environment that I have been showing. My job as a natural history filmmaker is to convey the reality of the environment so that people will recognise its intrinsic value, its interest, its intrinsic merit and feel some responsibility for it. After that has been done, then the various pressure groups can get at them through their own channels and ask them to send a donation to, let us say, the World Wildlife Fund. (Attenborough, quoted in Burgess and Unwin, 1984: 105–106)

Attenborough's legacy, as a result of the global reach of the landmark format and the programmes he voiced for *Wildlife on One* (1977–2005), is to have communicated the diversity and uniqueness of wild animals and plants around the globe to countless millions of viewers. "How we treat others," film critic Richard Dyer has pointed out, "is based on how we see them" (1993: 1). In this respect, the idea that an ethic of environmental concern might be distilled from beautiful imagery seems reasonable. It speaks of the role of romanticism and nostalgia in public understandings of nature, and a broader public appreciation for the environment. Yet, given the urgency that surrounds the current ecological crisis, this perspective has become more difficult to defend.

Nevertheless, in *Life on Earth* Attenborough began the tradition, continued in subsequent landmark series, of addressing human impacts and broader environmental issues in his final to-camera statements. The ethos of environmentalism in these statements tempered the near total avoidance of environmental politics in these series, which prior to *The State of the Planet* (2000) gave very little airtime to controversial and politically challenging topics.[6] Stripped of environmental concerns, save for those voiced in Attenborough's final statements, these series demonstrate the tension between the representation of environmental issues and the desire to reach large international audiences. This tension has arguably become more pronounced since the BBC's joint-venture partnership with Discovery, first brokered in 1997, which means that Discovery is now the dominant co-producer of BBC wildlife programming, with considerable editorial clout. *The Life of Mammals* (2002), with a budget of £8 million, is a case in point. Vanessa Berlowitz, who produced the last episode, revealed that executives at Discovery objected to Attenborough's final remarks in the series, in which he focused on the need to control the human population:

> Perhaps the time has now come to put that process into reverse. Instead of controlling the environment for the benefit of the population, perhaps it's time we control the population to allow the survival of the environment. (Attenborough, *The Life of Mammals*, 2002)

Fearing that a veiled reference to contraception might alienate viewers in the American Midwest, the Discovery producers asked for Attenborough's narration to be altered in the US version of the series (Berlowitz, 2012). But Attenborough and senior producers at the BBC steadfastly refused, and his remarks on population control remained intact. This example, amusing as it is, demonstrates the pressures that international co-producers now exert on the content of wildlife programmes, in this case, seeking to eradicate even the smallest hint of environmental politics from the narration. Alongside the wildlife genre's predilection for uncontroversial science, the need for landmark series to appeal to global television audiences is a key factor shaping the representation of environmental issues.

GREENING WILDLIFE DOCUMENTARY: FROM BLUE CHIPS TO GREEN CHIPS

The turning point in the treatment of environmental issues in BBC wildlife documentary came in 2000 when Attenborough presented *The State of the Planet*, a three-part landmark series on the environment. This was a shorter landmark series, entirely financed by the BBC, which used large portions of stock footage and broke with generic convention to feature interviews with scientists and environmentalists. In spite of the use of a *Horizon*-style "talking head" format, it still cleaved to many of the traditional characteristics of landmark wildlife series, including the use of spectacular imagery of untouched natural environments. In a marked departure, however, these more conventional images were interspersed with far more unsettling visions. Images of denuded forests, entire landscapes taken over by the cross-hatched fields of industrial farming, oceans swirling with plastic debris and the smoke-haze of polluted city skylines sat alongside beautifully choreographed footage of gorillas and tigers, whose populations Attenborough now informed us were in crisis.

The State of the Planet looked comprehensively at global environmental issues such as introduced species, over-harvesting, destruction of habitats, islandisation and pollution. It also touched briefly on global warming and climate change, but stopped short of ruling definitively on the extent of the threat. In his narration, for example, Attenborough argued in no uncertain terms that human action was responsible for global warming, but he was careful to use caveats in reference to its wider impacts: "There is one kind of pollution, however, that *could* have worldwide consequences—that is the global warming that results from human activities that pump carbon dioxide into the atmosphere". Six years passed before Attenborough went on the BBC's *Ten O'Clock News* in 2006 to talk about the devastating impacts of human-induced climate change. Commenting on his reluctance to speak publicly about global warming, he cited BBC impartiality and his sense of himself as a non-expert.

> I'm not a climatologist. I am a reporter and my views, whatever other people might attribute to me, but I always make it absolutely clear, they're second hand. I haven't analysed all those ice cores, I am just reporting. I am reporting when there is enough academic support for you to be able to report that opinion. (Attenborough, 2012)

This comment underlines the difficulties of exactitude in science and the wildlife genre's romance with safe science.

Green chip programming began to proliferate within prime-time, big-budget wildlife programming in 2006. BBC series such as *The Truth About Climate Change* (2006), *Planet Earth: The Future* (2006) and *Frozen Planet* (2011) finally dealt conclusively with climate change, while series like *Saving Planet Earth* (2007) and *Last Chance to See* (2009) took the form of a celebrity quest to highlight conservation issues. This key turning point in wildlife documentary's treatment of green issues, as I argued earlier, was linked with the larger transformation in the way that international news media embraced climate change as a global threat (Cottle, 2009a; Lester and Cottle, 2009). It was also shaped by the rise of international cable and satellite channels specialising in wildlife programming, such as the suites of channels operated by Discovery Communications and National Geographic, which saw the dominance of the blue chip format begin to wane (see also Cottle, 2004; Chris, 2006). New technologies and story-telling practices, combined with the resurgence of wildlife documentary as a profitable niche television market, helped to pave the way for a more inclusive approach to environmental issues.

In the two-part series *The Truth About Climate Change* Attenborough used a similar format to *The State of the Planet* to examine the consequences of climate change. Neil Nightingale, a senior NHU producer who helped to devise this series, argued that it coincided with the strengthening of the scientific consensus around climate change.

> I was delighted that we could do the two programmes with David Attenborough. Climate change was such a big topic and there was a lot of misinformation around at the time. It was brilliant to be able to do something which clarified things more than anything else and that David felt very comfortable doing. The science was solid. (Nightingale, 2012)

Yet, as the science behind anthropogenic global warming became more solid, allowing BBC producers to present this information unequivocally, it was accompanied by a subtle shift in how environmental issues were framed. In a contemporary rendering of the slogan "think globally, act locally", the second episode focused on how audiences could effect change by engaging in waste-reduction schemes and reducing their carbon footprints. This series helped to pave the way for a new style of wildlife programming that focused on climate change and other complex environmental issues not as "doom and gloom" scenarios, but as problems that could be solved though concerted local, national and global action. Problems, in other words, that could be recast as upbeat, feel-good solutions.

A similar approach was used in *Saving Planet Earth*, in which Attenborough and a host of British and Irish celebrities focused on the success of individual conservation projects—Will Young on gorillas, Graham Norton on wolves, Jack Osborne on elephants, and, in a slightly bizarre choice, Carol Thatcher, daughter of Margaret, on the albatross in the Falklands. In many ways these programmes were an extension of the "green crusade" films featuring environmental activists in the 1980s, which, as Luis Vivanco argues, were popular because they offered "carefully crafted win-win visions of conservation and sustainable development" (Vivanco, 2002: 1202).[7] Far from being a condemnatory lecture, *Saving Planet Earth* tended to be more upbeat and inclusive. In the first episode Attenborough issued the following invitation, "Some scientists suggest that up to a quarter of animal species could be extinct by 2050. But it's not too late—you can be involved in *Saving Planet Earth*". Each programme explored the work of different conservation projects before appealing for public donations to the BBC Wildlife Fund, a charity formed to coincide with the launch of the series. *Last Chance to See* (2009) is another notable series in this tradition.

The renewed focus on environmental issues in BBC landmark wildlife series followed two distinct strategies. The first, typified by *Planet Earth* (2006), saw the landmark format return to the familiar convention of avoiding environmental issues altogether until Attenborough's final statements. However, *Planet Earth* was also accompanied by a separate three-part series, *Planet Earth: The Future* (2006), which used interviews with scientists and conservationists to highlight conservation issues surrounding the species and environments featured in *Planet Earth*. In the UK, this series was broadcast on BBC4 just after the last three episodes of *Planet Earth*, where it reached a much smaller audience than that attained by the spectacular high-definition landscapes featured in *Planet Earth* on BBC1. This strategy of creating two separate series represents a desire to maintain the historical separation of pristine wilderness and environmental concern in the wildlife genre, perhaps with revenues from international television sales and DVDs in mind, whilst still adopting an ethic of environmental concern.

The second approach, exemplified by *Frozen Planet* (2011), was more inclusive. Following the precedent set by *The Living Planet* in 1984, in which the last episode had focused on the destruction of ecosystems, an entire episode of *Frozen Planet* was devoted to the exploration of the effects of climate change on the Polar Regions. Alastair Fothergill, the executive producer of *Frozen Planet*, argued that the choice to include an episode on climate change and to get David Attenborough to author it had been made from the outset.

> We worked very hard to make it feel like it was part of the main series. What does that mean? That means that visually it was as glossy as the rest of the series, so it looked fantastic. The other important thing was to get David to author it, because he is enormously trusted by people. There is a great deal of respect for David as a person, so when he tells you things you tend to believe them. (Fothergill, 2012)

Commenting on the fact that the "climate change" episode rated just as highly as other episodes in the series, which attracted the exceptionally high average audience figure of 8.2 million, Fothergill argued:

> I think people watched the programme more readily because they really cared about the place by then. If it had gone out as a single film on its own, it never would have got that audience. But as part of a series that had by then grown enormous momentum—everybody was watching it. (Fothergill, 2012)

The episode predictably ignited controversy in the press, where it elicited criticism from British climate change skeptics, most notably Nigel Lawson (Porritt and Lawson, 2011). However, it was the accusation that the BBC had given international channels the option to drop the last episode of the series (dubbed "the climate change episode") to help the show sell better in international markets that highlighted another growing source of concern. *Frozen Planet* had been offered as a six-part series, with the option to include the climate change episode and a behind-the-scenes episode as "optional extras". More than thirty networks bought the series, but a third of them rejected the additional two episodes. It was rumoured that Discovery, the largest co-producer of the series, was planning not to air the climate change episode due to a "scheduling issue". Instead, producers at Discovery planned to incorporate elements from this programme into their final show (Bloxham, 2011). In effect, Discovery's proposal meant that Attenborough's nuanced take on climate change would not be broadcast in the US, where the largest population of climate change deniers resides. Discovery later backtracked on their decision and opted instead to broadcast all seven episodes, including the one on climate change (Hough, 2011).[8]

The exclusion of environmental issues in wildlife documentary is a feature of the generic constraints of the wildlife genre, where audience expectations and the appetites, both perceived and actual, of American co-producers, as well as the BBC's public service values of balance and impartiality, combine to ensure that controversial issues are suppressed. There is, however, another facet to the birth of green chip programming. By putting a positive spin on conservation projects, utilising celebrity endorsement or neatly corralling environmental issues into separate programmes and even separate series, the producers of wildlife programmes have succeeded in making concern for the environment more palatable to local and global television audiences. This change represents a significant shift in how environmental issues are produced and framed in wildlife documentaries. Nevertheless, when viewed in a cynical light, these programmes can also be understood as cheaper offshoots of more profitable wildlife series, riding on the popularity of eco-consciousness while landmark series like *Planet Earth* (2006) continue to present, in Stephen Mill's words, "period-piece fantasies of the natural world" (Mills, 1997). Ultimately, it is not just whether environmental issues are excluded from wildlife documentary that matters. It is how they are financed, produced, represented and broadcast (in all their versions) when they do get airtime.

ENDNOTES

1 "Landmarks" are multi-part documentary series focusing on academic subjects and authored by a single knowledgeable on-screen presenter. Attenborough's landmarks began with *Life on Earth* (1979) and continued to *Frozen Planet* (2011). Other BBC series such as *The Blue Planet* (2001) and *Planet Earth* (2006) are variations on the landmark format, with Attenborough acting only as a narrator rather than on-screen presenter and writer. This trend continued in *Frozen Planet* (2011), in which Attenborough appeared on-screen only in the final episode.

2 Environmental issues have been part of television wildlife documentaries almost since the genre's inception, featuring regularly in the BBC's *Life* (1965–1968) and *Nature* (1983–1994) series, and in environmental or conservation films like National Geographic's *Save the Panda* (1983) and Bullfrog Films' *Blow Pipes and Bulldozers* (1988). The trend for using celebrities to present environmental programmes, which was pioneered by Tigress Productions' *In the Wild* series

(1992–2002), broadcast intermittently on ITV and PBS, allowed environmental programming to briefly break into the mainstream.

3 Bousé cites three major categories of proto-wildlife films—"Safari Films, Scientific-Educational Films, and Narrative Adventures" (2000:46). However, I would like to distinguish ethology films, made by professional scientists such as Niko Tinbergen and Konrad Lorenz, from scientific-educational films, as they subtly differ in their roots and modes of address.

4 *The Living Desert* (1953), for example, was produced for roughly $300,000 and is reputed to have earned between $4 and $5 million in its first domestic cinematic release. In the following year, 1954, *The Vanishing Prairie* earned $1.8 million, or around fifteen times its production costs (Chris, 2006: 35).

5 There are exceptions to the wildlife genre's avoidance of "talking head" formats as a vehicle for controversial issues. *Warnings from the Wild: The Price of Salmon* (2001), for example, used this format to highlight the catastrophic environmental impacts of fish farming on wild salmon populations. However, the use of counter-posed "talking head" interviews remains an under-used device in natural history programming.

6 *The Living Planet* (1984) provides a notable exception. In this twelve-part series on the world's ecosystems, the final episode was devoted entirely to the destruction of ecosystems.

7 Luis Vivanco outlines two common narrative strategies used in environmental filmmaking. The first portrays conservationists as "green crusaders", heroic activists struggling "to save species from the ignorance, greed, and overpopulation of local people", while the second strategy, exemplified by *Blow Pipes and Bulldozers* (1988), relies on depictions of the "noble savage living in Harmony with nature" (Vivanco, 2002: 1199–2000). There are, of course, exceptions to these simplistic formulas. Both Vivanco and Dan Brockington (2009) cite *The Shaman's Apprentice* (2001) as evocative of a new kind of environmental film that attempts to chart the more complex environmental, cultural and economic politics behind conservation projects by addressing issues such as social justice for local communities. Another major trend is the increasing use of celebrities to endorse conservation projects and present wildlife documentaries (Brockington, 2008, 2009; Cottle, 2004; Vivanco, 2004).

8 The first six episodes of the US version of *Frozen Planet* (2012) were narrated by the actor Alex Baldwin; however, Attenborough's narration and his on-screen appearances remained intact in the controversial programme on climate change.

▪ CHAPTER TWELVE

Whither the "Moral Imperative"?

The Focus and Framing of Political Rhetoric in the Climate Change Debate in Australia

MYRA GURNEY

Like most complex issues, climate change has many dimensions with competing and conflicting values and agendas. These are debated using frames which filter perspectives through the lens of existing ideological beliefs and worldviews (Goffman, 1974). Frames, according to Nisbet and Mooney (2007), use language to "organise central ideas, defining a controversy to resonate with core values and assumptions". Unpacking how an issue is framed is important in order to expose the underlying assumptions and power relationships upon which knowledge production in any particular discourse is founded (O'Brien et al., 2010b).

The media play a pivotal role in both the production and reproduction of different frames, or what Gamson and Modigliani (1989: 3) call "interpretative packages". These make suggested meanings available for the attentive public through linguistic devices such as metaphors, catchphrases, historical examples and visual images. They also act to construct reasoning devices such as appeals to causal relationships and to principles such as moral arguments (Gamson and Modigliani, 1989: 3–4). Lakoff (2008) contends that because repeated frames are cognitively reinforced, the discourse becomes "internalized," as Foucault (1972) would argue, making it difficult to think outside the boundaries of those recurring frames.

Understanding the nature and direction of the debate also requires considering the way people use media to inform their opinions, the power of media as a gatekeeper and framer of different narratives (Couldry, 2000), and the importance of language in this process. In the case of climate change in particular, the shift from political bipartisanship to ideological divide on environmental issues has forced individuals to seek out what Brulle et al. (2012: 8) describe as "elite cues" wherein "individuals use media coverage to gauge the positions of elites and interpret the information based on their party and ideological identification". This is particularly cogent in the case of politics, which Castells (2009) has argued is itself fundamentally framed by "the inherent logic of the media system" (cited by Lester and Hutchins, 2009: 590), resulting in the privileging of some frames over others. As a result, political parties, and even activist organizations like Greenpeace, are often forced to operate within, and therefore unwittingly legitimize and reproduce, dominant and powerful frames in order to attract media attention and garner public legitimacy, a strategy that, it is argued, "limits their political capacity to spur genuine political transformation" (Lester and Hutchins, 2009: 582).

Numerous researchers believe that more fundamental issues lie at the heart of the general lack of engagement in Australia and elsewhere (see Leiserowitz, Maibach, Roser-Renouf and Smith, 2011) with the urgency for unilateral action to reduce carbon pollution. For some (e.g., Hamilton, 2009, 2010; Merchant, 1993; Lovelock, 2006; Douglas and Wildavsky, 1982), the explanation lies in the human species' historically fraught cultural, ideological and psychological relationship with nature and climate. Chakrabarty (2009), for example, notes that historically, humankind has viewed itself as a "prisoner of climate", a "biological agent", forced to operate within and to be constrained by boundaries set by nature. Climate change, however, has required a reconceptualising of humans as "geological agents" (Oreskes, 2004: 93) whose actions have altered the power dynamic to fundamentally shift the natural order. This has collapsed the traditional humanist distinction between "natural history" and "human history", requiring social theorists to re-examine their traditional assumptions about the relationship between humans and the planet and hence their moral and ethical responsibility for its long-term welfare.

These shifting paradigms and peculiar dynamics of the problem (Moser and Dilling, 2004) make our ability to consider the phenomenon of climate change, and the threats it poses, much more difficult. Despite the best efforts of those lobbying for urgent action, arguments that challenge traditional ways of thinking about the relationship between humans and the planet are continually derailed because they are "fought" within the dominant frames of those in whose interests it is to maintain the status quo or who frame the debate as "a choice between progress and planet"

(Charlton, 2011: 65). The discursive focus on individual self-interest and repeated foregrounding of the spectre of dire economic consequences makes engagement with the bigger picture more difficult, doubt easier to fuel, and substantive action something that can be deferred. The temporal lag and spatial dispersion of climate change means the connection, particularly in media coverage, between cause (emissions) and effect (weather) (Howard-Williams, 2009) is lacking, making the "moral" argument about our responsibility for future consequences and present culpability more difficult to popularly conceptualise.

Nisbet (2009: 18) provides a useful overview of some of the competing frames in the climate change debate, examples of which are present in the Australian context. But it is the moral and ethical frame, which became so rhetorically significant in 2007 when highlighted by (then) Prime Minister Kevin Rudd, in which this chapter is interested.

THE POLITICAL CONTEXT AND THE "MORAL IMPERATIVE"

Prior to the December 2007 Australian federal election, Kevin Rudd, then leader of the opposition Australian Labor Party (ALP), tapped into the prevailing zeitgeist of climate change concern when he famously amplified the political rhetoric by describing the need for international action on greenhouse gas abatement as "the great moral challenge of a generation" (Rudd, 2007). This widely cited rhetorical trope became a metaphorical "call to arms" with an almost religious dimension, and after 12 years of conservative government, Rudd led the ALP to a resounding victory. With an impressive mandate to act on climate change abatement, he made the introduction of an emissions trading scheme, or Carbon Pollution Reduction Scheme (CPRS), as it was renamed, a key policy platform of his government's first term. Despite initial public enthusiasm, the enactment of the policy became mired in complexity, poor communication and political brawling, which both confused and rapidly sapped the patience of the electorate, especially after the failure of the much anticipated Copenhagen Climate Conference in December 2009.

The conservative opposition Liberal/National Party also had its own share of climate change casualties, most notably Malcolm Turnbull, who was narrowly defeated as leader in December 2009 by climate change skeptic Tony Abbott after a party room revolt at Turnbull's agreed support of the CPRS. Turnbull's demise spelled the end of bipartisan support for the CPRS, and eventually was the death knell of Rudd's own leadership. After failing twice to pass the legislation through the upper house, he announced in April 2010 that his government would postpone until 2014 further attempts at enacting climate change policy. This announcement sparked a dramatic decline in Rudd's public popularity and became one element in his party's decision to replace him in June 2010 with his deputy Julia Gillard.

Since that time, Gillard has fared no better. In the wake of the August 2010 election she was forced to negotiate with the Australian Greens and three independents to form minority government, with climate change abatement one of the agreed conditions. Unfortunately, Gillard had stated during the election that "I don't rule out the possibility of legislating a Carbon Pollution Reduction Scheme, a market-based mechanism, I rule out a carbon tax" (Kelly and Shanahan, 2010). It is the last part of this statement upon which the Liberal-National Party coalition and large portions of the conservative media have since focused, casting aspersions on the Prime Minister's integrity and significantly (some say terminally) damaging her political credibility. While the Clean Energy Act (generally referred to as the "carbon tax") passed through Parliament in November 2011, opposition leader Tony Abbott, whose own positions on climate change have vacillated over time (see Keane, 2011), has been spectacularly successful in maintaining media focus on both the broken promise and the supposed economic impacts of the carbon tax. He has dramatically pledged a "blood oath" to repeal the legislation should he become prime minister.

The slow pace of action and the lack of political bipartisanship would appear to have impacted both the Australian public's belief in the anthropogenic drivers of climate change and their preferred government policy (Leviston, 2011; Connor and Stefanova, 2012). But what role has the framing and tenor of the debate since 2007 had in shaping the current state of public opinion? Why has the "moral challenge" appeal, a rhetorical trope which captured the Australian national imagination in 2007, become subjugated by other discourses and other frames? What did Rudd mean by the "moral challenge", and where does climate change as a "moral challenge" fit in relation to the broader frames of the debate more broadly, in particular those strategically encapsulated by important political speeches?

The climate change debate has a number of moral or ethical perspectives. Nisbet's (2009:18) typology defines this frame as "a matter of right or wrong; or of respect or disrespect for the limits, thresholds, or boundaries". Some are overtly articulated while others are implied or assumed. Many are esoteric, most are difficult to quantify and all are shaped by divergent worldviews and by differing cultural and ideological perspectives (Hulme, 2009). The "concept" of climate, as a range of scholars have argued, is about more than merely science, meteorology, weather patterns, ecosystems and the environment; it also has reverberations for societies, cultures and worldviews. As human life, like that of other species, develops within and is dependent upon climate, the notion is culturally central. From a moral perspective, responses to climate change incorporate considerations of equity, security and fundamental human rights, as well as the rights of non-human species.

A major moral or ethical issue is that of intergenerational equity—the needs and rights of future generations to a habitable planet. But how "quality of life" is measured, how "costs" are calculated and how "responsibilities" for damage and future actions are apportioned are matters of dispute (Gardiner, 2004). Should rich, industrialised countries shoulder the "burden" of action when poorer, developing nations are excused even if their emissions rise as they too aspire to Western-like living standards powered by cheap fossil fuels? The IPCC (cited by Gardiner, 2004: 579) noted that while developed nations are historically responsible for the current level of emissions and have been the ones to largely benefit, the impacts are expected to fall disproportionately on poorer nations who are least able to adapt. Additionally, the magnitude of change, both environmentally and economically, is compounded the longer unilateral actions are delayed, resulting in higher costs and more severe consequences for future generations.

Another moral issue is that of stewardship of the environment in such a way that we do not contribute to climatic changes which will impact on future quality of life for a diverse number of societies and cultures, including the potential intrinsic and aesthetic loss of ecosystems and non-human "capital". Many religious leaders see the planet as a gift from God which we have a duty to preserve (Hulme, 2009: 149; Rudd, 2006). Others (Donner, 2007; Chakrabarty, 2009) see the difficulty in the challenge to entrenched religious and historical paradigms which traditionally posit climate as what Donner terms "the domain of the gods", an infinite and transcendent system beyond human control, distinct and separate from humans and their earthly activities. Donner argues that accepting human culpability for climate change necessitates a reframing of this power relationship, requiring a difficult philosophical mind shift. Climate change generally, according to Jamieson (1992: 290), poses fundamental questions about "how we ought to live, what kinds of societies we want, and how we should relate to nature and other forms of life".

From a secular perspective, Dunlap (2004) points to three important parallels between some religious traditions and environmentalism, including the focus on symbiotic relationships between humans and nature, the recognition of the spiritual or transcendent dimensions of the natural world and the view that while science is powerful, it is never totally able to give us access to all that is truly important. Similarly, Lovelock (2006) contends that we are ethically bound to protect "Gaia", the metaphor he invokes for the Earth as a living organism, because it has value of and for itself, not merely as a resource upon which we humans feed for our own ends.

A major difficulty for those prosecuting the moral or ethical arguments of the debate is that as a substantially deferred phenomenon (that is, we are currently living with the results of past emissions), long-term impacts of climate change are difficult to imagine, making short-term pain and costs more unpalatable. Politically, other short-term priorities like improving economic growth and maintaining living standards become more cogent, tangible and urgent. Many politicians and economists consider that disregard for the disruptive economic consequences of policies which may impede the level of economic growth to which the Western world has become accustomed and to which the developing world aspires is itself immoral and unethical (Charlton, 2011). Others (O'Brien et al., 2010a) argue that a singular focus on environmental perspectives ignores the potential threat to human security posed by climate change. The practical issue for international policy makers has created what Gardiner labels "a perfect moral storm" (2006).

Brown et al. (2006) contend that scientific and public policy instrumentalities such as the IPCC raise serious but rarely considered ethical issues, and that the assumptions underpinning definitions of values and even methodologies are not adequately interrogated in policy formulations. Widely used indices of economic modeling such as cost-benefit analysis reflect unquestioned value judgments based on a paternalistic Western paradigm and assume a level of cultural and political homogeneity which does not necessarily exist (Gardiner, 2006; Jamieson, 1992; Nelson, 2008; Spash, 2007; Morgan et al., 1999; Neumayer, 2007; Toman, 2006). Charlton (2011) argues that this was central to the failure of the 2010 Copenhagen Climate Conference. Gardiner (2006: 408) notes that the tendency of political actors to emphasise considerations such as economics and scientific uncertainty and to focus selectively on individual powerlessness and national self-interest serves to "problematise action", whereas a focus on intergenerational ethics demands action—a position which is uncomfortable to those with a vested interest in "business as usual". Yet, insufficient consideration of the ethical implications and moral underpinnings of policy may mean that proposed solutions are potentially unfair and unjust to those most at risk.

APPROACH

This study is interested in the extent to which the various moral and ethical themes outlined are present in the Australian climate change debate, as evidenced by the language and discourse of key speeches of the major political actors between 2007 and 2011. These speeches have been chosen as illustrative only, and the choices do not imply that these are the only positions or frames used by these speakers. Political speeches have an important strategic place in political discourse (Glover, 2007) because they are explicitly written to set the agenda and vocabulary for debate and often become a key source for political commentary in both print and electronic media. In the case of Kevin Rudd's speech, his rhetorical evocation of the "great moral challenge of our generation" not only set the Australian climate change debate in motion, but also has been the phrase around which much of the ensuring debate has pivoted.

The speeches analysed were Kevin Rudd's opening remarks as opposition leader to the National Climate Change Summit in Canberra (Rudd, 2007), current opposition leader Tony Abbott's speech at the David Davies Memorial Dinner (Abbott, 2009), and current prime minister Julia Gillard's address to the National Press Club in July 2011, delivered immediately prior to the introduction of the Clean Energy Act (carbon tax bill) into the Australian parliament.

The speeches were analysed using a qualitative critical discourse analytic (CDA) framework, a methodology concerned with examining the nature of power relationships within language and the manner in which these contribute to, or inhibit, the enactment of social or political change (Van Dijk, 2002; Fairclough, 2001). The symbolic importance of language in political debate is recognised, especially as the insatiable needs of the 24-hour modern news cycle means that political comments become key sources of news content, often forensically analysed for nuance of meaning and intent (Tanner, 2011; Jones, 2012). More broadly, as language, discourse and frames reflect the social, political, cultural and economic milieu within which they operate, and from which they are spawned, they are important sources of insight.

RESULTS

KEVIN RUDD AND "THE GREAT MORAL CHALLENGE"

It was this speech that set the benchmark motif of the climate change debate in Australia. The opening salvos are somewhat like a "call to arms" on climate change, an issue that Labor strategists believed was politically significant and one that Rudd needed to muster effectively to distinguish himself from incumbent prime minister John Howard, who, along with the then US president George W. Bush, had steadfastly refused to ratify the Kyoto Protocol. At the time, Rudd was working hard to develop his image as someone with a vision for Australia who also was concerned about ordinary "working families", as distinct from Howard, who was being framed as a prime minister desperately clinging to power, whose policies and ideologies had had their time.

Consistent with Rudd's personal style and developing public persona, the speech is almost evangelical in tone. The word "challenge" is repeated—*moral* challenge, *environmental* challenge, *economic* challenge, *security* challenge, a *challenge* of massive dimensions—all of which creates a tenor of religious fervor. He thanks the participants for "making the pilgrimage to Canberra" and acknowledges a possible diversity of views, saying that he doesn't expect the participants to be all "singing from the one hymn sheet".

Strategically, Rudd is positioning himself as someone who is up for the "challenge" in many arenas—diplomatically, economically, technically—a "can-do" prospective prime minister, consultative and consensus-driven. There are numerous references to Australia's international role and the regard in which we are held, our predilection for being "sensible", "practical" and well liked, our position as a global leader punching above our weight—characteristics popularly considered uniquely Australian. With the repeated use of the generic "we", Rudd is inviting all the attendant captains of industry, trade unions and government to be part of this new consensus. Helping the global "challenge" on climate change is part of his "mission".

While the "great moral challenge" is mentioned *first* as a rhetorical anchor and opening flourish, it is the *last* of five core reasons for action that Rudd details. In order, these are: science and risks of unpredictable climate variations; the costs to the economy and jobs; the global nature of the problem requiring global participation; and the diplomatic advantage to Australia to lead. The fifth and final reason is that of intergenerational justice. Of this he says:

> How you sustain a proposition which says that when the evidence is in and the scientific evidence is in, the economic data is accumulating that when that is presented to us in the year 2007 and we fail to act, how can we look towards the interests of the generation which comes after us and say, "I'm sorry, it was too difficult to act". For me, that is a compelling argument as well. (Rudd, 2007, para. 23)

The last line flags Rudd as signaling that he may be alone in considering the moral and ethical dimensions of climate change. The "as well" can be read as an indication that this is something of an afterthought, that this perspective is not the most important, nor perhaps one that holds much weight among the assembled congregation to whom he is making what is a strategic political pitch. Rudd's referencing of the moral challenge is not entirely strategic however, but reiterates the position articulated in his essay "Faith in Politics" (2006) and is therefore consistent with his personal beliefs and "evangelical" fervor, as well as with the public persona he was working hard to establish (Marr, 2010; Stuart, 2010).

However, as this speech proceeds, there is little further exploration of the moral imperative. He possibly believed that other arguments were more likely to resonate with those present and hence were more pragmatically translatable into political support. Strategically, Rudd was also keen to establish his management "credentials", having earlier described himself as a "fiscal conservative" in order to deflect the usual conservative framing of Labor policies as fiscally reckless, evoking the spectre of the excesses of Gough Whitlam's 1972–1975 Labor government. For this reason, the choice of the word "challenge" itself is telling, as it is an example of the euphemistic managerial style of language of which Rudd was extremely fond.

Towards the end, he does return briefly to the arena of ethics and morals by alluding to questions of "responsibility":

> … what are the best policy settings for what might be described as personal responsibility, corporate responsibility, community responsibility agenda? How do we individually act as citizens, engage with the great challenge of climate change to do our bit to reduce our own carbon footprints? (para. 35)

Overall, however, more time and emphasis is given to the economics frame. For example, in the paragraphs about the economic implications, he revisits the conclusions of the Stern Review (Stern, 2007). Later, he concurs with Stern when he says: "climate change does represent significant market failure, that's where Governments have to enter the field …" (para. 43). So therefore, it seems that while the moral imperative is important for *him*, it is the concern for the Australian economy and the possible price "it"—the economy, the all-powerful "market"—will have to pay without climate change abatement that remains the most politically strategic and cogent argument for action.

TONY ABBOTT AND "A REALIST'S APPROACH TO CLIMATE CHANGE"

For Tony Abbott, the man who reportedly said "[climate change] is absolute crap" (Rintoul, 2009) and whose ascension to the leadership of the opposition Liberal Party spelled the defeat of Rudd's CPRS, it was difficult to find a speech to analyse. The one discussed here was delivered in July 2009 and sits mid-way between Rudd's "moral challenge" speech and Gillard's "Clean Energy Future" speech. Given Abbott's shifting positions on climate change science and policy, it makes interesting reading, particularly in the current Australian political context.

The lecture is intriguingly titled "A Realist's Approach to Climate Change". Abbott, "the realist" of the title, presumably, opens with an appeal to pragmatism: facts that can be "measured and considered" should be what govern any "rational discussion", rather than people's beliefs (which presumably are not necessarily rational; an interesting point from a publicly devout Catholic). Arguments should be separate from ideology and, in a dig at environmentalists and pro-CPRS supporters, should not "become a basis for heresy hunting" (Abbott, 2009, para. 3). The moral imperative, he argues, should not be used as an emotional weapon against those who question the science when the rationale and outcomes are uncertain. He diminishes the moral aspects of the environmental argument by an appeal to common sense with the comment that

> Every sensible person understands that we have to protect the environment because it's the only one we have. This is why the debate over climate change shouldn't be couched in morally loaded terms such as believers versus deniers. (para. 3)

Environmentalists, or "eco-fundamentalists", do not have exclusive claim to the moral high ground. Protecting the planet should be assumed to be the ultimate aim of everyone, not just those who label themselves "environmentalists". The descriptor "eco-fundamentalists" diminishes their status and rhetorical position with respect to the "realists", who legitimately question the science and who demand certainty before supporting policy which may not have the desired environmental outcomes.

In terms of policy, adaptation—here Abbott cited the "skeptical environmentalist" Bjorn Lomborg (2007)—might make more practical sense.

> What we can say, though, is that we should try to make as little difference as possible to the natural world. As well, prudent people take reasonable precautions against foreseeable contingencies. It's the insurance principle. The premium we are prepared to pay, though, should relate to the extent of risk and the magnitude of the possible loss. If carbon dioxide might be contributing to harmful climate change and emissions can effectively be reduced at reasonable cost, it certainly makes sense to do so. Of course, what we shouldn't do is embark on a cure that turns out to be worse than the disease. (para. 7)

This paragraph is interesting for the way it mixes the discourses of risk and economics, positioning them as rational. While Abbott begins with a disclaimer about the need to make as little difference as possible to the natural world, there is no sense that humans have a broader ethical obligation beyond what is "prudent". He remains skeptical about the science ("*if* carbon dioxide *might* be contributing to *harmful* climate change") and about the impact (climate change is part of the natural ecological processes). The words "reasonable precautions", "foreseeable contingencies", "insurance", are economic terms, all with an implied set of values against which action should be measured in order to be "rational". Policy decisions therefore should be neither emotively nor ideologically driven. The ideological underpinnings of Abbott's own position are conveniently ignored, and the assumptions about what constitutes "reasonable cost", how that cost is measured, and who must pay that cost are not interrogated. His position is assumed to be value-free, and therefore inherently more "rational". Because the need for "rational" consideration is framed in terms of economic costs and benefits, the moral dimension can be assumed and is not necessarily a primary argument in itself.

Ironically, given his current position (see Liberal Party of Australia, 2010), at the time of this speech Abbott was not against either an emissions trading scheme or a carbon tax, with the caveat that it must be "necessary". These solutions, however, are couched in the language of doubt: "should" is an oft repeated word. An emissions trading scheme, he observes, is in some ways inherently dishonest because it creates a perception that "... it's a cost-less way to avoid climate catastrophe". "Cost" again is framed as economic.

On the subject of Australia's international obligations, Abbott plays the line that compared to China and India, Australia has only a small impact on global emissions (despite the fact that per capita, Australia's contribution is one of the highest), so therefore a go-it-alone scheme will only hurt our economy without any impact on global emissions. He goes on to say:

> As a good global citizen and on the right issues, Australia should be prepared to take a lead but there are normally limits to unilateral action even in the best of causes. (para. 11)

Presumably, this is not one of the "right issues" or the "best of causes"; Australia's economic interests should be paramount, and we should not be taking a lead should there be a danger that our economy may suffer. The economy is framed as something separate from issues of global environmental responsibility, the implication being that Australia can remain aloof from its international obligations.

Finally, Abbott argues that the political contingencies, rather than the moral ones, must be considered in how the coalition votes. He chides Rudd for acting out of political expediency rather than genuine commitment (para. 22). He says:

> As long as people are thinking about the possible dangers of climate change, they are unlikely to be worrying about the more imminent and more certain dangers of economic change. (para. 24)

Economic change is the primary issue, and the moral argument is merely an emotional and political distraction, a concern of "eco-fundamentalists" who are framed as "non-realists" and a danger to the economy in their ideological quest for a pseudo-religious but uncertain cause.

JULIA GILLARD AND THE "CLEAN ENERGY FUTURE"

Prime Minister Gillard's speech, delivered in July 2011 prior to the introduction of the Clean Energy Act (carbon tax bill) into Parliament, is quite different in style from Rudd's evangelical approach. The central trope is the "future"—"clean energy future" is the catchphrase. Ironically, however, Gillard's arguments are framed in the context of the past, as a continuation of major Labor structural reforms of the 1980s Hawke-Keating era, universally lauded as having been instrumental in Australia's economic success in a rapidly evolving global economy. Gillard seeks to position herself as a reformer in this Labor tradition, with fairness and equity as the central principles. The carbon tax is couched within the paradigm of such major structural changes. Carbon pricing, she says, "is this Government's biggest reform yet", a reform which has long-term implications, achieving "a change of far greater structural significance: decoupling the growth of carbon pollution from the growth of our economy". In other words, it will encourage continuing growth on the back of structural changes in energy emission. Growth is the mantra, propelled by transformative new technologies which will enable new opportunities. In paragraphs 31–34, the reform frame is well and truly reinforced, but it is economic reform first and environmental reform second. The moral imperative to drive the reform is absent from the argument.

Outlining how the carbon price and the permit system will work, she notes that it is predicated on the ultimately "rational" basis of

> ... evidence-based emissions targets, abatement at the lowest economic cost. A new bottom line, where polluters pay. (paras. 61–62)
>
> ...
>
> The inherent logic of the market is central. Costs are decentralized with the government mopping up the excess with tax cuts and household assistance, tax cuts which ultimately will provide extra incentives to create work. Understanding the environmental imperative is coupled with understanding the economics. (para. 77)

In the final parts of the speech, Gillard reiterates her reform agenda with the repetition of "walking the reform road". The future is down this road—a road, and a future, defined by economic progress and continued growth (paras. 94–96). The moral to which she refers is encompassed by equity and fairness in economic policy rather than by any broader, altruistic imperative.

Finally, the proposed reform is framed as a matter of faith: "faith in [Australia's] capacity to reform", "realistic faith", "determined faith", "faith shared between a creative and confident people". Despite being a declared atheist, Gillard, like Rudd, is evangelising, exhorting the "true believers" (to use the traditional ALP descriptor) to put their faith in the mystical historical propensity of Labor for fairness and reform, as well as in the "inherent logic of the market". Like that of Abbott, Gillard's faith is parochially rather than globally focused: on Australia rather than the world, on individual needs rather than global responsibilities, predicated on the "natural" inclinations of the market rather than any acknowledgment of a broader moral responsibility to the global environment.

CONCLUSION

For each of these Australian political actors, climate change has had significant political ramifications, their political fortunes and credibility pivoting around the strategies with which they have engaged with the issue. The three speeches reflect the differing styles of the speakers and their divergent political positions, but also the stages in the debate in which they were delivered. While Rudd flagged the "moral challenge" as a central trope of his sermon, the dominant frame was economic. Abbott positioned himself as a calm, conservative rationalist holding back a sea of hysteria from uncertain science and rampaging "loony" environmentalists hell-bent on wrecking the economy for their own ideological ends. Gillard extolled the virtues of economic reform with an environmental purpose, but it is reform which makes no reference of either the individual's or nation's moral or ethical responsibilities to the global environment.

To what extent did these speeches address the moral and ethical aspects of climate change? What underlying values does the discourse reflect, what are the dominant frames, and how do the speeches work rhetorically to construct the problem as a moral and ethical one? In general, despite Rudd's "call to arms" about the moral imperative for unilateral global action, his arguments are overwhelmingly economic. The moral is acknowledged, but its function in this speech is fundamentally rhetorical. Abbott's cursory reference to the larger question of moral responsibility to the planet implies that his focus on economic consequences encompasses those concerns anyway, and that any other argument is ideologically driven emotional blackmail rather than policy based on reason and evidence. The "clean energy future" extolled by Gillard is essentially business as usual via a different energy route. The "future" about which she speaks is an economic, not ecological one. In each case, the central argument is that the only way to protect natural resources is to price them, to treat them like a product or commodity and trade them. The "costs" are economic and political. The moral imperative is cursorily acknowledged, but is mostly either assumed or subsumed to more urgent and important economic imperatives.

Both Rudd's and Gillard's speeches are strategic political "acts", designed more to bolster their short-term political currency than to reflect a genuine ideological narrative. The general political inability to engage with the ethical or moral tenets of the debate, Gardiner (2006: 400) argues, reflects a broader problem of "institutional inadequacy" in Western political structures, where the power of media and of public opinion generally hampers political action. Always the political pragmatist (Coorey, 2011), Abbott in his speech attempts to further seed the doubts on the reality of climate change by casting suspicion on those claiming the moral high ground. Doing so serves to frame the skeptical position as value-neutral and "rational" in comparison with the "eco-fundamentalists" of the pro–climate change lobby.

The UK Stern Review (Stern, 2007) and the Australian Garnaut Review (Garnaut, 2008) which focused on the economic costs of inaction on climate change both have shaped the language and policy agendas in the UK and Australia. They equally reflect what Paton (2011: 355) refers to as a "market trance" and empower "the intellectual imperialism of mainstream economics" (Hamilton, 2010: 59) which has become so normalized in media discourse generally, and pervasive in climate change discourses in particular. The "market" has been psychologically and discursively conferred a level of animism acting on some intrinsic, natural organising force or "'natural' properties and truths" (Doyle, 2010: 158) to the extent that the focus on the welfare of individuals as "consumers" has weakened the power of sovereign governments to speak about, let along act on, climate change in any way which challenges the prevailing powerful neo-liberal narrative. While the "moral imperative" trope served Kevin Rudd politically in 2007, allowing him to raise the national consciousness on the importance of global action on climate change abatement, this chapter has argued that Rudd's rhetoric was hollow, and subjugated to the more strategically powerful economic frame. The question remains as to the extent to which this is detrimental to the long-term objectives of this important debate.

▪ CHAPTER THIRTEEN

As Fukushima Unfolds

Media Meltdown and Public Empowerment

KUMI KATO

The aftershock of the 3.11 East Japan earthquake was felt globally—most significantly as a re-think on nuclear power. In Japan there is a strong public movement towards a nuclear-free and more sustainable and democratic society. Several issues have fuelled this public mobilization, including the realization that, as a small nation of just over 120 million people, Japan has lived with 54 nuclear power reactors introduced progressively with minimal public consultation, and also that media reports on Fukushima merely reiterated the official government line with feeble "expert" commentaries. The more than one million relief volunteers who contributed to the reconstruction effort are also now recognizing their social and personal power; and inevitably, some may say, a truly independent alternative media network and a range of NGOs continue to provide credible information, nationally and internationally. The general public gradually became empowered and found a voice as the meltdown of Fukushima unfolded in the mainstream media. This chapter attempts to account for the lack of critical media debate while, at the same time, the general public is making a major turn-around over the nuclear issue.

June 16, 2011, was the forty-third nuclear-free day in Japan since operation of all nuclear power plants was suspended. At more than 100 venues around the nation, crowds gathered to watch a documentary, *The 4th Revolution: Energy Autonomy*, which advocates the possibility of a 100-percent renewable future, using examples of successful innovations around the world. The audience, feeling uplifted by the positive prospect presented in the film, were met with the deflating news that the central government had decided to resume operation of two of the four reactors at the Oi Nuclear Power Plant as agreement had been reached with its local government, Fukui Prefecture.

The catastrophe of the 3.11 East Japan earthquake, followed by the massive tsunami and nuclear plant accident, was beyond the imagination of all but a few enlightened observers. Cities and villages along the 500 km coastline of Tohoku turned to massive piles of rubble, empty parks and sunken land inundated with seawater. The scar left by the earthquake and tsunami is grave; the consequences for Fukushima are invisible and immeasurable. *Fukushima* no longer signifies simply a place in south Tohoku, but also a devastating accident equally as grave as Chernobyl, radioactive homelands to which residents cannot return for years to come, and the truth about nuclear power that had been supplying nearly 30 percent of Japan's electricity. What shocked Japan's general public in the aftermath of 3.11 was the fact that in this small nation, we have been living with 54 nuclear power reactors progressively introduced since the mid-1960s with little public consultation. The nation woke up to the fact, sadly recognizing its own ignorance and indifference: how could we have forgotten the history of Hiroshima and Nagasaki (1945) or the Bikini Atoll disaster (1954)? Indeed, we talk about *genpatsu* (nuclear power) as if it is completely distinct from *genbaku* (nuclear bomb) or *kaku* (nuclear), as Nishi claims:

> Since bombs fell on Hiroshima and Nagasaki, Japan has cultivated a religion that condemns nuclear arms. Along the way, however, Japan metamorphosed into a strange creature that felt immune to things nuclear. (Nishi, 2012; see also Kawamura, 2011)

It is time to face the question: will this realization drive Japan to a new kind of future, or is the society simply not capable of such change? While decisions by Germany, Italy and Switzerland to move towards a nuclear-free future provide hope that a positive lesson can be learned from Fukushima, will the sacrifices of the victims be ignored in their own country? Will Japan's leadership betray its own people—160,000 displaced in Fukushima who will not return home for a long time, if ever, and 200,000 people who surround the prime minister's residence in protest against nuclear power? And what role did media play in this public realization and mobilization? This chapter attempts to address these questions.

My perspectives are from ecohumanities and the wider social context of contemporary Japan. They are also informed by my experience of participating in the relief work as a volunteer and a university academic working with student volunteers in 2011 and 2012. I have visited Tohoku eight times, mainly in Rikuzentakata, where nearly 10 percent of its population of 20,000 was lost.

This chapter has three sections. After briefly reviewing 3.11, I give an overview of the domestic and international media reports that immediately followed, focusing on the nature of information in the domestic mainstream news, which was unmasked by international and (domestic) alternative media which gradually alerted the general public to its own ignorance and lack of criticality. This lack of criticality is discussed in the second section; I believe this led to the empowerment and mobilization of the public as discussed in the last section. As mentioned above, the term *Fukushima* here signifies nuclear power issues broadly, as well as the specific accident at Fukushima Daiichi Genpatsu (the nuclear power plant owned by TEPCO, Tokyo Electric Power Company).

EAST JAPAN EARTHQUAKE

The East Japan earthquake devastated the east coast of north Japan (Tohoku), claiming nearly 20,000 lives and destroying 43 cities and towns along 500km of the coastline. Over 22,000 fishing boats at 300 ports were damaged or swept away; some were later found as far away as the west coast of the US and Canada. A total of 390,000 houses were completely or partially destroyed, and a further 700,000 sustained some damage and 35,000 were flooded. More than 23,000ha of agricultural land was affected. Estimated damage exceeds 20 trillion yen, and 18 months after the disaster, only about 5 percent of the 2.2 million tons of rubble had been processed.

In Japan, an earthquake-prone country, small tremors are daily occurrences, but the term *shinsai* (earthquake devastation) now refers specifically to the 3.11 East Japan earthquake. Travelling along the Sanriku northern east coast, signs marking the "Estimated Tsunami Inundation Area" have been in place for many years, based on accumulated data, but the tsunami of March 2011 surpassed many of them. There are also more than 200 stone monuments warning about the risk of tsunami, some of which specifically warn not to build houses below the marked point.

MEDIA ON FUKUSHIMA

Since the moment of the earthquake at 14:46 on 11 March 2011, graphic images were beamed throughout the nation and around the world. An immediate tsunami warning was issued all along the east coast of Japan alerting the whole nation, including those who had not felt the quake, indicating that a major earthquake had occurred somewhere in the country.

There are now innumerable on-line archival reports and footage of the devastation—an Internet search for "Japan earthquake 2011," for example, produces 117,000 clips capturing the tsunami, victims in evacuation, rescue workers, and devastated areas at various stages of recovery. At the moment of the earthquake, however, clearly there was no time to receive information through media, apart from warnings and messages between families and friends if mobile coverage was available.[1] The Fukushima accident, however, was a different situation. Media information did play a critical role as the disaster unfolded. The following section discusses the nature of the domestic mainstream TV news, analyzed from domestic alternative media and international newspaper reports.

DOMESTIC REPORTS

"No immediate health effect to human body"[2] (Tadachini jintaini eikyo-wa nai) is perhaps one of the most frequently heard phrases in the TV news reports on Fukushima. Now it symbolizes the inadequacy of the government response and mainstream media reports, which was gradually revealed by the domestic alternative media and international reports that people turned to for more objective and diverse information and commentary through live broadcasts and video clips on U-stream and YouTube, as well as active discussion through SNS.[3] One of the active alternative media channels, *Our Planet TV* (OPT)[4] broadcast a 3-hour program, *How the TV Reported the Nuclear Accident,* which examined how the five major TV stations—*NHK,*[5] *TBS, Nihon TV* (and its regional affiliate *Fukushima Chuo TV*), *Fuji TV* and *TV Asahi*—dealt with Fukushima during the first six days after the earthquake (14:46, 11–16 March). The program had been scheduled to be aired on 22 March on a cable news station, Asahi Newstar, as a 30-minute program, but it was cancelled because it was potentially controversial. OPT instead broadcast the full 3-hour program.

In the program, three journalists who provided commentaries had each published their own analyses of the media reports on Fukushima: *How Did the TV Broadcast the Nuclear Accident?* (Ito, 2012), *NHK—Benefits and Misuses of Domination* (Odagiri, 2012), and *Verifying the Nuclear Disaster Report: What Was Told Then* (Hirokawa, 2012).[6] The discussion and analysis revealed a number of critical issues, and for the current discussion, the following five points are pertinent: Fukushima, evacuation order, explosion of the reactors, radiation level, and meltdown. They raise serious questions about the nature of the mainstream news media in Japan, which, as I discuss later, prompted the general public to question the credibility of media information and, further, the democracy of the nation. The five points are discussed below.

FUKUSHIMA

On the coasts of Fukushima, Miyagi, and Iwate the first wave of the tsunami was detected within 10 minutes of the earthquake (14:46–14:55), but the largest tsunami hit the coasts 20 to 40 minutes later (15:15–15:50).[7] In terms of the nuclear power plants, initial attention was given to Onagawa nuclear power plant because of its proximity to the epicentre (70 km off Miyagi coast), with three stations (Nihon, TBS, Asahi) reporting on it from 14:59 to 15:08. TBS broadcast live the tsunami reaching Fukushima at 15:15 with a picture from the weather cam, captioning the picture as "Tomioka Coast" and reporting that "the wave was reaching the truss tower (*tetto*)." Fuji (15:31) also mentioned the "tsunami hitting the *tetto* and [that] a part of the cliff was collapsing," but neither station verified that the picture was of the Fukushima Daiichi nuclear power plant. More curiously, NHK reported on Fukushima only at 16:30, but used a picture of Fukushima with a sunny blue sky and calm ocean. When it was established that the plant had lost power (16:47), TEPCO immediately issued a statement that there was no immediate danger and no radiation problem, which was followed by then prime minister Kan's assurance of "no radiation risk" (16:55).

EVACUATION ORDER (11 MARCH, 21:53)

After a series of statements declaring no radiation leak, an evacuation order was issued at 21:53 on 11 March for the area within a 3 km radius of Fukushima, as well as an order that those within a 10 km zone were to "stay inside". None of the stations questioned the reason for setting the 3 km and 10 km zones. Further, 20 km and 30 km zones (to stay inside) were announced on 15 March, affecting 136,000 people (Nippon Hoso Kyokai [NHK], 2011). Two hours earlier (19:46), a "nuclear energy emergency" had been declared, but Minister of Economy, Trade and Industry, Edano Yukio, stated that this was "just so that we prepare for the worst case scenario", rather than advising people to prepare and be ready for the expected long-term evacuation. Stations also repeated the reassurances—e.g., "no radiation leak," "just in case," "just to be on a cautious side," "just to ensure safety"—and also reminded people "not to be misled by wrong information." Many people in the evacuation zones believed that they were going to evacuate "just for an overnight and return home tomorrow." Instead, many did not see their home for months.

EXPLOSION OF THE NO. 1 REACTOR (12 MARCH, 15:36)

The explosion of the No. 1 Reactor was reported immediately by two stations (TBS at 15:39, Fukushima Chuo at 15:40) as "white smoke that appears to be vapour is rising", but all other stations reported it more than an hour later, almost simultaneously (Nihon, 16:50; Fuji, 16:51; NHK, 16:52, Asahi, 17:05). The later reports mentioned an "explosion-like sound" and "vapour-like smoke," and expert commentators explained that it was a "vent to let out explosive gas" or a "controlled explosion" and that the gas was "filtered," implying the benign nature of the gas. Only Fukushima

Chuo (Nippon affiliate) mentioned that the wind was "blowing the smoke northward," implying the risky nature of the smoke.

RADIATION LEVEL

The radiation that "had not leaked outside" before the explosion suddenly appeared as a topic on the news—not that it actually *had* leaked, contrary to earlier reports, but that "its level is within the acceptable dose." On 13 March all stations explained the detected radiation level with the identical "expert" commentary, comparing it with an X-ray (0.1mSv) and a return flight between Tokyo and New York (0.2mSv). The reports suggested that radiation exists naturally in our living environment, and therefore we are naturally exposed to a degree of radiation, and also that the dose indicator was per hour, so we wouldn't be exposed to the amount unless we stayed in the environment for the whole hour. All these reports implied that one should not be overly concerned about exposure, and none explained that all the examples of safe levels of exposure exclude infants and children, who are more vulnerable than adults; that natural and artificial radiation are not comparable; that internal exposure is as serious as external exposure; and that low-level exposure over a long time cannot be dismissed as a 'safe dose.'

On 13 March an "expert" went so far as to state that *hibaku* (radiation exposure) was not to be confused with *osen* (pollution), and that (in the current situation) "you'd only need to brush off the pollutant from your clothes" because the situation was "unlike the case of the Three Mile Island accident, where substantial radioactivity was detected" (TBS, 13 March 17:09). After the series of explosions over four days, on 16 March (NHK, 17:55) Edano used the phrase "no immediate health risk," which has come to symbolize the absence of accurate information, and the inadequacy and incompetence of the central government and the mainstream TV news.

MELTDOWN (ROSHIN YOYU)

The first mention of the word *roshin yoyu* (meltdown) was made 6 hours after the earthquake by an academic commentating on the state of Fukushima (20:06, 11 March, Fuji TV). It had been established that an alternative cooling system was needed because the electric cooling system had been destroyed, and thus the expert contemplated the danger of "hydrogen explosion and meltdown." As cesium was being detected around the plant, the possibility of meltdown was also suggested by an expert just before the explosion (NHK, 14:15; Asahi, 14:38). In the following days, the word *roshin yoyu* disappeared altogether from all stations, and experts started to dismiss it as an "unfounded rumour" and said "there are many definitions and degrees of meltdown," implying that this case was not dangerous. As a series of explosions were occurring,[8] the "meltdown" dominated reporting outside Japan, while inside Japan the word became taboo, and the sense of emergency was suppressed. On 13 May TEPCO issued a statement that a meltdown did occur at the No. 1 Reactor of the Fukushima Daiichi Genpatsu 4 hours after it had been overwhelmed by the tsunami, and that the same "seemed to have occurred at the No. 2 and No. 3".[9] This did not surprise the public, who had gathered enough information by then to assume that all reactors were in a critical state. By then, the general public was turning their attention to other information sources flourishing on the Internet.

INTERNATIONAL REPORTS

In contrast to the domestic media, international reports[10] informed the public (and the domestic alternative media) of more diverse, objective and critical views on Fukushima, shining a light on the nature of the domestic mainstream media.[11] The first report of the disaster in the *Boston Globe* on 12 March, for example, already used the phrase "nuclear meltdown," along with 47 photos of the devastation: "Japan raced to avert a nuclear meltdown today by flooding a nuclear reactor with seawater after Friday's massive earthquake left more than 600 people dead and thousands more missing..." (*Boston Globe*, 12 March 2011). A survey of four major newspapers in the US, UK, France and China for one month following 3.11 (Yokouchi et al., 2012) shows that in all four countries, the major focus was on the nuclear threat—the scale of devastation and consequences of the radiation contamination, and its political, economic, social and cultural effects. The four papers examined were the *New York Times* (USA), the *Shanghai Wen Wei-po* (China), *The Times* (UK) and *Le Monde* (France). The articles published for four weeks starting from 3.12 were sorted into four categories: nuclear power plants, earthquake, tsunami and other. Articles totalled 206 (*NYT*), 225 (*WWP*), 77 (*TT*) and 155 (*LM*), of which 113, 113, 54 and 92, respectively, were categorized "nuclear plants." This analysis is noteworthy because it reveals not only the nature and extent of Japan's bilateral relationships with each country, but also how the current state and its policy regarding nuclear energy are reflected in the media.

Features in the *New York Times* on the earthquake in Japan indicate close bilateral relationships in politics, economics, and social and cultural aspects. While commending Japan's earthquake-proof architecture and evacuation procedures learned from history, the reports tend to downplay the nuclear risk because the US, with its 104 reactors, has been promoting nuclear as "clean energy." The *Shanghai Wen Wei-po* focused on domestic issues because radiation contamination is a real threat to China due to its geographical proximity to Japan, as well as on effects on food imports from Japan. Radiation level sampling started on 3.12, and the newspaper expressed concerns for Chinese students and travellers in Japan. One of *Le Monde*'s reports, "Seisme au Japon," focused on effects on Europe and the world, and compared the Three Mile Island and Chernobyl cases. Demonstrating France's weaker economic ties to Japan, few of *Le Monde*'s reports focused on economic effects. *The Times* special section *News Japan* (16–20 March) showed concerns for the effects on shares, insurance and banking, and focused on the inappropriate response from government and inadequate facilities rather than the nuclear risk itself.

The correspondence between the nature of the reports on Fukushima and each country's nuclear policy is revealing. The percentage of the nation's total electricity supplied by nuclear power generation and numbers of reactors in each country are: France (75.2 percent, 58 reactors), USA (20.2 percent, 104 reactors), UK (18 percent, 17 reactors) and China (1.9 percent, 14 reactors) (World Nuclear Association, 2012). Presidents Sarkozy, Obama and Hu and Prime Minister Cameron had expressed support for nuclear power. Cameron, on his first visit to Japan in April 2012, signed an agreement that opened up UK's nuclear market to "Japanese companies' technical expertise in new plant design and construction" in return for UK experts helping Japan with decommissioning and clean-up (Reuters, 2012). Earlier, in February 2012, Cameron agreed a joint nuclear development with Sarkozy, signalling a "shared commitment to civil nuclear power" (Press Association, 2012). This commitment, however, took a significant turn when Sarkozy was voted out in May 2012 in favor of François Hollande, who stated publicly that he intends to "reduce France's dependence on nuclear power to 50 percent by 2025" (Leone, 2012). This seems to suggest that the media information on nuclear energy can be heavily influenced by government policy, and thus, as Yamamura (2012) suggests, that stronger opposition is likely to be expressed in nations free of nuclear energy compared with those that either are dependent on it or have plans to expand it.

International media expressed critical views of the competence of the Japanese government. While all four papers praise the Japanese people, using words such as *self-sacrificing, unselfish, orderly, patient* and *resilient,* they are critical of the government's response as *slow, inadequate, ambiguous, dishonest* and *inconsiderate of its people, especially nuclear plant workers.* The praise for the Japanese people, is in a sense, an expression of sympathy for their suffering caused by the disaster, but also by the incompetence of the government. This was also demonstrated by the selection of Minami-Soma mayor Sakurai Katsunobu as one of *Time* magazine's "100 Most Influential People in the World" (2011). Sakurai posted his desperate plea on YouTube on 24 April, reaching some 200,000 viewers as his town's southern half was declared within the 20 km and 30 km exclusion zones. More than 20,000 people were advised to "stay inside," but many were left in isolation because food delivery to the 30 km zone was suspended due to safety concerns. The city was left to deal with the fear of aftershocks, further exclusion, and lack of basic support. Sakurai's recognition by *Time* is largely unknown in Japan.

On 25 June 2012, the Australian ABC's *7:30* current affairs program said of the current state of the No. 4 reactor at Fukushima Daiichi that the damaged structure contains "spent fuel that contains roughly 10 times more C137 than released by Chernobyl, according to experts," and that it is stored 30 m above ground and easily damaged by earthquake, typhoon or tornado, according to Robert Alvarez (former advisor to the US Secretary for National Security and the Environment) and a Kyoto University academic, Koide Hiroaki. Koide explained that the nuclear fuel is 2.5 times more than what's needed in the reactor core. It contains 5,000 times more cesium than was released by the Hiroshima bomb, and "the pool is just hanging there". Suzuki Tomohiko, a journalist who went into Fukushima with a hidden camera, was told by a staff member who worked reinforcing the walls that he was shocked to discover the instability of the wall. On the same day, NHK reported on the bulging of the wall containing the spent fuel, saying that the inspection found a 3.3 cm and 4.6 cm lean of the 13 m wall, but that this is "within the architectural safety standard and will withstand a large earthquake." Tokyo Electric Power (TEPCO) affirmed that the bulging wall can withstand a large earthquake, and that a crane would be installed to remove the fuel sometime next year.

This analysis reveals the lack of independence in mainstream Japanese news stations, and thus their incapacity to communicate urgency and risk by providing objective and credible information. Instead, they continued to repeat the government lines and reinforced their agenda "not to panic the public". Some critical information was withheld, or its risk downplayed, to the extent that any information alerting the public to radiation risk was criticized for provoking panic and fear and challenging the authorities, of which the mainstream media are a part. NHK, in particular, is regarded as a credible information source, and its reassurance of safety sounds like a "gold-seal" guarantee. Its ethical responsibility should be questioned even more strongly than that of other commercial stations (discussed below). The grave "human

Figure 1: Miss Kitty being tested for radiation level.[12] Illustration by Taylor Jones for *Hoover Digest.*

error" that Yamamura (2012) and many others refer to is the way Fukushima was handled by the government and the media, but also, more seriously, how critical questions are not asked by the mainstream media and how, so far, the general public has accepted that.

The Japanese government, and the mainstream news media, failed to practice the "assumption of state responsibility" (Klein, in Pantti et al., 2012: 139) when the people's resources, mobility and, most importantly, information were restricted. State responsibility and media ethics (Pantti et al., 2012), which are taken for granted and needed by the public, both failed. The uncritical nature of the mainstream media, especially the publicly funded station, reflects the lack of critique and social debates thus far, as the general public come to terms with the situation. One aspect of "un-criticality" is discussed below.

POLITICAL APATHY AND THE SOCIAL PHENOMENON OF 'INDIRECTNESS'

Both Japanese society and the Japanese language have been described as "indirect" (e.g., Bratt Paulston et al., 2012; Hodgson et al., 2000). Indirectness in the extreme can lead to avoiding discussion of critical issues or diminishing the seriousness of social matters. One strange phenomenon of social indirectness in contemporary Japan is the rising popularity of characters (*kyara*) created to represent a place, event, product, organization or even governmental department. A *kyara* is an imaginative figure (human or animal), often originating from manga, anime, novels, or computer games. Hello Kitty, or "Kitty-chan", may be the best known Japanese character, as famous as Mickey Mouse, Snoopy, Miffy or Tom and Jerry. These characters are not just for children, but are used in all social contexts.

In recent years, such characters—sometimes called *yuru-kyara, yuru* (lovable, somewhat clumsy, friendly and affectionate characters)—have been created for events, organizations (ministries, local governments, banks), places and products. Their names are associated with the products or organizations, and they are used to soften, brighten and make accessible the organizations or products they represent. More than 40 universities, 17 banks, 43 central government agencies, 48 police departments, 25 fire and emergency departments and more than 1,000 local governmental units have their own characters. For example, *Soramame-kun* (Broad-bean boy) created for the Atmospheric Environmental Regional Observation System (AEROS) for the Ministry of the Environment "watches the sky frequently and attentively." Another character, *Shinkyu-san,* used by the Ministry of the Environment, specifically works for the 25 percent reduction of CO2 in the Challenge 25 Campaign.

Tetsuwan Atomu (Mighty atom, or Astro boy), a well-known character, was initially featured in manga (1956–1968) and TV anime (1963–1966). As evident in his name, he represented the period when it was believed that science and technology, and especially nuclear power, would usher in a new era. A series of adventure stories about Atomu promoted scientific advancement and achievement, and their popularity extended to the US, UK, Ireland and Australia. Following this example, several scientific figures were created, and some specifically promoted nuclear issues. The most controversial was *Puluto-kun* (Pluto-boy), who appeared in the promotional materials of the Japan Atomic Energy Agency (JAEA).[13] In the video "A Reliable Friend, Puluto-kun" (1993), Puluto-kun explains the safe nature of plutonium and its potential for peaceful use, emphasizing that plutonium is "not like the highly toxic and potentially fatal potassium cyanide; even ingested with water, it will be simply eliminated from the human body. No cases of cancer caused by plutonium have been reported." The 11-minute video was highly controversial, particularly among residents near the Monju reactor, and it was removed from the JAEA website and YouTube, but Pluto-kun is still used in JAEA's promotional material. Another character, *Pellet-kun* (Fuel pellet), appears on the website of Nuclear Fuel Industries. Pellet-kun takes you through a quiz on nuclear power plants and industries' role in developing fuel, which specifically mentions the safety of various measures taken by industries and how nuclear energy contributes to "counteracting climate change." The friendly character here causes serious harm by trivializing a most critical issue.

These examples give a glimpse of Japan's current social context, with its lack of criticality, social and personal vigour, and robustness.[14] Active discussion and negotiation of diverse opinions, ideas, and ways of life help to develop initiative, leadership, critical thinking, and open and direct communication. Perhaps 3.11 and Fukushima—the devastation itself and the unsatisfactory response by media and government—have moved the general public to voice their opinions and take action.

What also empowered the general public is their participation in various forms of volunteer work. By June 2012, one million volunteer workers had visited Tohoku. This number represents only those who registered at the Department of Social Welfare (Shakyo). Initially, volunteers were advised to be self-sufficient in terms of food, water, accommodation and access to the area, but in April 2011 "volunteer buses" started to operate from various parts of Japan. From Wakayama, for example, 12 "volunteer bus trips" were organized from April to August 2011, with 233 volunteers participating.[15]

EMPOWERING NEW VOICES

On 16 July 2012, 170,000 people gathered in Central Tokyo Park to express opposition to nuclear power plants, following the restarting of the Oi nuclear power plant on 4 July despite overwhelming opposition from the general public, experts, and a third of the parliamentarians. Since the announcement of the resumption of operation, a protest crowd gathers every Friday night in front of the prime minister's residence. The public is angered by the government decision, which ignored "80 percent of the public voice and the 160,000 displaced people in Fukushima, not to mention the current and future suffering of children."[16] Kansai Denryoku (KEPCO) issued a notice of the possibility of *keikaku-teiden* (scheduled power cut) for the "expected power shortage during the peak in summer", but within a week of restarting the Oi nuclear plant, KEPCO announced a plan to close eight fossil-fuel power plants, which are more costly to operate than nuclear power plants, exposing the truth behind the power shortage claim. The majority of the general public has claimed that they'd "go without electricity rather than live under nuclear threats".[17] In fact, it was reported that even on the hottest day in Tokyo (22 July) the electricity consumption was more than 20 percent less than the previous year (Tanaka, 2012).

The first anti-nuclear gathering after 3.11 was held on 14 March 2011 in a central park in Tokyo (Koenji), when 15,000 gathered to call for the closing of all nuclear power plants. *Genpatsu Zero Action*, a demonstration on 9.11, the six-month anniversary of the disaster, was held all around Japan, and now a group is campaigning for the 10 March 2013 referendum on nuclear power. There is a rising civil protest against nuclear power, and many are taking to the streets in major cities (Tokyo, Kyoto, Osaka). NGOs such as FOE (Friends of the Earth), Greenpeace, WWF–Japan, and ISEP (Institute for Sustainable Energy Policies) are calling for a serious re-think on energy policy and a transition to renewable energy (see, for example, Greenpeace International, 2011a, 2011b; Greenpeace Japan, 2012). Some call it the "third major turning-point Japan is facing, after the Meiji Restoration and WW2" (ISEP), and the current mass protest has been dubbed the "hydrangea revolution" after its beginning in the June hydrangea blossoming season.

The Japanese government's decision to resume the power plant operation was claimed to have been based on "local consensus," which refers only to the small local community of 6,500 people whose jobs depend on the plant and associated industries. Nuclear responsibility, however, clearly extends globally beyond time and space, and thus calls for an international discussion and responsibility.[18] The global view of 3.11 certainly mobilized the general public in Japan, many of whom see this as an opportunity to create a new kind of future. The vision to rebuild from the 3.11 devastation will, I hope, lead the nation to a future that is sustainable and democratic, equipped with new media and global networks, and with a global perspective and a deep local appreciation.

ACKNOWLEDGMENT

The author acknowledges the helpful comments provided by Simon Wearne and the editors.

ENDNOTES

1 Mobile phone networks experienced major congestion. The author, in Wakayama where no major quake was felt, was unable to use her mobile phone for nearly 3 hours.

2 (FRYING DUTCHMAN "humanERROR", 2012). This has become a symbolic song among the anti-nuclear movement, translated into English, French and German.

3 e.g., *Citizens' Nuclear Information Centre* (http://cnic.jp/), *Our Planet TV* (http://www.our-planet-tv.org/), *IWJ* (*Independent Web Journal*), *NPJ* (*News for the People in Japan*, http://www.news-pj.net/), *Japan Real Time* (*The Wall Street Journal*, Japan, http://jp.wsj.com/japanreal-time/blog/archives/12206/) and *E-Wave Tokyo* (http://eritokyo.jp/independent/today-column-ewave.htm).

4 Established in 2001, and funded entirely by membership fees, workshops and production fees (no corporate sponsorship).

5 NHK Nippon Hoso Kyokai (Japan Broadcasting Cooperation).

6 Titles are the author's translation.

7 Japan Meteorological Agency, 11 March 2011. http://www.jma.go.jp/jp/tsunami/observation_04_20110312193944.html.

8 Explosions occurred at No. 3, 14 March, 11:01; No.2, 15 March, 6:10; No. 4, 15 March, 6:14.

9 AFP BB News, 24 May 2011. http://www.afpbb.com/article/disaster-accidents-crime/disaster/2802249/7257386.

10 Available on-line or as a print publication. More than 30 foreign newspapers are also available in Japanese, 10 Japanese papers have English versions, and the *Japan Times* is an English-language newspaper.

11 A diverse range of reports include Brumfiel and Fuyuno (2012), Coghlan (2011), Cyranoski (2012), Fackler (2011), MacKenzie (2011), Nakamoto (2011), Schiermeier (2011), Smith (2011), and Soble (2011).

12 Nishi, T. (2012). On the cesium road. http://www.hoover.org/publications/hoover-digest/article/113111.

13 Former Power Reactor and Nuclear Fuel Development Corporation (1967–1998).

14 "Super-ageing" also plays a part in this (Muramatsu and Akiyama, 2011; Hirata, 2004; Fukuwara, 2007; Ono, 2005; Ministry of Land, Infrastructure, Transport and Tourism, 2006).

15 All trips were over weekends, leaving on Friday afternoon and returning early Monday. The overnight bus trip took about 16 hours, and the second day was at a local guest house. The trip cost for each volunteer was around 5,000 yen (for guest house accommodation). Wakayama had to divert the relief effort to its own disaster—five days of continuous rain caused flood and landslides in the southern part of the prefecture in early September, claiming nearly 100 lives.

16 Comment by a protester at Oi Nuclear Power Plant re-opening (30 June 2012). Author's field notes.

17 Comment by an organizer of Stop Nuclear gathering in Tokyo (22 June 2012). Author's field notes.

18 This includes the nations that promote uranium mining and exporting. For example, the Australian indigenous community issued their strong opposition to the Olympic Dam mine; a large amount of the uranium mined is exported to Japan. "Many of our food sources, traditional plants and trees are gone because of this mine. We worry for our water: it's our main source of life. The mine causes many safety risks to our roads—transporting the uranium from the mine. It has stopped us from accessing our sacred sites and destroyed others. These can never be replaced. BHP never consulted me or my families, they select who they consult with. Many of our people have not had a voice. We want the mine stopped now, because it's not good for anything." (Eileen Wingfield, Kokatha elder, http://lizardsrevenge.net/5-2/).

▪ CHAPTER FOURTEEN

Public Communication, Environmental Crises and Nuclear Disasters

A Comparative Approach

CLIO KENTERELIDOU

Environmental crises and nuclear disasters are unexpected and irreversible. They are an immediate natural and/or human-induced disaster of a large magnitude causing large-scale threat to human health and/or the environment. Such events lead to severe health problems and environmental degradation (Fink, 1986), and affecting the parties involved as well as their publics, products/services, or good name (Fearn-Banks, 1996; Ho and Hallahan, 2004; Kenterelidou and Panagiotou, 2006).

Environmental crises and nuclear disasters do not respect borders; they become global almost instantly with the help of the media. The news media transmit environmental crises information to audiences, deliver news about the crises, issues and policies, and distribute information about pertinent issues lying outside people's immediate realm of experience (Cox, 2010; Iyengar, 1991; Scheufele, 2002). Environmental crises and nuclear disasters stick in people's minds and frequently have a much wider public impact than any other kinds of crises. Crises also have a strong (inter)linkage with powerful environmental lobbies that achieve a lot of publicity by publishing stories about these events. The newsworthiness of environmental crises and nuclear disasters can be explained by the following:

a) almost everyone has an interest in the environment, which as a word and idea evokes different emotions and responses for different people. Images of dying wildlife and humans have a huge visual and mental impact on all generations;
b) the crises involve drama, tragedy, and the presence of political elites or celebrities and their comments about the accident;
c) they can be personalized; and
d) they entail conflict due to questions arising around transparency and decision-making. (Perko, 2011)

As social issues with global impact, environmental crises and nuclear disasters represent "socio-technical" problems requiring complex remedies and a synthesis of practices (Granot, 1998). They inevitably receive wide media attention, and consequently attract the public's attention (Gonzales-Herrero and Pratt, 1998: 85). Every move of the actors involved attracts the attention of the global media, and therefore this is the time for actors to show their capacity to make and execute decisions, and overcome difficulties and solve problems. This is also the time, however, where links between means and ends are not widely understood, as the stakeholders' framed agendas may be highly persuasive, or at least provide political cover for actions that would otherwise be discredited as special-interest politics. All actors concerned well understand the political logic of such crises moments, as they are a public issue concerning everyday citizenship and a political issue for policy implementation and initiative, acts and pressure. The result of this combination is that the environment becomes an issue in the media agenda. Generating public awareness and getting people to engage in the behavior required during a crisis and environment-related issue or condition entails making sure communication works together with publicity. Otherwise, shockwaves can shake public trust.

Managing communication efficiently and effectively means that a crisis communication strategy focuses on helping the general public, through public information, to express their concerns and fears. The crucial part of public communication during a crisis, especially in environmental and nuclear disasters, is not to restrict communication only to facts. Crisis is inherently pragmatic in nature, but information can fail to account for the general public's or individual's knowledge (or lack thereof), their perception of risks, and their relative inexperience (Perko, 2011). Indeed, there are examples of "targeting" and "tailoring" public information, and of strategic framing of crisis communication from the stakeholders concerned, that resulted in a failed public communication strategy and a general erosion of public trust (Ho and Hallahan, 2004; Kenterelidou and Panagiotou, 2006; Seon-Kyoung and Gower, 2009). These examples include the accident at the Paks nuclear power plant in Hungary (2003), the Tokaimura nuclear accidents in Japan (1999 and 2006), the Goiânia accident in Brazil (1987), the Soviet submarine K-431 accident at Chazhma Bay in Vladivostok, Russia (1985), the accident at Stationary Low-Power Reactor Number One (SL-1) in Idaho, United States (1961), the Kyshtym disaster at the Mayak nuclear plant in the Soviet Union (1957), and the atomic bombings of Hiroshima and Nagasaki during World War II (1945).

A perception of risk brought forth by crises can also be created. It can be produced by the media and may differ from reality, or may become magnified, intensified, weakened or filtered with respect to its attributes and importance (Kasperson et al., 1988; Pidgeon et al., 2003). An incident in one country can rapidly become an international crisis fuelled by 24-hour news media coverage. Actors are then involved in a crisis and are faced with the exponential and unrelenting expansion of communications media activity, and the proliferation of mass media attention.

As news is regarded as a form of social construction (Gitlin, 1980), the way environmental crises and nuclear disasters are presented in the news media can have a significant impact on the public's understanding of the nature of the crisis and environmental problem itself and the proposed remedies or solutions, as well as on its crisis and environment-related behaviors and perceptions. In this context, this study of the "ecosystem" that surrounds environmental crises and nuclear accidents and their public communication helps highlight the mechanics involved and mechanisms (operations, actions) employed by the political communication actors concerned (see also Pantti et al., 2012). Under this optic, public communication is considered to be more than a practice of informing the public; it is a means of viewing and perceiving the modern mechanisms of political communication actions and operations of the actors in a society. When these actions are not related to openness, transparency, visibility and candor, then the concept of public interest within the action of informing appears to be fluid, and public trust is at stake (Schedler et al., 1998). Moreover, public communication is described as a seesaw, as information and risk evaluation flows back and forth between political communication actors (i.e., governments/states, governmental officials, the media, business enterprises, other political-economic-societal entities, academic experts, NGOs, institutions, regulatory practitioners, interest groups, and the general public; Leiss, 1996) in their effort to name the world's parts, and to certify "reality" as reality. Crisis can be defined in different ways because these conditions inform perceptions held by the highest-level decision makers of the actors involved (Giddens, 1991). Under this lens, public trust again is foregrounded whenever effective communication and a constructive dialogue is sought between all those involved in a particular debate about the crisis and its risks.

At this point, it should be noted that public trust is established and strengthened when it interacts with communication. Every public information and communication act is socially and politically charged, and is registered in people's minds positively when something is actually being done for the management of the crisis, and negatively when nothing is being done and nobody deals with the crisis situation (Kenterelidou, 2009). In other words, in the short run, effective public communication increases awareness and understanding of protective actions and improves the public's response, and in the long run, it facilitates the remediation process and the return to normal life.

The purpose of this study is to explore modern public information tools and the ways that democratic societies respond during such crises in the era of globalization. Specifically, this study is a comparative approach of the ways in which actors define, set principles, and provide public information to citizens via their public communication and the resonant communication mechanisms that they draw upon in their crisis management. By exploring the actions of stakeholders in environmental and nuclear events and identifying their crisis communication strategies, the study examines how those directly involved with a crisis engage with those who can influence or have an interest in the incident.

This study considers the entire sets of relations, actions and public information processes and institutional settings within which information was formed and deployed. This approach encompasses the political communication actors and issues involved in a nuclear accident, including the history, sociopolitical and economic setting, media scenery, power relations, and workings of the governments and of other political, economic and societal entities within the context of a mega-crisis (Schmid, 2011). The design of the study is a brief comparative analysis of three environmental mega-crises case paradigms: (1) the Fukushima Nuclear crisis (Japan, 2011); (2) the Chernobyl disaster (Ukraine, 1986); and (3) the Three Mile Island accident (Middleton, Pennsylvania, United States, 1979). These crises are explored using eight criteria designed to assess responses and attempts to reach the goal of effective public communication. These criteria have been developed from a comprehensive reading of the crises communication literature (notably, Coombs, 2006; Seon-Kyoung and Gower, 2009; Kenterelidou and Panagiotou, 2006; Sandman and Lanard, 2004; Ferguson, 1999). The reasons for choosing these cases are:

a) all three crises were experienced at a local level but possess a global dimension;
b) the damage, the devastation and chaos caused, and the disruption to people's lives were considerable in all three;
c) the crises focused the attention of a wide audience due to their newsworthiness;

d) they prompted the assembly of a large spectrum of political communication actors (i.e., the general population, media, government agencies, business enterprises, academic researchers and experts);
e) all three crises presented relevant scientific problems suitable for in-depth study; and
g) they all involved nuclear power plants, and there was consistency in the reaction to the disasters-accidents, including the reaction regarding the dissemination of public information, the use of public communication strategies and the variations of public trust.

THREE MILE ISLAND (PENNSYLVANIA, UNITED STATES, 1979)

The Three Mile Island (TMI) incident on 28 March 1979 was a nuclear accident in a nuclear power plant run by Metropolitan Edison Company, based in Middletown, Dauphin County, Pennsylvania, United States. The nuclear accident was rated level 5 —"accident with off-site risk"—on the International Nuclear and Radiological Event Scale (INES), with 7 being the highest and 0 the lowest. It remains the worst nuclear accident to date in the United States. TMI was a pressurized water reactor that experienced a partial core meltdown in reactor block 2 in which a third of the reactor block melted. The accident was, in short, a combination of malfunctioning construction and human error. As a result, radioactive steam was released into the atmosphere and gallons of contaminated water were released into the Susquehanna River. Yet, experts say the resulting radiation exposure was never enough to cause a detectable health effect in the general population because the radioactive material never actually escaped the containment vessel. No deaths were linked to the event, but there remains some controversy over subsequent fetal deaths (World Nuclear Association, 2012; Walker, 2004).

CHERNOBYL (UKRAINE, 1986)

The Chernobyl nuclear disaster happened on 26 April 1986 near Pripyat, Ukraine S.S.R.. It became the worst and most devastating recorded nuclear catastrophe in history. It was rated as level 7—the highest level—which refers to a "major accident" on the INES. A low-power engineering test of the Unit 4 reactor at the Chernobyl nuclear power station resulted in a sudden increase in heat production and loss of cooling water, causing a catastrophic steam explosion that completely destroyed the reactor core. The explosions and subsequent meltdown of the reactor core, attributed to numerous human errors, resulted in a fire and the release of radioactivity into the atmosphere, which was then spread by the winds and fell as radioactive rain for days over Eastern Europe. Over a 10-day period, winds carried radioactivity that caused severe contamination in vast territories of republics of the former Soviet Union—Russia (Brjansk), Ukraine (Chernigov) and Belarus (Gomel)—as well as over parts of Europe and even China, Canada and the United States (International Atomic Energy Agency, 2005; World Nuclear Association, 2012; Mould, 2000; Belyakov, 1999; United Nations, 1998; Abbotti et al., 2006).

FUKUSHIMA (JAPAN, 2011)

Fukushima was a series of incidents that occurred on 11 March 2011. A natural disaster—the Tohuku earthquake (magnitude 9) and the subsequent powerful tsunami—devastated the northeastern coast of Japan and caused a loss of power at Japan's Fukushima Daiichi nuclear power plant. The plant's owner is a private sector company, Tokyo Electric Power Company (TEPCO), the largest utility in Asia and the fourth largest in the world. According to the INES, the nuclear accident was rated, finally, as level 7. The loss of power in the plant led to a series of equipment failures and the partial melting of three cores, with the backup power supplies flooded by the tsunami which knocked out electricity at the facility, preventing emergency cooling. As a result, pressure explosions occurred in the reactors hosting the buildings, deliberate venting was done to avoid further damage, and water was dumped into the ocean. Radioactive material was released. Further, weeks after the initial nuclear accident, it became necessary for TEPCO to dump thousands of tons of highly contaminated radioactive water into the ocean. In other words, radioactivity was released again into the environment. Moreover, human errors and the design of the plant exacerbated the severity of the accident (World Nuclear Association, 2012; Tertrais, 2011; Cooper, 2011).

RESEARCH FINDINGS AND DISCUSSION

Criteria 1—Rapid and Continuous Communication to the General Public and Affected Groups

In TMI, the first announcement of the accident was by a radio news reporter who stated simply that a problem had occurred at TMI and that the company offered assurance that the public was in no danger. Later on, an Associated Press release filled the media with the story. In Chernobyl, there was an information vacuum, as the accident was made known only 36 hours after it had happened. In Fukushima, the news took a long time to emerge. The response to the crisis was flawed by poor communication, delays in releasing data, and a constant chorus of misinformation from all of the stakeholders involved (see also Kato, this volume). Official authorities took a month to change the initial estimation of level 5 and admit the level 7 accident assessment. Therefore, the public turned to alternative, real-time information sources on the Internet, which offered not only online encyclopedic knowledge that enabled public understanding, but also misinformation. Citizen journalists unfettered by rules or ethical considerations were able to gather, filter and distribute news as fast as established news organizations. In summary, it can be observed that in TMI, the company's and the local authorities' communications were confusing, and the Chernobyl disaster revealed a secretive government operating at the expense of human lives, whereas at Fukushima, the officials controlled the release of critical information to suppress political cost and preserve economic interests.

Because all three crises involved nuclear power plants, one might assume that there was some consistency among them in the dissemination of public information and the employed public communication strategies. Ultimately, this did not happen; they all mishandled their communication with the public and failed in fundamental principles of good crisis communication, because the actions taken ran counter to its core elements. Underplaying the seriousness of the crises gave a (false) sense of security, but in the long term that translated into a loss of confidence in the stakeholders.

Criteria 2—Express Empathy and Address People's Concerns about Risks

In TMI, there was false reassurance of the public's concerns and intentional withholding of information that continued to appear even throughout the following years. Continuous scientific disputes among experts weakened the trustworthiness of both the utility and the officials. A deterioration of the public's confidence predominated because doubts about manipulations of information remained. In Chernobyl, the plight of the radiation victims was ignored, especially by those who were financially or politically dependent on nuclear activities (Bertell, 2008). The tactic of "keep it low and trouble will be avoided" was applied by all levels of officials. In Fukushima, initially, the public was assured that the radiation released was "safe and poses no health risk" (J. Rosenthal, 2011). Later, the situation was characterized as "no immediate danger" because, as it was said, resultant cancers would take decades to develop. Next, the evacuation of the area was ordered. Days later, the public once again was assured that the radiation would be "diluted by the ocean" and everything would be fine because "no one is allowed to fish in the offshore evacuation area"—as if fish don't migrate and ocean currents don't circulate the planet (J. Rosenthal, 2011). In all cases, there were worldwide communicators of a plethora of emotions, and an establishment that repeatedly played down the risks and suppressed vital information.

Criteria 3—Provide Information About How People Can Protect Themselves

In TMI, the only information was provided at some demonstrations that were organized by activists. In Chernobyl, there were conflicting announcements, findings and recommendations. There was a lack of general awareness, and a disregard of the effects of the accident. In Fukushima, in an effort to educate the public, some journalists and experts recognized the necessity of working together to ensure that the most accurate information was delivered to the people. As the technology was complex, they published erudite analysis and opinion pieces. Moreover, the new/social media enabled rapid public access to a plethora of online information that shed light on the situation and enabled public understanding. Overall, however, it appears that in all three crises, the information was conveyed in fragments, and there was lack of clarity, obfuscation, sensationalized rumors, and inconsistent, misleading comments.

Criteria 4—Designate Crisis Spokespersons, Formal Channels and Methods of Communication

In TMI, the government sent a mixed message. While stating that things were under control and there was no danger to public health, they concluded their statement by adding that there was radiation emitting into the environment. However, credible public media sources (Walter Cronkite, CBS Evening News) publicized the conflicting statements about radiation releases made by all involved, and the confusion of the communications began to be noticeable. In Chernobyl, the incompetence or lack of experience of the officials and experts resulted in conflicting recommendations and communication practices. The plethora of conflicting statements made them useless and misleading. In Fukushima, there was a lack of specialized personnel able to respond appropriately during the crisis. This, in turn, resulted in the appearance of a plethora of message senders and voices, causing communication chaos (Chang, 2011). The common characteristics in all three cases were slow communication with contradictory comments, lack of coordination and unity of messaging, and an unclear sense of strategy.

Criteria 5—Make Sure that Communicators Have a Good Understanding of the Crisis Circumstances and Potential Outcomes

In TMI, the local authorities' communications were confusing because they involved too many senders and messages revealing different definitions of whether there was a crisis and what the subsequent consequences could be. In Chernobyl, there was a lack of clear lines of responsibility and inadequacy of communication among all actors involved. There was also incompetence or lack of experience on the part of the officials and experts, which explains why there were no answers. In Fukushima, there were no specialized personnel with a good understanding of the crisis circumstances. There were also harried officials incapable of making decisions, a loss of trust among the major actors, and conflicts that produced confused flows of sometimes contradictory information. Moreover, the operating company was poorly regulated and did not know what was occurring, and the Japanese government was slow to accept assistance from international nuclear experts (Chang, 2011).

Criteria 6—Admit Uncertainties

In TMI, the utility refused to accept full responsibility for the accident, stating that the reactor manufacturer and the government were also at fault. In Chernobyl, all actors involved downplayed the seriousness of the accident. There was an information vacuum, a lack of general awareness and a disregard for the effects of the accident, even many years after the accident. The government and the nuclear industry elite misled and avoided accountability for mistakes and failures. There were also misconceptions, scarcity of reliable information, apathy, lies, and secrecy shrouding the accident, which is known as the "secret disaster." Chernobyl combines the unseen, unheard, unfelt and unsmelt (Barnett, 2007). In Fukushima, this is still an ongoing process. However, only a year after the accident, it was revealed that data and vital information sent to official entities were not shared with others, remained unread and were deleted afterwards. They also were not used because the disaster countermeasure office regarded the data as "useless because the predicted amount of released radiation is unrealistic" (*Mainichi Shimbun*, 2012).

There is a common denominator in all three cases: the apparent reluctance to admit a mistake or a construction error on the part of the plant's owner, the nuclear industry and the government officials. The reason for this reluctance is risk of loss—business and financial loss due to regulatory constraints, organizational changes, litigation, physical risk, sabotage or terrorism, and loss of confidence in institutions (Kasperson et al., 1988; Slovic, 2000; Pidgeon et al., 2003).

Criteria 7—Attempts for Remedy

In TMI, in an attempt to improve the government's credibility, the Environmental Protection Agency recruited in March 1980 a number of people in the area to educate the public by teaching them to measure radiation levels. Moreover, in October 1979 the Nuclear Regulatory Commission recommended what was (up to then) the largest fine in its history—US$155,000—against the utility for 17 violations associated with the accident (Walsh, 1983). In 1983, however, the United States government considered restarting the TMI reactor. In Chernobyl, in an attempt to mitigate the situation, without imposing major restrictions on information flow, Soviet officials invited journalists from abroad to report on the decontamination work. Nevertheless, Western journalistic norms and practices as well as the concept of newsworthiness differed from the Soviets'. It was the first time that journalists were able to report directly from the Soviet countryside, and understandably, the journalists' interests moved far beyond the territory of the power plant; their reporting included interviews and information from people who until then had had no voice. This approach contradicted the image that had been crafted by the Soviet officials who, in the end, reintroduced controls on information in order to preserve their view of the incident and the image that the state was taking care of their civilian population (Schmid, 2011). Additionally, there were some governmental decisions on restrictions of nuclear energy use. In Fukushima, it is still an ongoing process. It is worth mentioning that the Japanese government has admitted that it did not keep records of key meetings during the crisis, even though such detailed notes are considered a key component of disaster management (BBC Online, 2012). In all cases, the close and often unclear relationships and interdependencies among actors served to circumvent rules and regulations or remedy initiatives. The result of this situation was the loss of public trust in government and all the parties involved.

Criteria 8—Aftermath

In TMI, there was confusion among the actors involving communication and misinformation. The already troubled United States nuclear power industry and regulatory agency faced even more serious problems. The media played a role in conveying information about the incident and countermeasures, but did not really have much impact in uncovering the mistakes. Public fears of nuclear accidents rose and public confidence about nuclear safety was shaken, bringing long-ignored reactor safety problems into sharp focus. A move to abolish nuclear power emerged. The public information and communication, or the lack of it, became a prime characteristic of how nuclear catastrophes were dealt with in society.

In Chernobyl, fear, rumor and a lack of honesty dominated the atmosphere (Rahu, 2003). Nuclear affairs were state secrets—everything was classified, and only successful accomplishments were announced (Schmid, 2011). Uncertainty took root in the public and that, in turn, caused the questioning of the ethics of risk management (Barnett, 2007; Oughton, 2011). The poor handling of the disaster and the fact that the parties involved downplayed the seriousness of the accident shattered the former Soviet Union's credibility and the "new" culture of openness and transparency (*glasnost*). The formation of interest groups, lobbying and NGOs for environment and public health followed. Decisions to restrict the use of nuclear energy were made by several European governments (Organisation for Economic Co-operation and Development Nuclear Energy Agency, 2002: 44). Chernobyl, a word and place unknown until then, became globally identifiable, researched and analyzed. New "nuclear-related" words entered the global vocabulary. Chernobyl had enormous social, economic and psychological consequences (Gray, 2002; Bay and Oughton, 2005; International Atomic Energy Agency, 2005) and taught humanity as much about industry and politics as about radiation (Sich, 1996). Chernobyl was not just an event—it is an ongoing process.

In Fukushima, the crisis continues. The crisis reinforced the public's concern about the potential for severe accidents and long-lasting harm to health and environment. Because the crisis happened on the twenty-fifth anniversary of the Chernobyl accident, it brought back memories and reignited skepticism and critique about nuclear power. It made clear that the liabilities of a nuclear accident of this magnitude can be larger than a private company can bear (Reynolds, 2011), and brought under the spotlight a re-examination of all aspects of nuclear power and the need for an energy portfolio to meet climate-change mitigation objectives (Glaser, 2011). Fukushima's economic consequences are greater than Chernobyl's economic impacts because the crisis affects a relatively more wealthy and productive economy that overlaps with another all-powerful economy, the United States (in the United States reactors use Japanese technologies like Fukushima's). The economic consequences are also greater because Japan is the world's top country in terms of managing its nuclear power plants, and a nation with a reputation for discipline, scientific knowledge, and engineering prowess. In summary, the after-effects were that TMI scared off investors and induced additional regulation of the industry, the Chernobyl disaster set many Europeans against nuclear power, and the radioactive releases at Fukushima following the earthquake and the tsunami are changing the policy landscape (Andrews, 2011).

Overall, the three disasters demonstrate the nuclear power industry's willingness to risk the health of humanity and environment, and to favor military needs and financial gain over public understanding. The outcome is that nuclear, political, environmental and public agendas will be reshaped and new policies in countries around the world will be introduced because safety is not only a public health or environmental issue, it is also an economic and, thus, a political issue (Cooper, 2011).

CONCLUSIONS

Communication with the public is an integral part of any crisis and must be woven into every aspect of the crisis response (Golan, 2003: 126), not only at the end phase with report findings, but also at the planning and problem-formulation phase. The fundamentals of an effective public communication during environmental crises and nuclear disasters are:

a) contents of communication (i.e., releasing information about the status of the nuclear plants, the extent of the damage, and the risks of further radioactive emissions);
b) scientifically correct answers;
c) rapid decision-making;
d) frequency and extent of communication; and
e) dialogue, social awareness, and some involvement of the affected populations and the general public in decision-making processes (Heath et al., 2007).

Because every public information and communication act during a crisis is socially and politically charged, there is an imperative for the nuclear industry to understand that bigger issues are attached to nuclear power—issues of public concern regarding environmental, political, public health and societal factors—which means that a public communication strategy should be focusing on providing information and regaining and amplifying trust (Ramana, 2011; Kenterelidou, 2009). The media are the link between the officials and the public, because public communication is clearly a two-way process (Giddens, 1991; Smith and McCloskey, 1998:47; Miller, 1999: 1249). The news media must be viewed as the conduit of information to the public, not as an afterthought. Environmental and nuclear crises must be represented in the media as public events (Granot, 1998). The media can play a positive role in providing information and dealing with risk by discrediting any inaccurate depictions or estimations of risk and facilitating remedies and the restoration of public trust (Smith and McCloskey, 1998; Trettin and Musham, 2000).

The contemporary approach to public communication during environmental crises and nuclear disasters should go beyond alerting or reassuring the public about potential risks. The approach should entail both policy and operational planning and the involvement of citizens in decision making by increasing public knowledge and social awareness. If the costs from such crises are socialized, the public pays the price. Additionally, there is a need to improve international cooperation and preparedness, a need to adopt a "whole-ecosystem" approach where the political communication actors are required to share a singular purpose: to inform the public honestly, promptly, openly, efficiently and consistently, and to work with the media to achieve this outcome (Fischhoff, 1995; Chess, Salomone, and Hance, 1995; Trettin and Musham, 2000). Trettin and Musham note that "creating a better and critically informed public entails obligations not only on the part of actors involved during the crisis but on the part of concerned citizenry as well" (2000: 417). An educated public triggers public mobilization and social action. The result of this strategy is the validation of the actors' reactions as immediate, all-embracing and effective (Kenterelidou and Panagiotou, 2006), and contributes to the formation of a new worldview on the environment.

▪ PART 4

Contested Claims

▪ CHAPTER FIFTEEN

Climate Change, Media Convergence and Public Uncertainty

ROBERT COX

While concerns about climate change have faded in the United States in recent years, actions to control greenhouse gas emissions have "rolled out apace" in other nations (E. Rosenthal, 2011: para. 7). Europe's carbon emissions trading is expanding; Australia has begun to tax carbon emissions and has committed to link its carbon price to emissions-trading schemes in Europe and New Zealand; India has placed a carbon tax on both domestic and imported coal; and China's latest five-year plan contains a "limited pilot cap-and-trade system, under which polluters pay for excess pollution" (Milliken, 2011: 68).

At the same time, the United States (referred to as U.S. throughout this chapter) has remained a conspicuous outlier in its halting attempts to address climate change. "Now that nearly every other nation accepts climate change as a pressing problem, America has turned agnostic on the issue" (E. Rosenthal, 2011: para. 2). This agnosticism has been due, in part, to a decline in the American public's concern about global warming and corresponding doubts either that it is happening or that human activities are a contributing cause (Ipsos Global @dvisor, 2011; Jones, 2011; Saad, 2011; Harris Poll, 2011; Pew Research Center, 2011a). As columnist Paul Krugman observed, climate scientists in the U.S., while increasingly confident in modeling climate change, have become frustrated by their "inability to get anyone to believe them" (2009: A21).

Why, then, has public doubt persisted in the U.S.? Two different but complementary explanations have been forwarded. By one account, scientists, educators, and many journalists have found climate change more difficult to communicate to the U.S. public than other environmental or health issues. In addition to the complexity of climate science, the impacts of climate change are often seen by many Americans as distant, subtle, or outside the personal experience of readers or viewers (Moser, 2010). As evidence, the Yale Project on Climate Change Communication found that most Americans remain seriously confused about the causes of climate change (Leiserowitz, Smith, and Marlon, 2010: 3).

Since the late 1990s, U.S. journalists have also documented the role of corporate advertising and the funding of contrarian reports and information campaigns that are intended to "manufacture uncertainty" or question the scientific basis of climate change (Michaels and Monforton, 2005: 362; see also Gillis and Kaufman, 2012; Shabecoff, 2000; Cushman, 1998). As a result, U.S. media "have been significantly more likely than media in other industrialized nations to portray global warming as a controversial issue characterized by scientific uncertainty" (Jacques, Dunlap, and Freedman, 2008: 356).

Both accounts—difficulties in communicating climate change and the deliberate manufacturing of uncertainty—assume a kind of information or democracy "deficit." Implicit is the belief that if only journalists and science educators would do a better job of communicating the facts of climate change, a literate public would know the truth and demand that public officials take action (Schwarze, 2011). While perhaps appealing, this thesis ignores a more basic challenge to the American public's understanding and support for actions addressing climate change.

Compounding these accounts, I suggest, is a more systemic and problematic transformation in key sites of production and dissemination of public knowledge about climate change and climate science in U.S. media. Beyond the contribution of individual journalists, this chapter considers the convergence of certain developments in legacy media (newspapers and network television) and new media that have begun to affect the sources, composition, and flows of public knowledge generally and climate science specifically. While some of these changes are not peculiar to the United States, I argue that climate skeptics in the U.S. (and to a lesser extent in the U.K.) have been particularly successful in exploiting the consequences of this media convergence and, as a result, have been able to sustain not only broad skepticism but also a partisan divide in the American public's understanding of both climate change and climate science.

In the analysis that follows, I consider two related sets of developments in U.S. legacy media and new media and their contributions to the production and distribution of public narratives of uncertainty about climate change and climate scientists. The chapter illustrates some of the results of this media convergence by tracing U.S. news coverage of the 2009 "Climategate" story and related reports, as well as surveys of Americans' attitudes about climate change and scientists during this period.

LEGACY MEDIA AND SCIENCE JOURNALISM

Understanding of climate phenomena and climate science can be difficult under any circumstance. Recent scientific organizations' reports on the impacts and accelerating pace of climate change,[1] however, have been occurring at a time of crisis in the business model of legacy media in many nations, and particularly in the U.S. The Pew Research Center's Project for Excellence in Journalism reported that, by 2010, "every news platform [in U.S. media] saw audiences either stall or decline—except for the Internet" (Pew Research Center, 2011b: para. 11). The structural economic problems facing U.S. newspapers, particularly, have been severe, with declines in advertising, rising production costs, and a falloff in subscriptions. As the Internet and cable television took off in the decade from 2000 to 2010, revenue for newspapers fell substantially, in what one study called the industry's worst financial crisis since the Great Depression (Kirchhoff, 2009).

The sharp drop in revenue in the latter half of the last decade, particularly, has left newspapers downsizing core aspects of their operations—daily circulation, newspaper size, news beats, space devoted to news, and the number of reporters. Major U.S. newspaper companies have filed for bankruptcy, while "many have laid off reporters and editors…or turned to Web-only publications" (Kirchhoff, 2009: i). Overall, U.S. newspapers have lost over 25 percent in daily circulation since 2000, and newsrooms have shrunk by a similar amount. The Pew Research Center estimated that with 1,000 to 1,500 newsroom jobs lost in 2010 alone, newspaper newsrooms are now 30 percent smaller than in 2000. This has left "the largest newsrooms in most American cities bruised and necessarily less ambitious than they were a decade ago" (2011c: para. 25).

As news media cut staff, there is inevitably a loss of science expertise. Some U.S. newspapers are eliminating entire beats. The *San Jose Mercury News* reported that, "two decades ago nearly 150 papers had a science section. Now fewer than 20 are left, and most usually dedicate their scarce column inches to lifestyle and health" (Daley, 2010: para. 16). Some analysts also worry that with fewer journalists, many with less training, there is an accompanying general devaluing or a "de-skilling" of the profession (Pew Research Center, 2011a: para. 17). One veteran journalist stated the concern bluntly: "The ranks of reporters best equipped to cover…major environmental and climate change stories at most news outlets, particularly in local markets, are being decimated" (Daley, 2010: para. 6).

Correspondingly, coverage of climate change in U.S. legacy media has steadily declined since a peak[2] in 2007. In one tracking survey, for example, the number of articles on climate change or global warming in the largest five U.S. newspapers[3] fell by more than 300 percent between 2007 and 2012 (Boykoff, 2012). (An exception occurred in late 2009, when U.S. and international media gave extensive coverage to the so-called "Climategate" scandal of hacked emails; I will return to this episode below.) While the reasons for this decline are many, including the competition from news coverage of financial crises, war, and unemployment, the attrition of experienced science reporters at major U.S. newspapers undoubtedly has been a contributing factor.

The downward trend is similar in radio and network television news. Overall, audiences for TV evening news have continued a 30-year decline, with negative consequences for newsrooms (Guskin, Rosenstiel, and Moore, 2011). In late 2008, for example, CNN eliminated its entire science and environment news team; at the same time, NBC cancelled The Weather Channel's climate program, *Forecast Earth* (Spencer, 2010). More broadly, the decline in news staff at radio and television stations has resulted in fewer science and climate-related news stories, or stories reported by less-qualified staff. A survey of U.S. radio and TV news directors by George Mason University's Center for Climate Change Communication found:

- Television news directors were interested in running science stories, but few had staff dedicated to this beat;
- Climate change was covered relatively infrequently on local TV news;
- Most news directors were comfortable with their weathercasters reporting on climate science.

Even among news directors who do run stories about climate change, 90 percent believed that coverage of climate change should reflect a balance of viewpoints, i.e., both scientists and climate skeptics (Maibach, Wilson, and Witte, 2010: 4–5).

Although most U.S. newspapers now maintain online sites, and some have migrated entirely online, these sites usually employ fewer news staff to produce content. When the *Seattle Post-Intelligencer*, for example, ended its print edition and began digital publishing, it slashed its newsroom of 165 reporters to only 20 (Yardley and Pérez-Pena, 2009). Other online sites report similar trends (Preston, 2011). Furthermore, some online legacy news sources are locking content behind pay walls. Ironically, featuring one of the best remaining science sections, the *New York Times* is the most prominent example of this trend. On the other hand, content on sites such as *Climate Depot* are free, furthering the imbalance online between legacy journalism and skeptical sources. The Pew Research Center concluded that even the best of the online news sites "still have limited ability to produce content," and doing so ultimately "will depend on finding a revenue model far larger than what exists today" (2010: para. 3).

As a consequence, online news sites, like legacy media, are increasingly relying on syndicated news feeds and content aggregators for science and environmental stories. More importantly, legacy sites are only a small portion of a wider ecosystem of sources and distribution networks that now produce content about environmental and climate-related concerns.

ALTERNATE SITES OF PRODUCTION AND DISTRIBUTION

While U.S. newsrooms shrink, the number of non-journalistic sources of climate and environmental news is growing rapidly. New media—blogs, dedicated websites, vlogs, social networks, and social media technologies—are shifting the tools of content generation and dissemination to those whom New York University journalism professor Jay Rosen has called the people "formerly known as the audience." As a result, "news is no longer gathered exclusively by reporters...but emerges from an ecosystem in which journalists, sources, readers, and viewers exchange information" (*Economist*, 2011: 10).

At the same time, search and content aggregators are driving traffic to a range of web content—news headlines, podcasts, videos, tweets, blog posts, and other links. Aggregators such as Google News, the *Drudge Report*, and Popurls also provide feeds to producers at cable television stations, talk radio, and other popular news platforms. Further, the ability to share or recommend content across the "horizontal media" of social networks, micro-blogging (e.g., Twitter), and other social media, without the involvement of traditional media organizations, is allowing users to act collectively as "a broadcast network" (*Economist*, 2011: 10). The Pew Research Center's annual "State of the News Media" summarized the impact of these developments: "In the digital realm, the news industry is no longer in control of its own destiny" (2011b: para. 4).

With each technological advance, a new set of players has entered the processes of generating and connecting content to audiences and advertisers. Of particular note has been the explosive growth of partisan or "self-interested information providers"—bloggers, social media users, talk radio hosts, and others—who, as legacy newsrooms get smaller, are increasingly able "to exploit a perceived opportunity in journalism's contraction" (Pew Research Center, 2010: para. 6). The Pew Research

Center reported that, as a result, these new information providers are increasingly able to vie for "the public's and the media's attention, and their resources, in contrast to those of traditional independent journalism, are growing" (2010, para. 6). The result has been that, as reportorial journalism has declined, "the commentary and discussion aspect of media...is growing—in cable, radio, social media, blogs and elsewhere.... Quantitatively, argument rather than expanding information is the growing share of media people are exposed to today" (2010, para. 4).

Climate change skeptics, particularly, have proven adept at exploiting the openings provided by new media platforms as well as the decline in science journalism as a news source. One of the most popular skeptic aggregators is the conservative *Drudge Report* (www.drudgereport.com). *Drudge* gets two to three million visits daily, globally, from news directors, reporters, other media producers and bloggers, as well as individuals (Quantcast, 2011). The website drives "more than double the traffic" to online media, blog posts, and partisan commentary than Facebook and Twitter combined (Donaldson, 2011: para. 3–4). Known for its provocative content, *Drudge* is also a prolific aggregator of sensational news about climate change and is "a popular... opinion source for American conservatives and man-made global warming deniers" (Source Watch, 2011: para. 1). Recent *Drudge* headlines have included: "Obama climate czar has socialist ties," "EXPERT: GLOBAL WARMING IS OVER," and "Global cooling? 30 years of warmer temperatures go poof!"

As a result, the *Drudge Report* is more often the source for a cynical story about climate change in popular media than a research report from a scientific journal. Consider this story, initially from *Drudge*, reported by Fox News on January 11, 2010: "30 Years of Global Cooling Are Coming, Leading Scientist Says." On air and at the Fox News website, the story ran:

> From Miami to Maine, Savannah to Seattle, America is caught in an icy grip [which] one of the U.N.'s top global warming *proponents* says could mark the beginning of a mini ice age.... It could be just the beginning of a decades-long deep freeze, says Professor Mojib Latif, one of the world's leading climate modelers.
>
> Latif thinks the cold snap Americans have been suffering through is only the beginning. He says we're in for 30 years of cooler temperatures—a mini ice age, he calls it, basing his theory on an analysis of natural cycles in water temperatures in the world's oceans. (FoxNews.com, 2010)

At the time of its report, the Fox News Channel (FNC) was the most-watched cable network in the U.S., with 3.2 million viewers in prime time. And while Fox accurately identified Dr. Latif (at the Leibniz Institute in Germany) as a leading climate scientist, the statements attributed to him were erroneous. When a reporter telephoned to confirm the Fox News account, Latif replied, "I don't know what to do. They just make these things up" (Romm, 2010: para. 1).

The source of the Fox News error was apparently its use of aggregator links to a conservative blogger at the *Mail Online*, whose post two days earlier was headlined, "The Mini Ice Age Starts Here" (Rose, 2010). The blogger had misquoted Dr. Latif,[4] but the post was picked up by the *Drudge Report* and other aggregators and quickly distributed to other bloggers and cable news producers, and embellished further by FoxNews.com.

Errors occur even in news stories by legacy reporters. My point is that the rapid multiplication of an erroneous story is more likely as news operations cut staff, including reporters, editors, and fact-checkers. As a result, misleading or ideologically driven stories more easily survive and are distributed more rapidly, particularly to popular media outlets such as cable TV and talk radio, and in syndicated op-ed columns.

To be fair, content aggregators serve science as well. The National Oceanic and Atmospheric Agency (NOAA), NASA's Goddard Institute for Space Studies, and other federal agencies also have websites which gather and distribute high-quality climate research to the public. And news aggregators like AOL and Google News occasionally link to stories of new climate research. Yet, for every NASA site, or story aggregated by Google News, there are far more numerous sites operated by climate skeptics, as well as platforms that aggregate and distribute skeptics' views to wider outlets. Skeptic websites such as *Climate Depot, Air Vent, Climate Audit,* and *Watts up with That,* and skeptical policy centers like the Heartland Institute function largely to aggregate and distribute reports of climate anomalies and skeptical opinion in the blogosphere and across media outlets.

Many of the skeptic websites, along with aggregators like the *Drudge Report,* serve as alternative news feeds for popular conservative news sources such as Fox News, talk radio hosts, and conservative columnists. In addition to the market share of Fox News and conservative radio hosts, the distribution of conservative commentary in syndicated "op-ed" columns can be considerable. Conservative columnists, for example, appear in newspapers with a combined total circulation of more than 152 million; and in three of four regions (South, West, and Midwest) of the U.S., conservative syndicated columnists reach more readers than do progressive columnists (Media Matters, 2011).

As a result, news and commentary about climate phenomena and climate science increasingly occur in a milieu of alternative media platforms, sources of knowledge claims, and competing ideological agendas. I now consider the news framing of "Climategate," one the most pernicious of the skeptic narratives about climate science.

THE NARRATIVE OF "CLIMATEGATE"

On 17 November 2009, just weeks before the United Nations climate summit in Copenhagen, hackers broke into a server at the University of East Anglia's Climactic Research Unit and downloaded over 1,000 personal emails from prominent climate scientists. David W. Norton describes what happened next:

> About one hour later, a link to stolen data—personal emails and research documents—was anonymously posted to the Climate Audit blog [a skeptic blog], but immediately removed by the site administrator. Two days passed before links to the stolen data were suddenly posted to two other conservative blogs: *The Air Vent* and *Watts up with That.* (2010: 6)

Within hours, commentators on the *Watts Up* blog jumped on the story, coining the term "Climategate" (Delingpole, 2009). One post suggested a narrative frame of the emails for news media: "The media are clueless. They need to be helped to understand the significance of—CLIMATEGATE! LEAK OF SECRET EMAILS SHOWS TOP CLIMATE SCIENTISTS ENGAGED IN MASSIVE FRAUD! GLOBAL WARMING WAS HOAX DESIGNED TO ENRICH POLITICIANS AND RESEARCHERS!" (Quoted in Norton, 2010: 6).

In the following days, the Climategate frame of scandal and hoax exploded in the blogosphere and on Twitter. The Twitter hashtag #climategate actually preceded "the growth of the catchphrase within all other [media] domains" (Norton, 2010: 10). While the frame circulated through horizontal media and the blogosphere, the *Drudge Report* featured it prominently on its website, helping to direct "millions of hits to websites reporting on the email scandal" (Rayner, 2011: para. 21). A week after the emails were posted, Google searches for the term "Climategate" (10,400,000 searches) had surpassed those for "global warming" (10,100,000 searches) (Watts, 2009: para. 5).

Fanned by social media and the *Drudge Report*, the Climategate frame went viral, influencing the coverage of mainstream media (Norton, 2010).[5] Gaining press attention across the U.S., the scandal also generated considerable coverage internationally, in major newspapers and scientific journals, and on television and radio networks. In the U.S., "approximately 1 out of 4, or 58 million American adults had both heard of and followed the Climategate story" (Leiserowitz, Maibach, Roser-Renouf, Smith, and Dawson, 2010: 5, 6).

In the U.S., after years of declining articles about climate change, coverage of the controversy spiked in major U.S. newspapers (Boykoff, 2012). Headlines in newspapers and cable TV news such as "Hiding Evidence of Global Cooling" (*Washington Post,* 24 November, 2009), "Climate Scientists Accused of Manipulating Global Warming Data" (Fox News, 23 November, 2009), and "Hacked E-Mail Is New Fodder for Climate Dispute" (*New York Times,* 21 November 2009) reinforced a sense of scandal and uncertainty in climate science. Prominent climatologists themselves were harassed. Dr. Phil Jones, director of the Climatic Research Unit at East Anglia, told a reporter that he was getting death threats from all over the world: "people said I should go and kill myself" (Hickman, 2010).

Receiving less attention were six independent investigations[6] in the U.S. and U.K. that within months cleared the accused scientists of charges that they had tampered with their research. The inquiries also found nothing in the emails that contradicted the underlying science itself (Gulledge, 2011). Despite these results, and perhaps because of the disparity in media coverage of the hacked emails and the later investigations (Rousey, 2010), "many people were left wondering whether climate change was really as much of a threat as it had been made out to be" (Rigg, 2011).

CLIMATE SCIENCE AND PUBLIC UNCERTAINTY

In 2012 the drumbeat of uncertainty about global warming continued online and in popular media in the U.S. While Climategate stories had died down, narratives of scandal, suspicion, and manipulation of data continued to appear in blogs and social media (Peach, 2011) and to be aggregated by sites such as *Climate Depot* (climatedepot.org) and the *Drudge Report*. In the years following Climategate, Americans' belief that global warming was occurring or was caused by human activities also continued to decline and/or remained at a low ebb, while at the federal level, the "political environment for climate science initiatives [had]…turned hostile" (Gillis, 2011: para. 2).

Surveys of American public opinion, particularly, have reflected this drumbeat of uncertainty. Three trends during this period are noteworthy: (1) A general *decline* in Americans' concern about climate change and their belief that global warming is happening, and that human activities are a major factor; (2) an *increase* in Americans' beliefs that news media exaggerate reports of climate change and that climate scientists are divided or cannot be trusted; and finally, (3) a *partisan divide* in Americans' beliefs about climate change and climate scientists.

It is perhaps not surprising that Americans' concern about climate change has declined both as public fatigue with global warming set in (Kloor, 2011) and as news of recession, unemployment, and other problems dominated news coverage. By 2011, surveys found that the number of Americans who worried about global warming had fallen nearly to an historic low (Agence France-Presse, 2011; see also *New York Times* & *CBS News*, 2011).

While worries about other problems may explain a decline in the *salience* of climate change, the disconnection between climate science and U.S. public opinion goes further. In the period considered by this chapter, opinion surveys also found that many Americans became more uncertain or cynical about the basic tenets of climate science itself. Both Gallup and Harris polls, the two longest-running surveys of Americans' views about global warming, reported sharp declines in the belief that warming was happening from 2007–2008 to 2011: 78 percent to 44 percent (Harris Poll, 2011) and 61 percent to 49 percent (Jones, 2011). Other surveys generally have found similar trends in U.S. public opinion (Pew Research Center, 2011a; *CNN World*, 2010).[7]

There also has been a similar decline in the belief that human activities are largely to blame for climate change. Harris, Gallup, CNN/ORC International, and Pew Research Center surveys, among other surveys, have reported steady declines in the belief that human activity is a contributing factor in global warming. A survey by Yale University and George Mason University, for example, found a 10-point drop from 2008 (57 percent), before Climategate, to 2011 (47 percent) among Americans who believed that global warming is caused mostly by human activities (Leiserowitz, Maibach, Roser-Renouf, Smith, and Hmielowski, 2011: 3).

A second major trend has been the *increase* in Americans' beliefs that news media exaggerate reports of climate change and that climate scientists themselves are divided over the issue and/or cannot be trusted. Among major surveys, Newport (Gallup) (2010) found that, following Climategate, 48 percent of Americans believed that the seriousness of global warming is "exaggerated" by media, the highest skepticism in Gallup's survey since it first asked this question in 1997, when 31 percent believed media exaggerated. Recent surveys also report an increase in those believing that there is "a lot of disagreement" among climate scientists (*Washington Post–ABC News*, 2009; Leiserowitz, Maibach, Roser-Renouf, Smith, and Hmielowski, 2011).

An increasing number of Americans believe not only that scientists can't agree about global warming but also that they cannot be trusted. Leiserowitz, Maibach, Roser-Renouf, Smith, and Dawson, for example, found a 9-point drop since 2008 in "public trust in scientists as a source of information about global warming" (2010: 3; also see Cohen and Agiesta, 2009).[8] Even more disturbing, a Rasmussen survey of American adults reported that 69 percent believed that "it's at least somewhat likely that some scientists have falsified research data in order to support their own theories and beliefs, including 40 percent who say this is very likely" (Fox Nation, 2011: para. 2).

The decline in the belief that global warming is happening and the corresponding increasing distrust of climate scientists may be due, in part, to the extensive coverage of Climategate in U.S. popular media. Leiserowitz, Maibach, Roser-Renouf, Smith, and Dawson, for example, found that 47 percent of those who had followed the Climategate stories at least a little said these stories "had made them somewhat (18 percent) or much more certain (29 percent) that global warming is *not* happening"

(2010: 6). The survey also found that over half of those who followed the Climategate stories had much less (29 percent) or somewhat less (24 percent) trust in scientists as a result of the news stories. Over all, Leiserowitz, et al. concluded that media coverage of Climategate "had a significant impact on overall public opinion" (2010: 7).

Finally, recent surveys of Americans' beliefs about global warming tend to reflect *a distinctly partisan divide* among the public. The Pew Research Center, for example, found that 77 percent of Democrats and 63 percent of Independents believed there is "solid evidence" of global warming, while only 43 percent of Republicans agreed (2011a; see also Leiserowitz, Maibach, Roser-Renouf, and Hmielowski, 2011: 4). Similarly, an Ipsos poll found a partisan divide over the causes of warming, with 37 percent of Democrats believing that it is the result "primarily of human action," while only 14 percent of Republicans believed this (Geman, 2011: para. 6). Political beliefs also played a role in Americans' views of the media, with conservative Republicans "three times as likely as liberal Democrats to think the media is exaggerating the severity of global warming" (Agence France-Presse, 2011: para. 8–10).

The impact of media coverage of Climategate, specifically, also appears to have had a sharply partisan influence. A survey designed to test the effects of such coverage found that "the loss of trust in scientists among those Americans who followed the Climategate scandal was primarily among Americans already predisposed, for ideological or cultural worldview reasons, to disbelieve climate science" (Leiserowitz, Maibach, Roser-Renouf, Smith, and Dawson, 2010: 11).

CONCLUSION

In retrospect, the decline in Americans' beliefs about climate change should not be surprising. Although climate-related stories have declined as science journalism has contracted in legacy media, skeptical "self-interested information providers" have been able to exploit platforms in new media to produce, distribute, and sustain narratives of uncertainty and cynicism about global warming and climate scientists. As a consequence, the sources of public knowledge about climate change have shifted significantly in this period sustaining not only broad skepticism, but also a partisan divide in the American public's understanding of the nature and causes of climate change.

There are, nevertheless, some countervailing initiatives by scientists, journalists, and others to engage the sites at which public knowledge about climate change is now produced or distributed to wider publics. ClimateCentral.org, an independent, non-profit journalism and research organization, for example, has launched a multi-media site to help mainstream Americans "understand how climate change connects to them" (climatecentral.org/about, 2012). Other initiatives have sought to engage media producers directly. The American Geophysical Union, perhaps the premier professional association for climate scientists, maintains an online question and answer service, with scientists available to answer journalists' questions about climate science. Other scientists maintain a Climate Science Rapid Response Team (climaterapidresponse.org) "committed to providing rapid, high-quality information to media and government officials." Both initiatives are intended to help fill the gap left by the cutbacks in science journalists at newspapers and other legacy media.

Still, the reach of these initiatives remains limited. Unlike legacy media, sites like ClimateCentral.org must compete within a media ecosystem that aggregates science and non-science posts, links, and stories alike. The challenge of such sites, therefore, has been to transform themselves from a broadcast communication model to a "networked model" (Clark and Slyke, 2011); that is, a model with sources that are able to interact with multiple, intersecting flows of horizontal media, continually finding, ranking, filtering, recommending, and distributing content.

With a decline in science journalism and systemic shifts in the sites of knowledge production and dissemination about climate change in media, it is, therefore, important to inquire about the composition and uses of public knowledge itself. Such shifts obviously raise questions not only about the sources of public knowledge, but also about the authentication and/or *warrant-ability* of particular knowledge claims; that is, their ability to justify a conclusion or action. In the context of climate science, layered with complexity, it is also important to ask, with Ulrich Beck, "What counts as 'proof' in a world where knowledge and non-knowledge of risks are inextricably fused and all knowledge is contested?" (Beck, 2008: 8).

ENDNOTES

1 For example, the National Research Council (National Academies, 2010) reaffirmed its call for actions to address accelerating evidence of anthropogenic climate change; see Rignot, Velicogna, van den Broeke, Monaghan, and Lenaerts (2011) for an example of such accelerating pace of warming.

2 Extensive media coverage followed two events in the U.S. in 2006 and 2007: the United Nations' Intergovernmental Panel on Climate Change report in 2007 that "most of the observed increase in the average temperature since the mid-20th century is very likely due to the observed increase in anthropogenic greenhouse gas concentrations" (2007a: 10), and the release a year earlier of the Academy Award-winning documentary film *An Inconvenient Truth*, about global warming.

3 Boykoff graph at: http://sciencepolicy.colorado.edu/media_coverage. (The newspapers included in this graph are the *Washington Post, Wall Street Journal, New York Times, USA Today*, and *Los Angeles Times*.)

4 Latif himself had previously objected to such misquoting. Two months earlier, he told National Public Radio that climate skeptics were misusing his work to suggest that we were headed for a

period of "global cooling." His research, he explained, had merely suggested a few years' "hold" in temperatures, when human-caused warming might be partly offset by ocean cycles; "these short-term changes," he said, "are much smaller than the long-term warming trends. So…we are not talking about a net cooling"; after this "hold," warming would accelerate again (Raz, 2009: para. 14, 10).

5 The implicit claim of an agenda-setting influence of social media on mainstream media is, of course, difficult to prove. For a discussion of one statistical rationale behind this inference in the case of media coverage of Climategate, see Norton (2010: 11, fn. 5).

6 The investigations were conducted by Pennsylvania State University, two reviews commissioned by the University of East Anglia, a U.K. Parliament inquiry, the National Oceanic and Atmospheric Administration Inspector General's office, and the National Science Foundation's Inspector General's office. The inquiries did find that some scientists had used intemperate language or had ridiculed climate skeptics in their email exchanges; the most serious finding was that some scientists may have been overly cautious, refusing to share their data with critics (Gulledge, 2011).

7 Similar trends can be found in the U.K. The number of Britons who are unconvinced that climate change is a reality has nearly doubled in recent years (ClickGreen, 2011: para. 1.) A Reuters/Ipsos poll (15 September 2011), however, is an outlier among these surveys, finding that a belief in global warming rose to 83 percent in 2011 from 75 percent in 2010.

8 The increase in distrust of scientists is relative. Scientists generally are still the most trusted source of information about climate change. Leiserowitz, Maibach, Roser-Renouf, and Smith found that 76 percent of Americans trust scientists "strongly" or "somewhat," far more than the mainstream news media, trusted by 38 percent of Americans (2011).

▪ CHAPTER SIXTEEN

"That Sinking Feeling"

Climate Change, Journalism and Small Island States

CHRIS NASH AND WENDY BACON

There is a poignant moment in Jared Diamond's widely read book *Collapse: How Societies Choose to Fail or Survive* when he ponders the final years of the last surviving original inhabitants of Henderson Island in remote Polynesia—alone, isolated and doomed (Diamond, 2005: 133–135). His argument is that the decline and extinction of human habitation on Henderson Island, and other southerly Pacific Islands including Easter Island, were caused by the failure of the inhabitants to understand the consequences for the local ecology of their cultivation and harvesting practices—especially of trees for canoe construction—which fatally compromised their inter-island transport, continuing productivity and therefore survival.

He makes congruent cases about a broad range of other historical and contemporary ecological crisis situations, canvassing the scholarly literature in each instance, and projects the likely consequences of ecological mismanagement into the contemporary case studies. The time scale is calibrated over centuries, and Diamond reminds his readers that the ill-fated Norse occupation of Greenland survived longer than English-speaking society has existed in North America (Diamond, 2005: 276).

Diamond warns that some contemporary ecological problems are global in potential impact, although always with a local specificity (Diamond, 2005: 486ff). Notable among these is anthropogenic climate change caused by increased emissions of greenhouse gases, whose effects will include rising sea levels (RSL) that will impact severely on low-lying littoral areas and islands (Diamond, 2005: 493). Since the late 1980s, the threat posed to low-lying islands by RSL and other consequences of climate change has been discussed in the climate science literature (for periodic summaries, see Intergovernmental Panel on Climate Change, 2001, 2007b; Connell, 2003; Barnett and Campbell, 2010) and has figured strongly in scientific reports to international bodies including the United Nations Framework Convention on Climate Change (UNFCCC), the international summits at Rio de Janeiro (1992) and Kyoto (1997), and since 1990 in successive reports of the Intergovernmental Panel on Climate Change. From early on, it has also been canvassed in the social sciences literature (Pernetta and Hughes, 1990; Roy and Connell, 1991; Connell and Lea, 1992; Connell, 1993; Barnett, 2001) and in the international news media.

Island nation governments were not slow to organize themselves once the RSL threat was identified. In 1991, specifically to address the problem, they formed the Alliance of Small Island States (AOSIS), which by 2012 had 42 member states and another four with observer status (Alliance of Small Island States, 2012; Barnett and Campbell, 2010: 101). AOSIS has no formal constitution, and its profile from time to time depends largely on the capacities and activities of the island government holding its rotating chair. AOSIS has been active in promoting international recognition of the threat its members face, and in demanding policy responses from both national governments and responsible international organizations (Ryan, 2010; Barnett and Campbell, 2010).

As Diamond clearly understood, the spectre of marooned islanders confronting the consequences of ecological mismanagement is intellectually and emotionally powerful, particularly when, as distinct from the situation he was describing, the mismanagement is not of their doing. The trope of "disappearing islands" as harbinger of generalized climate catastrophe—as the "canary in the mine", as it were (Chambers and Chambers, 2001)—has been picked up by policymakers, activists, the media and concerned publics (Ryan, 2010: 194). Overall, the amount of media coverage of the issue is small, and while the social sciences literature on the relationship of the news media to small islands and climate change is similarly sparse, it is strongly critical of the media's performance.

Connell (2003), while agreeing with the broad climate science consensus on the anthropogenesis and dangerous consequences of contemporary climate change, contrasts a nuanced and complex "scientific perspective" with a putatively uni-dimensional, ignorant and misrepresentative "media intervention" that he claims has resulted in "garbage can anarchy". He has borrowed this term to describe an approach where "[n]ew problems have been grafted onto old ones and given a single cause [and] once isolated phenomena become systematically inter-related" (Connell, 2003: 98). In Connell's view the journalistic perspective is uniform, even monolithic, and wrong to the point of fraudulent. In his view it has resulted in a failure by the people and government of Tuvalu, at the forefront of the AOSIS campaign, to engage with eminently viable strategies of local adaptation and mitigation; instead, they seek compensation and migration opportunities from Australia, New Zealand and elsewhere. To a certain extent Connell is criticizing the journalists for not being scientists, and doesn't recognize that political and journalistic discourses have different purposes and characteristics to scientific discourse. But he is also effectively reducing the journalistic coverage to a rhetorical caricature.

Farbotko presents a discourse analysis of the *Sydney Morning Herald*'s coverage of climate change over the period 1990 to 2005, and finds a uniformity of representation where "the publication of sensational climate change stories privileges expressions of disempowerment and desperation by Tuvaluan leaders ... in opposition to perceptions of Australia as a place of safety where environmental refugees could possibly shelter" (Farbotko, 2005: 286–287). Farbotko acknowledges that this perception is not "singular or uncontested", but similarly to Connell, concludes that "[b]y focusing on vulnerability, alternative discourses of adaptation for Tuvaluans are marginalized. Constructing Tuvaluan identity in terms of vulnerability can also operate to silence alternative identities that emphasize more empowering qualities of resilience and resourcefulness" (Farbotko, 2005: 286, 289). In subsequent publications Farbotko presents a further series of binary oppositions in discussing media and popular representations of climate change in the Pacific: compassion and voyeurism (Farbotko, 2010a), disappearing islands and cosmopolitan experimentation (2010b), and science versus emotion in climate change politics (Farbotko and MacGregor, 2010).

Cameron (2011) deploys the conceptual framework of governmentalities to argue a similar case: that "the populations of small island states are deployed as objects of compassion through the media, orchestrated to mobilize an ecological citizenry into action, with Islanders conscripted to operate as proof of the global climate change crisis" (Cameron, 2011: 880). She concludes that this discursive deployment is ineffective, and that "the complexities ... suggest that propagating a cosmopolitan ethico-political horizon in relation to global action to mitigate climate change is too simplistic" (Cameron, 2011: 883).

Notably, all of these authors are based in Australia, and each of them presents challenging perspectives on the Western journalistic discourse about the vulnerabilities of small island states to anthropogenic climate change. However, in dealing with the media, they share some common limitations. Firstly, they essentialize the media discourse into either an effectively uniform position or, at best, a series of discursive polarities; secondly, they present the media as a sovereign discursive agent without recognizing media production as itself a site of struggle among interested parties and media professionals; and thirdly, they tend to impute an efficacy to Western media discourse such that it produces a broader passivity and disempowerment among island populations and their agents in policy-making forums, and reduces them to discursive artefacts of the Western imagination.

There was a huge global media focus on climate change in the lead up to and during COP15. In Australia, no other issue had more coverage in 2009 (Chubb and Bacon, 2010). In this chapter we present a detailed case study of media coverage of climate change with specific reference to small island states before, during and after the COP15 international conference in Copenhagen in December 2009 (formally known as the Fifteenth Conference of the Parties to the UNFCCC). This will address the first issue of imputed uniformity in media coverage, and the second issue of contestation in media production. To address the last issue—of alleged disempowerment and imposed passivity—would require a detailed empirical analysis of the mutual impact between activities in the Western media field and activities in fields such as domestic island politics, economics and international law and diplomacy. We are not attempting such an analysis here, although we do note the research that suggests the island governments and their citizens are variously active on the issue, and have been for some decades, both domestically and internationally (Ryan, 2010; Barnett and Campbell, 2010; Korauaba, 2012).

Regarding the impact of the media coverage, recent research confirms that media discourse on climate change is of primarily domestic political interest in its location of production and consumption (Eide, Kunelius and Kumpu, 2010). Likewise, there is no evidence that we're aware of to support a view that the populations or governments of AOSIS member states invest significant attention and resources to influence Western media coverage of their issues: their focus is on the detailed processes of international agreements, diplomatic forums and government-to-government negotiations, and on local adaptation and mitigation strategies (Ryan, 2010). Indeed, far from being rendered passive in the face of the challenge, the Pacific Island countries are acting vigorously on adaptation and mitigation strategies, as Barnett and Campbell document extensively (Barnett and Campbell, 2010: 111–137).

The aim of our broadly Bourdieusian field analysis is to identify the structure of the field of Australian media coverage of the linkage between small islands and climate change.[1] (For examples of this approach and an elaboration of its theoretical provenance, see Bacon and Nash, 2003, 2012; Roberts and Nash, 2009; Nash, 2010.) This analysis covers the content of ten major Australian newspapers and *ABC News* and *ABC Current Affairs* programs on radio and television, generating a total of 282 individual reports, over the period from 1 October 2009 to 30 November 2011. These dates were chosen to cover the lead-up to COP15 in Copenhagen (7–18 December) and its aftermath, and the COP16 conference the following December in Cancun and its aftermath up until the COP17 conference in Durban in 2011. COP15 is the starting point because that was the conference where AOSIS was particularly effective: led by the Tuvalu delegation and supported (with differences) by the Maldives, they caused major negotiations to be adjourned for two days in protest against the terms of a draft conference agreement (known as "The Danish Text" and subsequently abandoned), and were able to attract significant levels of international media attention. For an introductory outline of the key controversial events at COP15 and various contested interpretations of their significance, the *Wikipedia* entry is a valuable starting point (*Wikipedia*, 2012), and Ryan offers an account of the policy and political differences among the small island states at Copenhagen (Ryan, 2010).

We restricted our data to Australian media coverage for a number of reasons. Firstly, we wanted to produce not a generalized overview but a detailed field analysis, and because our initial scan of international English-language coverage revealed that the major sources were domestic to the national media outlets; therefore a detailed analysis would require a specific geographical focus. Secondly, the Australian Government (with New Zealand) reportedly was the "fiercest" contestant against the AOSIS position at Copenhagen (Ryan, 2010: 195, 197ff) and plays a significant role in Pacific affairs as a major aid donor, as a former colonial power in the Pacific and a leading member of the South Pacific Forum of Heads of Government, and as the proposed defendant along with the USA in a threatened suit to be brought by Tuvalu over climate change impacts. Thirdly, the major social sciences literature on the media and Pacific climate change is by Australian scholars reflecting on Australian media. Fourthly, in light of the above reasons, there was no countervailing argument why any other national media should be privileged. COP15 was significant for the failure of the European and North American delegations to assert hegemony over the proceedings, for the assertiveness of the so-called Group of 77 developing states as well as AOSIS, and for the relative intransigence of the BRICS countries (Brazil, Russia, India, China and South Africa) in the face of US and European pressure. A major comparative investigation of coverage by a comprehensive range of international media in relation to the respective national politics of the issue would be a valuable but different research project.

For Bourdieu, the structure of a field is determined by relationships among "objective" positions occupied by important participants in the field (Bourdieu and Wacquant, 1992: 97), but because social relations as such can never be directly observed (Harvey, 2006: 141) their existence and attributes must be induced (or abduced) from observable characteristics, patterns of behaviour and events. Methodologically then, to identify the structure of the field of media coverage of this topic, we have sought to identify patterns in the coverage over a 26-month period spanning the lead-up and aftermath of two COP meetings, and thereby identify the major players in the media field and their characteristic terms of intervention and representation by journalists. This analysis reveals the structure of the media field on this issue as a product of conflictual social relations, within which the dynamic process of contestation occurs, which for Bourdieu is the engine of reproduction but also of modification and development of objective structures over time (Bourdieu and Waquant, 1992: 7–9; Nash, 2010: 62 ff).

We are aware of a number of independent video and radio documentaries and niche text publications that address issues of climate change in small island states. They are not our concern here because we are interested in the larger socio-political field. However, it is important to acknowledge that such niche media play a significant role among constituencies for particular policy and political positions.

We conducted a database search of selected publications, which comprised all major metropolitan general dailies in each Australian capital city (Brisbane, Sydney, Melbourne, Hobart, Darwin, Adelaide, Perth), *ABC News* (national and state) and *Current Affairs*, plus two small-circulation but prominent online publications (*Crikey* and *New Matilda*). The Boolean search terms were ((carteret*) OR (Kiribati*) OR (fiji*) OR (maldives*) OR (marshall*) OR (PNG*) OR (((pacific) OR (small) OR (low NEAR lying)) OR islands OR nations) OR (Tokelau*) OR (Tuvalu*) OR (Samoa*) OR (Solomon*)) AND ((climate NEAR change) OR (global NEAR warming)). There were 282 reports identified by these search terms. We coded (with validation checks for consistency) the form and content of all reports against a set of criteria that identify the genre and date of publication, the sources quoted and the relationship of the sources to the identified theme, and the authorial perspective on the content of the report.

The ABC programs (together) and the national, Sydney and Melbourne broadsheets (*The Australian*, the *Sydney Morning Herald* (*SMH*) and *The Age*, respectively) had comparable amounts of coverage: between 43 and 50 reports each over the 26-month period. The targeted audience/readership for these outlets is generally educated and at the more affluent end of the socio-economic spectrum. Of the remaining newspapers, which all include less affluent and less educated demographics in their target readerships (although not necessarily exclusively[2]), the Melbourne, Sydney and Brisbane tabloids had between 10 and 22 reports (half or less of the totals in the broadsheets), while the Darwin and Hobart tabloids had 6 and 7, respectively. These results illustrate the highly differentiated spatiality of news media flows and therefore the relative in/visibility of certain issues and events to spatially defined populations; in this finding, they concur with other studies of Australian media coverage of international conflict and humanitarian issues (Bacon and Nash, 2003) and the carbon economy (Bacon and Nash, 2012).

Temporally, the news coverage was spread unevenly across the 26-month period. Half of the reports appeared in the three months leading up to and including the COP15 event in December 2009, with three-eighths of them appearing in December. This indicates several things. Firstly, the situation of small island states was a significant component of the media's coverage of COP15. Secondly, the issue was effectively inserted into the media's agenda for the conference in the preceding two months, and the coverage was not just a product of the AOSIS activities at the conference itself. Thirdly, after the COP15 conference, there was a steady, low-level stream of coverage averaging six reports a month across all outlets, suggesting that it had entered the media's conceptual framework or "map of meaning" (Hall et al., 1978: 54) for reporting on climate change and/or small island states. Fourthly, the peaks in coverage of 10 or more reports a month accompanied either major diplomatic events (COP16 in Cancun in December 2010—10 reports; Pacific Islands Forum in Auckland in September 2011—15 reports) or a flurry of controversy over whether scientific evidence showed that Tuvalu was not sinking but expanding in size (13 reports in June 2010), which we will consider in one of our case studies below.

The spatio-temporal spread of reports reveals profound gaps in the availability of reports to specific geographic communities. For example, there was no reporting at all on small islands and climate change in *The Northern Territory News* in our period until 11 months after Copenhagen, there was a 10-month gap in coverage in *The Courier Mail* after January 2010, and a 20-month gap from December 2009 in *The Daily Telegraph's* coverage.

The 282 reports were spread across a range of journalistic genres which are not hard and fast categories, but rather blend at their interfaces. There was a high proportion of features, comment and editorial in print/online publications (55 percent) and non-news programs in ABC coverage (66 percent), which indicates that editorially, the issues were seen to require amplification, explication and discussion. However, the more extended non-news coverage was not distributed evenly across outlets: of the features, in which one would normally expect to find more depth, sources and analysis, *The Australian* (14) and the *SMH* (15) published the most, with *The Australian*, the *SMH*, *The Age* and *Crikey* together accounting for more than 80 percent (41) of the total features. All of these publications target affluent, educated readerships.

There were more comment pieces (59) than features (50), which may be taken as evidence of both public controversy on the issue and editorial determination to participate in that controversy. *The Daily Telegraph* in Sydney ran four news items but no features, and six comment pieces including two by noted climate change skeptics Andrew Bolt and George Pell, the Catholic archbishop of Sydney. *The Herald Sun* ran five comment pieces, all by skeptic Andrew Bolt, compared to two features; *The Hobart Mercury* ran one feature and two comment pieces, one by skeptic Piers

Akerman and one by climate change endorser Peter Boyer. On the other side of the controversy, *The Age*, strongly campaigning in support of action on climate change, ran fourteen comment pieces and five editorials, compared to eight features.

What flows from the analysis so far is that there is extreme variability in the formal dimensions—time, space, genre and point of view—of reporting on the issue. The spatio-temporal variability considered on its own may arguably be the result of disinterested decisions about news interest to targeted local audiences, but when put together with the patterns of genre above and the points of view identified in Table 3 below, this interpretation is not persuasive. It is more likely that the patterns of in/visibility are the result of identifiable editorial attitudes towards the issues, and a desire to promote specific points of view.

We will now turn to the content of the journalism, specifically the identity and location of journalists and story focus, the selection and usage of sources, and the thematic priorities and treatment in the reports. Of the 282 reports, 62 or almost one quarter did not identify the journalist, and were typically brief news stories. Of the identified journalists, Adam Morton, environmental reporter for *The Age*, had a total of 18 reports (four co-authored with Daniel Flitton), followed by Rowan Callick (9) for *The Australian*, Andrew Bolt (8) for *The Herald Sun* and other News Ltd publications, Lenore Taylor (7) for *The Australian* and Marian Wilkinson (6) for the *SMH*, and then Emma Alberici (5) and Sarah Clarke (4), both for ABC. There were a further four journalists who had three stories each, twenty journalists with two stories each, and 102 journalists with one story each. This indicates that a very limited number of journalists, working for the national Sydney and Melbourne broadsheets and ABC, gave substantial coverage to the issue over the period, and otherwise a huge number of journalists and commentators gave minimal coverage to the issue over the period. In other words, substantial coverage depended on the interests and capacities of a handful of journalists and outlets (in particular Adam Morton at *The Age* and Rowan Callick at *The Australian*), and for most other journalists and mastheads it was a story meriting occasional mention or examination.

Thirty-nine of the stories were date-lined from COP15 at Copenhagen, 37 from Australia, 11 from AOSIS member states, 2 from COP16 at Cancun, 24 from other locations and another 167 from unidentified locations. The stories were evenly split on the question of whether or not they were focussed on events or situations at the location from which they were reported or at other locations, and we explore this divergence below. While the issue made a significant impact in Copenhagen, the press gave it minuscule coverage a year later in Cancun; and only 4 percent of the stories mentioning small island states in the context of climate change were reported from those states, so by and large, the journalists were not travelling to the affected locations to explore the issues, but reporting their fates from afar. Of course, information about the situation was forcefully presented in other locations, especially Copenhagen, where the modes of presentation became part of the story (e.g., demonstrations, speeches, walkouts, etc.). Nonetheless, along with temporality, the spatiality of information sources and flows is an important dimension of their character, and these figures indicate that far from the situation presented by Connell (2003) of troupes of Western journalists visiting the islands with pre-conceived agendas and one-dimensional storylines, the overwhelming bulk of stories during our period of analysis were sourced from non-island locations where the analysis and interpretation could be, and was, deeply contested and subject to the vicissitudes of journalistic interest.

Regarding the locations of interest (as distinct from location of production above), they were spread over a wide range of island states and other countries. Tuvalu, with 104 mentions, was almost twice as often referenced as the next most common—the Maldives (56), Kiribati (50) and Papua New Guinea (45)—and all together there were some 60 distinct locations of vulnerability mentioned, more than half of them less than three times in total. The four high-profile states with most mentions were taking prominent but differentiated positions in proceedings at Copenhagen and afterwards, and so the coverage presented a range of perspectives and interpretations, and not a monolithic or bi-polar position as reported by Connell and Farbotko.

This brings us to the thematic concerns and information content of the reports. We coded all reports to identify whether the mention of the small island states with respect to climate change was incidental or significant to the main topic of the report. It was found that the spread was roughly equal across outlets (137 to 145, respectively) and over time, confirming that the issue was being treated seriously as an object of investigation by some outlets but also that it had been included as an standard, recognizable element in the "map of meaning" that journalists use to locate the elements of a story. Fascinatingly, mention of the small island states with respect to climate change in *The Northern Territory News* was incidental in all instances (albeit totalling only six over the 2-year period), suggesting that the journalists and editor believed that the linkage didn't require further exploration, either as a serious issue or in terms of sense-making for readers—it was just part of the context and implicitly sensible. This further indicates that news outlets do not necessarily see themselves (perhaps justifiably so in the case of the *NT News*) as a comprehensive source of information on important topics, but rather as position-taking within a pre-existing field by virtue of their contribution to the information flows that structure the field (Bourdieu and Wacquant, 1992: 7–9).

The overwhelming majority of reports (86 percent) accepted that climate change was a significant issue for small island states, while only 9 percent rejected or questioned that assessment (and for 5 percent of reports, the issue didn't arise). However, there was considerable variety in the thematic frames of the reports, as Table 1 shows.

Table 1: Thematic Focus of Reports

Population vulnerability	31%	Impact on fauna/flora	4%
Superpower interest	22%	Domestic politics	3%
Diplomatic negotiations	21%	Poor vs. poor conflict	2%
Developed vs. developing world	15%	Other	2%

Furthermore, in about 40 percent of reports there was a range of complementary factors that were integrated into the main thematic frame, as Table 2 illustrates.

Table 2: Complementary Issues

Rising sea levels	23%	Scientific issue	5%
Adaptation and mitigation	22%	Food and water	4%
Financing	14%	Other	4%
Extreme weather	10%	Energy	3%
Migration and relocation	8%	Increased overpopulation	1%
Environmental degradation	6%	Land expansion	1%

Interestingly, only 8 percent of stories linked the issue of migration and relocation with climate change for small island states—roughly a third of the proportion of stories that addressed adaptation and mitigation issues (22 percent)—and this low proportion accords with the rejection of this linkage by island governments at Copenhagen (Ryan, 2010:196). It also contrasts strongly with Connell's (2003) and Farbotko's (2005) characterization of media coverage outlined above.

In discussing the frames and issues, the reports took a range of interpretive perspectives, as set out in Table 3.

Table 3: Interpretive Perspectives

Unified (either for or against climate change)	58%
Conflicted (but supporting climate change)	20%
Balanced	17%
Conflicted (but against climate change)	5%

Fifty-eight percent of the stories were unified, implicitly or explicitly, in the perspective that climate change did or did not constitute a threat to the situation of small island states. The remaining reports presented conflicting perspectives on whether climate change was a threat, of which 17 percent were balanced in their presentation, 20 percent presented climate change as a threat against the opposing arguments, and 5 percent disagreed that climate change was a threat. In presenting these interpretative perspectives, there was only a limited appeal to (or disputation of) specific climate science evidence: only 26 out of 282 reports, or less than 10 percent, referred to debates or findings in specific peer-reviewed scientific literature.

Indeed, the reports followed the normal journalistic practice of contextualizing the issues within the field of agents or players in a position to respond to them. In the context of international diplomatic summits, the geo-politics of the issue are clearly relevant, and were evident at Copenhagen (Ryan, 2010). Throughout the 282 reports there were 842 references to 67 identified geo-political agents—individual states or named aggregations of states such as the UN or AOSIS. Table 4 ranks the named geo-political agents with more than 10 references each, and notably none of them are small island states (Indonesia had 9, Tuvalu had 8 and Bangladesh 6 mentions). Predictably, given that we are analysing the Australian media, Australia had by far the highest number of mentions.

Table 4: Mentions of Geo-Political Players

Australia	167	New Zealand	28
UN	124	G77	27
China	79	Developing countries	17
US	78	UK	17
Others	51	Denmark	16
India	41	Japan	16
Brazil	28	South Africa	16
Europe (misc.)	28	Developed countries	11

From this we can say that the "maps of meaning" with which the news media operate do not represent the small island states themselves as major forces or players in addressing the issue, even when they are the acknowledged harbingers of future threats, and they do not feature even in the 50 percent of reports where this issue is central to the report's narrative. At best, what news sources speaking on behalf of those states can hope for is that their perspective might be congruent with the perspectives articulated by the major players, or that they might be given a "right of reply" to opposing perspectives. It thoroughly validates the approach referred to above by small island states not to vest significant resources in influencing Western media coverage of their issues, and to prioritize activities in diplomatic forums.

Journalists reference authoritative individuals and organizations as the verifiable sources of their truth claims (Tuchman, 1978; Hall et al., 1978; Ericson et al., 1989). Table 5 aggregates sources by occupation. Unsurprisingly, political sources constitute the overwhelming majority (64 percent) of named sources, which accords with earlier studies of news coverage of international humanitarian issues (Bacon and Nash, 2003). It also reflects the geo-political approach to the significance of issues.

Table 5: Quoted Sources in Reports, by Occupation

Political	186	Religious	6
Academic	24	Business	4
NGO	18	Other	4
Scientist	18	Celebrity	2
Public	12	Government	2
Think-tank	8	Activist	1
Media figures	7		

Journalistic narratives are complex, and not all sources are quoted favourably, nor are they necessarily quoted in support of the preferred interpretation of the issue being promoted by the journalist. However, sources are used to define the terms of the issue being reported and any debates about those terms. Table 6 displays our analysis of whether the source is deployed in the narrative to define the preferred terms and perspective of the report, or as a counter-definer. Typically, the first source quoted will define the preferred interpretation, and subsequent sources will explore the range of perspectives on the issue, but when statements by that first source have authoritative and independent news value (e.g., when the source is a prime minister or president of a state), then the journalist may deploy second and third sources to position the first source as a counter-definer against their preferred interpretation.

These figures show that occasionally journalists do use a counter-definer to lead off a story, but overwhelmingly, first, second and third sources are quoted by journalists to verify the preferred definition of the key terms and issues (which may well include points of disagreement and opposing interpretations) and so to present a "unified" interpretation of the issue, as summarized in Table 3 previously.

To summarize, our analysis of the structure of coverage of this issue over the period shows considerable complexity and variation, and it would be misrepresentative to reduce it to simple one-dimensional descriptors or binary polarities. That is not to say that there are no patterns to the coverage. Indeed, the patterns are

Table 6: Sources as Thematic Definers

	1st source	2nd & 3rd sources
Defining	128	105
Counter-defining	9	31
Unassigned	4	14

stark. Firstly, unless the news provider is the ABC or one of the three broadsheets (*The Australian, The Age* and the *SMH*), Australians are likely to be seriously ill-informed on this issue, which means that geographically, most non-ABC listeners/viewers outside the south-eastern corner of Australia are ill-informed. Similarly, if they missed the four periods of peak coverage over the 2-year period they also will be ill-informed, unless they access alternative, non-mainstream sources of information.

Secondly, there was considerable controversy about the issues, as manifested by a large proportion of comment pieces and features in the mix of coverage. The coverage and controversy ranged over a broad set of primary and secondary aspects of the issues, and again cannot be reduced to single issues such as migration and relocation or financial compensation. Given that the media controversy largely concerns the validity of climate science, it was interesting that peer-reviewed science was referred to in less than 10 percent of stories.

Thirdly, the Australian media largely analysed the issues in geo-political terms within a global framework, privileging Australian sources over international sources, reporting from Australia and large international gatherings and only occasionally from the small island states themselves. The geo-political level is where the bulk of both coverage and controversy is located, and within that framework the small island states are cited both as harbingers for the rest of the world of future consequences of unmitigated climate change, and also in terms of their own specific circumstances, issues and policy negotiations.

CONCLUSION

Contrary to a prominent observation in the social sciences literature, Australian media coverage of climate change with respect to small island states is neither monolithic nor bi-polar, but deeply variegated by multiple factors including spatiality, temporality, genre, attitude towards climate science and the use of sources. It cannot be readily reduced to single or binary categories. This variegation is not random but deeply structured by the factors listed, and this complex structure constitutes the dynamic field of media reporting on this issue within which participants act.

What is striking about the reporting is the extent to which it is contested within the field. As Table 3 shows, 17 percent of reporting does not take an implicit or explicit stance in respect of the fundamental question of whether climate change poses a threat to small island states, which means that conversely, more than four fifths of the reporting does take such a stance. On the one hand, this may be taken as a lack of balance in the reporting; on the other hand, it may be taken as inconsistent with the scientific consensus (of 97 percent of climate scientists) that climate change does pose such a threat (Oreskes, 2004). Depending on the perspective, a 17 percent proportion of balanced reporting may be interpreted as outrageously low or outrageously high.

The differences in stance on this fundamental question manifest themselves in the ways journalists construct their reporting activities and the resultant publications, or put differently, in the ways journalists act within the structured and contested field to represent the issues and events. Our subsequent case studies examine the reportage of the two most individually prolific journalists in our sample: Adam Morton reporting for *The Age*, and Rowan Callick for *The Australian*, who took contrary positions on the fundamental question about the validity of the climate science consensus with respect to small island states. They conducted themselves quite differently on the factors we examined—spatiality, temporality and use of sources—and produced radically opposed perspectives. However, both journalists did partake of the journalistic consensus that the fundamental and overwhelming frame for the reporting of the issue is the geo-politics of the interests involved. This suggests that while journalism depends on the identification and deployment of relevant and verifiable evidence ("facts") as its core contribution to knowledge, the imperative concern is always with the political state of relations amongst the powerful positions and participants in the field. In the case of climate change, there are powerful positions in both the affirming and skeptic camps, and therefore the journalism engages with both those positions, and as a corollary, therefore engages with the question of which of these positions the scientific evidence supports.

It is both the fortune and misfortune of the small island states to be a crucible of this evidentiary test and geo-political power play, and the plight of their people to add a potential and readily identifiable human tragedy to the threat. This is what Diamond understood when he imagined the last Henderson Islanders looking out to sea, searching for the canoes that never came.

ENDNOTES

1 A subsequent publication will elucidate via case studies the approach of certain key players among the journalists and their sources within that structure.

2 There is local variation in target demographics; for example, *The Courier-Mail* in Brisbane presents itself as the authoritative interpreter for Queensland elites of local political and economic matters, and *The Hobart Mercury,* while resolutely blue-collar in its voice and idiom (Hall et al., 1978), has the only regular climate change columnist of any Australian newspaper, perhaps in recognition of the approximately 20 percent share the Greens get in Tasmanian elections.

▪ CHAPTER SEVENTEEN

"Skeptics" and "Believers"

The Anti-Elite Rhetoric of Climate Change Skepticism in the Media

ALANNA MYERS

In a 2000 article about the sociological barriers to widespread public understanding of climate change, Sheldon Ungar begins by admitting that after a decade of clipping articles from *Science* and *Nature*, a layperson's sense that climate change is real "ultimately boils down to picking the experts you think you can trust" (2000: 297). Not only is the problem neglected in the mass media, Ungar argues, but those individuals who do seek a deeper understanding through books, films and the Internet "remain utterly dependent on experts for evaluating the global circulation models on which the whole game is predicated" (2000: 297). Acknowledging the necessity of non-scientists deferring to experts on questions of scientific complexity, Ungar simultaneously underlines the problems that such a one-way relationship entails: the intimation that, even among experts, there is disagreement; and, more significantly, the inherent social divide that arises between those who have scientific knowledge and those who do not.

On the one hand, much has changed since Ungar's article was published: mainstream media coverage of climate change has increased dramatically, and a subsequent decade's worth of articles in *Science* and *Nature* indicate that there is now very little disagreement among scientists that current global warming trends are caused by man-made greenhouse gas emissions (Anderegg et al., 2010; Doran and Zimmerman, 2009). On the other hand, Ungar's concern over the disconnect between expert and lay understandings of climate change science is as relevant as ever. In Australia and the US in particular, public opinion on whether climate change is caused by human activities is far from settled (Leviston et al., 2011: 2–5), and climate change skepticism—once thought to be in decline following its prominence in the media throughout the 1990s and early 2000s (Russell, 2008: 47)—has risen again in the wake of the 2009 United Nations climate change conference in Copenhagen (COP15). In Australia, for instance, the federal Labor government's efforts to introduce a carbon tax law in 2011 attracted both strong support and heated opposition, with those opposed condemning not only the tax, but also the idea that the science of anthropogenic climate change was "settled". Framed by the media, climate politics often resembles not a debate over policy, but a bitter conflict between "skeptics" and "believers" as each side lays claim to the right to define what climate change is and what it means for the world (Anderson, 2009: 166).

This chapter contends that the two problems brought to light in Ungar's article—the first regarding the degree of uncertainty that defines the scientific discipline; the second tending towards a Foucauldian analysis of power-knowledge matrices—are key to understanding the resurgence of climate change skepticism in the Australian media in recent years. Accordingly, the chapter begins by considering the role of expertise in the Foucauldian concept of "governmentality". It then discusses how social critiques of expertise have been leveraged by conservative political movements to foster resentment against and rejection of policy recommendations issuing from so-called "elites", contextualizing this anti-elite discourse in relation to its Australian manifestations and, in particular, its appropriation by the national broadsheet *The Australian*. The way in which a populist anti-elite ideology has been applied to undermine public trust in climate change science and urge delay on action to mitigate the effects of global warming is then examined by means of a critical discourse analysis of news and opinion articles and editorials published in *The Australian* newspaper during COP15. The analysis seeks to demonstrate how the bifocalism of "skeptics" and "believers" was overlaid on top of an existing cultural battlefield of "us" versus "them", "elites" versus "the people", in which the consensus position on climate change was described as an "orthodoxy" of green ideologues, politicians, business leaders and scientists. The consequence of this oppositional framing of climate change science is the reduction of a tangible environmental problem to a question of social and political values, where delay can be fostered indefinitely in the name of debate.

GOVERNMENTALITY AND THE RISE OF EXPERTISE

Expertise is central to a Foucauldian understanding of modern governance as "governmentality." Under governmentality, Foucault (1991) proposes, power is decentralized from the state and spread across various discrete and specialized sites of expert authority who are charged with making recommendations to lay populations on how best to live their lives on the basis of what is "good, healthy, normal, virtuous, efficient, or profitable" (Rose and Miller, 1992: 175). Importantly, as Rose and Miller (1992: 174, 175) summarize, governmentality is concerned not so much with "imposing constraints upon citizens" or with weaving "an all-pervasive web of 'social control'" as with "'making up' citizens capable of bearing a kind of regulated freedom". The media functions within this system by acting as a conduit or relay between these expert authorities and the general public, providing them with the information they need in order to be self-governing (Nolan, 2008: 111). However, the media, too, functions as a site of expertise, charged with translating complex technical information into plain language for lay audiences as well as with keeping expert institutions accountable to their publics by exposing corruption and balancing contradictory claims both within and between disciplines (Hallin, 2000; Harrison, 2000: 109; Nolan, 2008: 111).

Though intended to grant citizens greater independence from the state, governmentality entails a tendency for experts to make recommendations for and on behalf of individuals in pursuit of a collective good, which may not align with all individuals or groups. Political scientist Timothy Luke, for example, argues in his book *Capitalism, Democracy and Ecology* that while the system of governance that experts uphold may have been intended to protect individuals from states (in the sense that they are no longer ruled over by a central sovereign power), "the proliferating panoply of entitlements and rights...has perhaps come to constrain real democratic choice, reducing communal self-governance to the individual ratification of expert opinions taken elsewhere" (1999: 10). In a similar vein, media studies scholar Anabela Carvalho has argued that the international diplomatic movement to address climate change is an example of governmentality that has effectively removed the individual's compulsion and capacity to act. While environmental activist groups have tried to return some of this agency by encouraging citizens to make changes in their own lives, Carvalho argues that reassurances from politicians and scientists about their ability to "solve" the problem using the tools of techno-science, sustainable development and ecological modernization, coupled with the institutionalization of responses in the UNFCCC and the IPCC, have left citizens feeling "confused and powerless" (2005: 12, 15). Simultaneously, Carvalho argues that media framing of the problem as a "global" issue that can only be solved through global approaches has further contributed to the sense that individuals have no meaningful contribution to make to climate change debates (2005: 11). This, Carvalho speculates, has concerning implications, for it may result in citizens losing interest and reverting to a strategy of denial.

Secondly, a critique of expertise has featured prominently in the ideology of populist anti-elitism, which has its origins in U.S. Cold War–era neo-liberalism but has been leveraged to great effect by various political and social movements in different parts of the world since (Greenfield and Williams, 2001: 37–38). In essence, populist anti-elitism imagines a new type of class system based on the possession of knowledge rather than the possession of property. In this new class system, those who rule are the "elites": privileged, university-educated intellectuals possessing highly specialized knowledge, whose views and recommendations are seen as being profoundly distant from—and hence irrelevant to—"ordinary" people. Greenfield and Williams provide the following useful explanation of how expertise is viewed in the rhetoric of elites:

> Attributed to the "caffe latte set," the "chardonnay socialists," the "chattering classes," the "politically correct," "the baby boomer collective," expertise is not a specific and technical acquisition but knowledge moralised and aestheticised. It is knowledge seen in the charismatic terms of an essence that is (unfairly) given to certain groups of people, and not others, and therefore enables those groups (the "elites") to unfairly manipulate and upstage others. It is knowledge that is not authorised by, runs counter to, is different from, the everyday knowledge of the "ordinary person." Defined from this perspective rather than by other, positive criteria—such as its institutional or professional protocols and its differentiated knowledge-outcomes—this understanding of expertise licenses the "out-of-touch" thesis: a view that somehow intellectual workers are only entitled to use their knowledge if they are in a mystical relation of representation with "ordinary people." (Greenfield and Williams, 2001: 40)

The consequence of framing complex social issues in this binary way hardly needs to be stated: the real causes of contention can be glossed over in favour of the disingenuous explanation that it is just another example of "elites" trying to impose their self-serving and out-of-touch policy recommendations on "the people," a view that enables policy decisions to be dismissed outright or delayed indefinitely.

In Australia, the rhetoric of populist anti-elitism was at its peak in the late 1990s (Scalmer and Goot, 2004: 141). Former Australian prime minister John Howard campaigned successfully to present himself as the defender of everyday "battlers" and "ordinary Australians" against privileged "elites"; One Nation politician Pauline Hanson framed her conservative immigration policies in comparable terms; and the mainstream media framed policy debates ranging from the referendum on whether Australia should become a republic to Indigenous policy as a conflict between "elites" and "non-elites" (Greenfield and Williams, 2001; Scalmer and Goot, 2004). Greenfield and Williams (2001: 37) contend that the dichotomous division of Australia into "elites" and "ordinary people" in the media became so routine during the 1990s that at times, "it became very hard to hear or even to allow oneself to think of different accounts of the Australian polity other than this bifurcated view."

The anti-elite discourse was invoked by media on both sides of the political spectrum during the 1990s—by News Limited's conservative national broadsheet *The Australian* and its high-selling tabloids the *Herald Sun* and *Daily Mail* as well as by their left-leaning rivals, Fairfax's *The Age* and the *Sydney Morning Herald* (Greenfield and Williams, 2001). However, studies suggest that News Limited newspapers—particularly *The Australian*—have employed the rhetoric of anti-elitism more consistently than most, with the language of "elites" versus "non-elites" carried into the 2000s and beyond in reference to a broad spectrum of political issues including policy on asylum seekers, the Iraq war, the "Stolen Generation" of Indigenous Australians and the Labor government's proposed mining tax (Scalmer and Goot, 2004; McKnight, 2012). *The Australian* is an important case for analysis for several other reasons. It is the flagship newspaper of Rupert Murdoch's Australian newspaper holdings, which make up 70 percent of the Australian newspaper market (McKnight, 2012: 7). Its readership is relatively small, but media scholars attest that its political impact is much greater owing to its "elite" audience of university-educated politicians, business people, professional and semi-professionals, and its perceived agenda-setting influence on the content of other Australian newspapers, radio programs and television news broadcasts (which is attributed to its having one of the best-resourced newsrooms in the country and an unrivalled presence in the federal parliamentary press gallery) (Manne, 2011: 4; McKnight, 2012: 8; Scalmer and Goot, 2004: 142). Most significantly, *The Australian* has an "unusually" ideological editorial line, a blend of neo-liberal economic policy and neo-conservative foreign policy coupled with an anti-elite ideology broadly though not definitively aligned with its owner, Rupert Murdoch, and News Corporation's suite of news media outlets worldwide (Manne, 2011: 3; McKnight, 2012: 16). As the analysis later in this chapter will demonstrate, this ideology was evident in *The Australian*'s coverage of COP15. Firstly, however, it is necessary to briefly outline the history of climate change skepticism in the media more generally.

POPULIST ANTI-ELITISM IN THE CLIMATE CHANGE DEBATE

Climate change skepticism in the media is nothing new. Indeed, studies suggest that throughout the 1990s and early 2000s, media reports commonly suggested that there was a greater degree of uncertainty about the existence and causes of climate change than actually existed in the scientific community (Antilla, 2005: 338; Boykoff and Boykoff, 2004, 2007; McCright and Dunlap, 2003; Zehr, 2000). Though this trend has been largely attributed to a highly successful campaign by the fossil fuel industry to foster doubt about the causes of climate change (Gelbspan, 2004; McCright and Dunlap, 2003; Monbiot, 2006; Pearse, 2009), Boykoff and Boykoff (2004) have shown how adherence to the journalistic norm of "balance" (the obligation to represent "both sides" of any significant issue or conflict in society) also contributed to the greater prominence given to skeptics in media reports during this time. Others have argued that the misrepresentation of climate change science in media reports was exacerbated by a lack of background knowledge on the part of journalists (Wilson, 2000) and poor communication on the part of scientists (Cole and Watrous, 2007). However, the number of skeptical voices gaining media coverage was found to decline from the mid-2000s onwards as public support for strong action on climate change grew, a trend often attributed to the release of former U.S. vice president Al Gore's climate change doco-drama *An Inconvenient Truth* in 2006 (Anderson, 2009: 169; Boykoff and Goodman, 2009: 398).

Then came Copenhagen. The Fifteenth Conference of the Parties (COP15) to the United Nations Framework Convention on Climate Change (UNFCCC) has come to be regarded as a significant turning point in international climate negotiations. After a period of intense media build-up and high public interest, the failure of world leaders to agree on a new global treaty to succeed the Kyoto Protocol precipitated a decline in public concern about climate change and the belief that human-induced climate change was occurring (Hanson, 2010, 2011). Coupled with the discovery of a minor factual error in the latest Assessment Report of the IPCC, as well as the so-called "Climategate" incident in which skeptics claimed that hacked emails from the Climatic Research Unit at the University of East Anglia proved anthropogenic global warming had been exaggerated, skepticism underwent a significant resurgence post-Copenhagen, and has continued to inform the political landscape in Australia. It is therefore worthwhile to consider how climate change skepticism was reported during this pivotal moment in *The Australian*, one of Australia's most influential newspapers.[1]

In partial support of McKnight's finding that *The Australian*'s position on climate change has been "erratic" (2012: 17), the analysis found that the newspaper's coverage of climate change science during COP15 was by no means uniformly opposed to the consensus or to the belief that some mitigating action should be taken on global warming. However, there was an editorial line evident across opinion pieces, editorials and, to a lesser extent, news articles that maintained that the science of anthropogenic climate change was far from "settled" and that, while the planet should be given "the benefit of the doubt", ongoing debate should be encouraged. This argument was predominantly made by establishing "debate" between two sides—on one side, the scientific consensus represented by an "orthodoxy" of scientists, green ideologues and left-wing politicians and business leaders; and on the other, those who refused to bow to the "alarmism" of the global warming "orthodoxy" and instead urged the canvassing of alternate views, represented by skeptical scientists, "Australians" and *The Australian* itself.

The strongest claims were made in *The Australian*'s editorials and directed against non-scientific elites—including politicians, business leaders, "green ideologues" (*Australian*, 2009e: 13) and other non-defined "experts"—who were attempting to silence the debate and oppress dissenting claims-makers with the "political refrain" that "'the science' is settled" (*Australian*, 2009c: 15). Sometimes, scientists were seen as the passive victims of these environmental extremists and political elites, their complex discoveries about the climate misrepresented and oversimplified by the hard and fast demands of politics. In one editorial, for example, *The Australian* argues that due to politicians' demands for certainty, the nuances of climate science have been overlooked and scientists have been pressured to "discourage critics within their own disciplines" (*Australian*, 2009c: 15). Simultaneously, however, scientists were seen to be more closely implicated in the global warming "orthodoxy" (*Australian*, 2009c: 15). In the same editorial, the hacking of emails from the Climatic Research Unit at the University of East Anglia is cited as evidence of "a lack of transparency about data and a culture keen to minimise dissent" (*Australian*, 2009c: 15), while one news article reports that scientists at the Met Office in Britain were being pressured to express their support for the scientists at the University of East Anglia (*Australian*, 2009a: 4). In the words of one unnamed scientist who is quoted in the article, "The Met Office is a major employer of scientists and has long had a policy of only appointing and working with those who subscribe to their views on man-made global warming" (*Australian*, 2009a: 4). Leaving aside the question of the accuracy of such claims, what begins to emerge from these reports and editorials is the sense of a divide between those vaguely defined politicians, scientists and institutions who collectively represent the consensus, and those individual scientists who stand against it.

Media institutions, too, were implicated in the global warming "orthodoxy", with *The Australian* engaging self-reflexively with the norm of balance to portray other media institutions as biased for not reporting theories that challenge the consensus. One news story is dedicated entirely to representing geologist and prominent Australian skeptic Ian Plimer's claims that the Australian press—with the exception of *The Australian* and its fellow Murdoch-owned paper, the *Herald Sun*—had ignored "a very large number of quite respectable scientists who have a different view" (Kerr, 2009: 7). Two days later, the satirical column *Cut-Paste* highlighted how "Climategate" had been under-reported in other newspapers, juxtaposing snippets from the *Sydney Morning Herald*'s coverage of the incident with ones from *The Australian*'s (2009b: 13). In the columnists' accompanying comments, the *Sydney Morning Herald* initially "ignores the story until November 25," then "buries it on page 10," and finally has to "explain Climategate to [its] readers" on 10 December, while *The Australian* carries out "no cover up" and reports the emails "on page one on November 23". Thus, while other media institutions are portrayed as being complicit in the global warming "orthodoxy", *The Australian* presents itself as the objective, impartial observer committed to representing the viewpoints and stories others are inclined to repress.

Those who supported radical action on climate change were dismissed in editorials as "extreme greens," "alarmists" and "doomsdayers" who avoided "inconvenient facts about natural cycles and the environment's ability to adapt" while pushing "politicised science and propaganda that [does not] line up with empirical evidence" (*Australian*, 2009e: 13). They were also guilty of "intellectual thuggery" for claiming that "'the science' is settled," because such a claim "bestows an unwarranted degree of certainty on a discipline that should, by its very nature, be contestable" (*Australian*, 2009c: 15). Protesters at COP15 were "slogan-chanting, empire-building activists, who adopt causes as careers in the way other people become teachers or accountants" (*Australian*, 2009e: 15), while the Copenhagen meeting itself was a "wealth-redistribution exercise" manipulated by NGOs and poor countries with a "cargo-cult mentality" to "attract aid dollars from the West" (*Australian*, 2009d: 13).

In contrast to the radical extremism of those who supported the consensus, the objective voice of reason was represented by "skeptical scientists" such as geologist Ian Plimer, whose work, *The Australian* attests in one editorial, has revealed "the fragility of much global warming science" (2009d: 13). A long paragraph is dedicated to carefully summarizing Plimer's argument that "rather than taking a 150-year time span to assess the problems of global warming, *we* need to look back several million years", which will in fact reveal that "far from heating up to dangerous levels, the planet is in a lull in an ice age that began 37 million years ago (*Australian*, 2009d: 13; italics added). The paragraph does not acknowledge that IPCC reports do, in fact, take into account long-term climatic changes and draw on experts from a range of scientific disciplines, not just the "atmospheric scientists" whom Plimer claims have "dominated the debate" (*Australian*, 2009d: 13). With this omission, what the piece achieves is a reasoned, matter-of-fact tone that presents Plimer's argument as the outcome of good common sense. The use of "we" includes *The Australian* along with its readers in the quest to break through the subterfuge of the global warming "orthodoxy".

Similarly, a series of articles about the health of the Great Barrier Reef pits marine biologist Ray Berkelmans against "doomsdayers" and "alarmists" who claim the reef will be destroyed under many warming scenarios predicted by the IPCC. Paraphrased in a front-page news article headlined "Scientists 'crying wolf' over coral", Berkelmans proclaims that far from dying, the reef is in "bloody brilliant shape", and scientists who suggest otherwise have a "credibility problem" (Walker, 2009: 1). The use of an Australian colloquialism ("bloody brilliant") appeals to an idealized version of the Australian national character that aligns Berkelmans with "ordinary Australians" rather than an elite class of scientific experts. The trend continues in an editorial that cites Berkelmans' research as "a reminder of the need to constantly test the claims made in the name of science" (*Australian*, 2009e: 13). It opines:

> International audiences are easily swayed by scare stories about the impact of global warming on the Great Barrier Reef. Australians who dive off the reef, year in, year out, are not so easily convinced that the end is nigh. They see first-hand what our coverage today shows—the Reef is more resilient than some might have predicted. (2009e: 13)

The implication is that "Australians" who dive off the reef "year in, year out" know more about its health than "green campaigners" and scientists whose knowledge derives from technical expertise acquired and applied in distant laboratories. Lay knowledge is thereby invoked to support science with "first-hand" evidence.

Having drawn the battleground between the climate change "orthodoxy" and ordinary Australians, *The Australian* proceeds to frame climate change as the latest fad of "green ideologues" who have seized on climate change science "to back up their arguments against development" (2009e: 13). Viewing climate change in this way—as the product of a left-leaning social movement, rather than a tangible environmental problem in its own right—enables fresh authority to be granted to skeptics who counter the consensus view (McKnight, 2012: 216). Meanwhile, the authority entrusted to certain sites of technical and scientific expertise is undermined on the basis of its alignment with a movement that is purportedly self-interested and politically motivated, as in the following paragraph:

> We live in an age where we commonly defer to experts. Getting to grips with the planet is, correctly, seen as a difficult task for the non-scientist. Yet this debate has revealed the shortcomings of relying on experts and allowing politicians to assert, rather than prove, the case for action. (*Australian*, 2009c: 15)

The explicit retreat from the authority of "experts" that is contained in this passage is symptomatic of the populist anti-elitist discourse that *The Australian* employed to cast doubt on the scientific consensus of anthropogenic climate change during COP15. Meanwhile, the inclusive language ("*We* live in a world where *we* commonly defer to experts") aligns *The Australian* firmly with its readers, those Australians who "deserve better" than the exaggerated claims that are purportedly presented by the elite class of politicians, experts, scientists, activists, business leaders and "green ideologues" calling for urgent action on global climate change. Against such overstatements, *The Australian* concedes that it is prepared to give the planet "the benefit of the doubt" on the need for action, but draws the line at "politicised science and propaganda that [does not] line up with empirical evidence" (2009e: 13). In doing so, however, it simultaneously puts the case for ongoing debate—and hence, delay—on meaningful action to mitigate and adapt to climate change.

DISCUSSION AND CONCLUSION

The preceding discussion has outlined the role that populist anti-elitism played in *The Australian*'s coverage of climate change during COP15. Far from being an isolated incident, longer-term content analyses of the coverage of climate change at *The Australian* from David McKnight (2008, 2012) and Robert Manne (2011) have similarly found that concern about climate change was commonly seen as "merely another claim from a liberal social movement" that was part of "a new orthodoxy representing a powerful 'new class' of intellectuals" (McKnight, 2012: 228–229). While to a certain degree this can be seen as a straightforward extension of the anti-elite tradition that has been mobilized so powerfully on other social and political issues, there are also important differences owing to the extent that science itself is a focus of the attack. As McKnight notes, climate change science, unlike feminism or environmentalism, is not a political or social movement, and to reject it is to resort to a "curiously 'post-modern' argument whereby the knowledge derived from climate science [is] regarded as socially constructed and [has] no special claim on our beliefs" (McKnight, 2012: 228–229). Thus, while *The Australian* rightly observes that, despite claims from politicians and business leaders, climate change science cannot and will never be truly "settled," what this criticism conceals is the degree to which such phrases are themselves a product of the translation of complex scientific information into media discourse and the construction of scientific risk in public life.

It is true that science is a discipline grounded in uncertainty, and that scientific theories should always be open to contestation. Yet, within that discipline there are theories which remain contentious and unresolved, and others which are supported by enough peer-reviewed scientific papers and which reflect agreement among such a majority of scientists working in the area that they are said to form a "consensus". The theory that the earth's climate is changing due to man-made greenhouse gas emissions belongs to the latter category. It is true that there are scientists actively publishing in the area who do not support that consensus, but their numbers are minimal compared to those who do support it. It is true, also, that these alternative views should gain a public hearing, but this should be in proportion to their representation in the scientific community. To argue that those who dispute the consensus should at all times be represented in public debates about climate change is to suggest that the two sides are equal, or that, rather than a scientific problem that can be proved to exist with very high degrees of confidence, climate change is a social issue on which input from a wide variety of claims makers is to be welcomed because everyone is entitled to their own opinion (McKnight, 2012: 216–217). It is desirable that public claims makers seek to avoid making technically inaccurate claims that the science of climate change is "settled" or "certain" for fear that, as *The Australian* argues (2009c: 15), publics will be made more suspicious when evidence of doubts do emerge (for example, as in the University of East Anglia incident). However, simplification is sometimes unavoidable in the quest to make lay audiences aware that on the question of the existence and causes of climate change science, there is no longer much dispute among scientists. If *The Australian* truly wishes to give the planet "the benefit of the doubt," it may direct its readers to the real debate that is currently taking place in climate change science and policy—over how much warming is now inevitable, and how drastic the cuts in carbon emissions must be over the next decade to ensure that the worst predicted effects of climate change are avoided.

While it is difficult to determine the impact a single media company's framing of climate change may have on broader public attitudes about the issue, the example discussed in this chapter can be viewed as one manifestation of a trend that has been highly successful at fostering ongoing public doubt about the existence and causes of global climate change, long after such doubts have been put to rest within the scientific community. The establishment of a demand and need for ongoing "debate" over climate change science is perhaps the greatest sign of the success of the populist anti-elite rhetoric to date. By framing the debate in terms of an unequal contest between an overbearing orthodoxy of social, political and scientific elites on the one side, and an oppressed minority of skeptical scientists and a misled public on the other, the argument that the two sides of the debate are equal (in terms of the quantity and quality of scientific evidence that each possesses) holds less weight than the argument that science must always be contestable and all theories have a right to be heard. Reframing the debate in these terms inevitably forces those who uphold the consensus onto the defensive, where they must seek to dispel accusations of orthodoxy by welcoming ongoing debate while, at the same time, struggling to make clear the weight and soundness of evidence that supports their position.

While it is easy to recommend that supporters of the consensus simply refuse to grant skeptics further legitimacy by denying them airtime or column space, the dangers of that approach are aptly seen when skeptics leverage cries of oppression. To simply dismiss skeptics' claims as opinion is to risk becoming entrapped in a politically correct discourse of openness and accountability where such dismissals look heavy-handed and autocratic. To engage with skeptics in the name of debate, on the other hand, is to enter into their project and counter-intuitively legitimize their right to representation in media reports. In both cases, the vehemence with which both sides defend their position fuels a debate conducted not over climate change, but over who has the right to participate in that debate. The more troubling outcome of this shift is that it in turn obscures the fact that the crucial questions at this time ought not to be whether anthropogenic climate change exists, but how severe its consequences will be, and what mitigation and adaptation steps must be taken in the present and immediate future to address it.

ENDNOTES

1 The discussion is based on a critical discourse analysis of news articles, opinion pieces and editorials published in *The Australian* during the two weeks of coverage of the proceedings of COP15 (8–21 December 2009). The Factiva search engine was used to identify all articles published during this period containing the words "climate change" or "global warming". The results were then manually narrowed to include only those articles in which climate change science was a core part of the story and which discussed or referred to climate skepticism, the "Climategate" controversy, or some other contested area of climate science.

▪ CHAPTER EIGHTEEN

Media, Civil Society and the Rise of a Green Public Sphere in China

GUOBIN YANG AND CRAIG CALHOUN

According to media reports, in April 2004 China's State Council halted the hydropower project planned for the Nu River in Yunnan Province. The decision came after months of intense public debates. China's premier, Wen Jiabao, reportedly cited "a high level of social concern" as an important reason for suspending the dam-building project (*Ming Pao Daily*, 2 April 2004). Such a reversal after public criticism was hardly typical of the Chinese government—nor was the nature of the public criticism typical. In the first place, the public debate addressed policy rather than a more common complaint such as the exposure of corruption or the suggestion that local officials deviated from the goals of the central Party and government. Second, a broad range of participants was involved in public discourse. This differentiated it from the "reportage" literature through which criticism flourished in the 1980s, for example, which typically required a strong individual personality, such as Liu Binyan, willing to focus on broader concerns in his or her writing.

How did the public debates about the Nu River happen? Who was involved? What media were used? We argue that the articulation of "a high level of social concern" depended on a public sphere of environmental discourse in China—a green public sphere. Communication and debates in the public sphere channeled citizen opinions to influence government policies.

A green public sphere fosters political debates and pluralistic views about environmental issues, and for this reason it is intrinsically valuable (Torgerson, 2000). The rise of a Chinese green public sphere commands special attention, however. First, it is exemplary of a variety of new forms of public engagement in contemporary China. These include, for example, feminist activism, cyber activism, HIV/AIDS activism, and rights activism more broadly. Second, with environmental issues as its central concern, the green public sphere represents the emergence of an issue-specific public. The differentiation of issue-specific publics is a relatively new development in China. Third, the transnational dimension of the Chinese green sphere indicates still another new trend, namely, its transnationalization (see Calhoun, 2004). In effect, then, a Chinese green public sphere is significant in terms of both content and form. With respect to its formal attributes, the green sphere is distinctive because it engages politics and public policy without being primarily political (Ho, 2007; Martens, 2006). Carving out a space for "nonpartisan" advocacy is a new development in China. It is also distinctive because of its reliance on a range of media and organizational forms, including traditional press, the Internet, "alternative media," as well as environmental NGOs.

This chapter[1] delineates the main features of the green public sphere as it emerged in China, analyzes the main factors that contributed to its emergence, and explores its functions. We argue that the emerging green sphere consisted of three basic elements: an environmental discourse, or "greenspeak"; publics that produced or consumed greenspeak; and media used for producing and circulating greenspeak. First, we show that one major indicator of the rise of a Chinese green public sphere was the proliferation of environmental discourse—a greenspeak. Contrary to an earlier Maoist and Marxist view of the human conquest of nature (Shapiro, 2001) this new discourse warned about the dangers of irresponsible human behavior toward nature and called for public action to protect the environment. Appearing in television programs, radio programs, newspapers, magazines, leaflets, flyers, posters, and on the Internet, this blossoming discourse represented a participatory conversational situation, one "of seven mouths and eight tongues" (*qizui bashe*), as a Chinese folk saying would have it. Second, we argue that environmental NGOs provided the pivotal organizational basis for the production and circulation of this greenspeak. A greenspeak that promotes a new environmental consciousness does not fall from heaven, but has its advocates and disseminators. We focus on environmental NGOs both because they are relatively new and because they play a central role in producing greenspeak. Third, we analyze the media of the green sphere. Distinguishing among mass media, the Internet, and "alternative media," we argue that because these different types of media differ in social organization, access, and technological features, they influence the green sphere differently. Finally, we return to the case of the Nu

River to illustrate the dynamics and functions of the fledgling green sphere. In the conclusion, we discuss the implications of our analysis for understanding the sources of political change in China.

A GREEN PUBLIC SPHERE AND GREENSPEAK IN CHINA

"Public sphere" is a controversial concept. Habermas initially defined it as "a domain of our social life in which such a thing as public opinion can be formed." Access to this domain is "open in principle to all citizens" who may "assemble and unite freely, and express and publicize their opinions freely" (Habermas, 1989a). Critics were quick to point out that Habermas's version of the public sphere was a bourgeois sphere that in reality excluded certain categories of people (such as women) and was fraught with problems of social, economic, cultural, even linguistic inequality (see, for example, Fraser, 1992). In response, Habermas later recognized the internal dynamics of the public sphere, the possibility of multiple public spheres, and the conflicts and interactions among them (1992).

In China studies, the concept of the public sphere has similarly been controversial. It was initially used to explain the rise of the student movement in 1989 (Calhoun, 1989). Then a symposium on "public sphere"/"civil society" in 1993 introduced an influential debate. Some scholars in the debate find civil society and the public sphere in late imperial China (Rankin, 1993; Rowe, 1993), the others argue that these concepts are too value-laden and historically specific for understanding Chinese realities (Wakeman, 1993). More recently, there has been a revival of interest in civil society and the public sphere among China scholars. For example, it has been argued that these categories are pertinent to China because they emerged out of experiences of modernity which transformed China no less than the West (Yang, 2004). Others use a relaxed notion of the public sphere, adopting more neutral terms such as "public space" or "social space," or focusing on publics rather than the public sphere (Yang, 2002; Lean, 2004). One reason why we continue to use the concept of the public sphere is that Chinese intellectuals themselves have come to embrace it. A 2005 Chinese book on green media, for example, focuses on the building of a "green space for public opinion" (*lüse gonggong yulun kongjian*), alluding directly to the Habermasian concept (Wang, 2005; see also Ma, 1994). Recognizing the historical baggage of the Habermasian concept, however, we maintain a broad conception of the public sphere as space for public discourse and communication. It consists of discourse, publics engaged in communication, and the media of communication. The fledgling green sphere in China thus has the following basic elements: an environmental discourse or greenspeak; publics that produce or consume greenspeak; and media used for producing and circulating greenspeak. "Public" is a broad and loose concept. By the publics of China's green public sphere, we refer specifically to individual citizens and environmental NGOs directly engaged in the production and consumption of greenspeak.

A main indicator of an emerging green public sphere in China is the proliferation of a greenspeak. Greenspeak refers to the whole gamut of linguistic and other symbolic means used for raising awareness of environmental issues (cf. Harre et al., 1999). The Chinese greenspeak includes recent neologisms in the Chinese language such as "sustainable consumption," "white pollution," "eco-centricism," "endangered species," "animal rights," "global warming," "desertification," "deforestation," "biodiversity," "bird watchers," and more. An entire dictionary of greenspeak can now be compiled. Different social actors use greenspeak for different purposes. Business corporations, for example, may use a green language as a way of "greenwashing" its interests (Greer and Bruno, 1996).

We focus on greenspeak produced by civil society actors. This civic greenspeak has several features. First, it is tacked onto the mainstream global discourse of sustainable development. The popularity of terms such as "one world," "common earth," "holistic approach," "global village," "Earth Day," and of course, "sustainable development" attests to the global dimension of this discourse. Following the 1992 Rio Earth Summit, moreover, the Chinese government published its strategies for sustainable development in a "China Agenda 21" white paper issued in March 1994, thus legitimating an official discourse of sustainable development in China.

Second, greenspeak expresses the tension between environmentalists and economic actors. Similar to new social movements elsewhere in the world, the environmental movement in China attracts some sections of the population and not others (see Calhoun, 1995). Its main constituency consists of students, intellectuals, journalists, professionals, and other types of urbanites. For example, university student environmental associations are a main part of the movement. A nonstudent member-based organization such as Friends of Nature (FON) also draws its membership mostly from these urban groups. Greenspeak gives these people, who tend to have more cultural capital than economic capital, a rhetoric for identifying themselves and their concerns in contrast to the dominance of a more crass economic rhetoric in society at large.

Third, in response to the ascendance of consumerism and materialism, greenspeak promotes new moral visions and practices. A central moral message is that environmentalism must be practiced as a new way of life. It promotes a new understanding of the relationship between humans and their natural environment, one that stresses human-nature harmony. A practical corollary of this view is that humans must treat nature and its flora and fauna with respect and kindness.[2] It also promotes the vision of a new personhood. Practicing a green consciousness must start with oneself. If each and every person lives an environmentally friendly life, then the earth might be saved. A common personal practice is to reject the use of disposable consumer products (such as disposable chopsticks and shopping bags). These practices convey some sense of religiosity. They embody a search for more spiritual "meanings" in life—again, something that tends to be opposed to sheer economism. In this sense, the green discourse continues an opening up to "expressive individualism" in China over the last 20 years.

Yet the relationship between economism and environmental protection is also a source of dilemmas and personal perplexity. Reflecting the growing richness and diversity of environmental discourse, these dilemmas are often openly discussed and shared. A good example comes from the publications of Green Camp, an unregistered NGO in Beijing. Each year since 2000, Green Camp has produced an informal publication featuring personal essays by participants in that year's green camp activities. These personal stories demonstrate how China's young environmentalists develop a deeper understanding of environmental issues by experiencing them first-hand and talking about their experience. One essay in the 2000 volume describes a small incident that happened to its author when he and a few other green campers were studying the feasibility of eco-tourism in Changbai Mountain. He heard the following conversation between a fellow camper and an employee of a local nature reserve protection station:

> CAMPER: What do you think is most needed for developing eco-tourism?
>
> LOCAL EMPLOYEE: Money.
>
> CAMPER: If you see people causing damage here, would you intervene?
>
> LOCAL EMPLOYEE: Yes!
>
> CAMPER: Why?
>
> LOCAL EMPLOYEE: Because this is my home.
>
> CAMPER: If you lose your job and have no income, would you go into the mountains to stealthily gather mountain stuff?[3]
>
> LOCAL EMPLOYEE: Yes!

After hearing this conversation, everyone became silent. It was a transformative moment for this individual. He realized that between reality and young environmentalists' idealism, there was a vast gap. He continues:

> After the field trip to Changbai Mountain, I came to a deeper understanding of the difficulties of environmental protection in China. To transform economy and environment into a virtuous relationship of mutual benefit and embark on a road of sustainable development—this is still a dream. It will take arduous efforts to turn the dream into reality. (Wei, 2000)

Part of the arduous efforts is to promote environmental consciousness and citizen participation. These efforts betray a fourth feature of the greenspeak—its political thrust. The Chinese greenspeak emphasizes participation and volunteerism. While recognizing that environmental problem solving depends on the joint efforts of government, citizens, and NGOs, greenspeak emphasizes the role of citizens and the importance of developing an NGO culture. In addition, greenspeak is a veiled way of talking about many other things, including government policies. One good example is a speech delivered by the representative of an environmental NGO from Qinghai Province at an NGO workshop in Beijing in October 2002. Referring to the central government's ambitious plan to develop the western regions, the speaker argued that in western minority regions such as Qinghai Province, the protection of the biodiversity of the natural environment should be integrated with the protection of cultural diversity, and that local communities should be involved in the decision-making process (Zhaxiduojie, 2002). Greenspeak thus can be political, though as Peter Ho (2007) suggests, this may be a depoliticized politics.

ENVIRONMENTAL NGOS: THE DISCOURSE-PRODUCING PUBLICS OF THE GREEN SPHERE

Michael Warner describes a public in the following terms:

> A public is a space of discourse organized by nothing other than discourse itself. It is autotelic; it exists only as the end for which books are published, shows broadcast, Web sites posted, speeches delivered, opinions produced. It exists *by virtue of being addressed.* (Warner, 2002, original emphasis)

Warner's description captures only half of what "a public" means. A public is not just an addressee, but also an addressor. It not only reads books, watches shows, reads Internet posts, listens to speeches, and receives opinions, but also publishes books, produces shows, writes blogs or responds to blogs, delivers speeches (or partakes in conversations, for that matter), and expresses opinions.

As mentioned previously, the publics of China's green sphere consist of all citizens and civil society organizations involved in the production and consumption of greenspeak. Here we concentrate on the discourse-producing publics. These are again diverse, and may include scientific communities, educational institutions, and a broad array of old and new social organizations. We focus on environmental NGOs because they are the most distinctive and novel organizational base for the green public sphere.

In one of the first systematic analyses of this topic, Peter Ho showed that environmental NGOs in China covered a wide spectrum, from more or less independent NGOs to government-organized NGOs (GONGOs), student environmental associations, unregistered voluntary organizations, and NGOs that were set up "in disguise" in order to bypass the registration requirements and hide their true nature from the government's view (Ho, 2001). A survey of university student environmental associations showed that as of April 2001, there were 184 student environmental associations (Lu, 2003). Nonstudent grassroots environmental NGOs numbered about 100 as of 2003, not including the numerous GONGOs (Yang, 2005). According to a Chinese news release, this number reached about 200 toward the end of 2006.[4]

During this period, environmental NGOs were engaged in a broad range of activities, from public education and community building to research and advocacy. In these activities, they resorted to all forms of media and public forums, including television, radio, newspapers, magazines, websites, exhibits, workshops, and salons. As a result, these organizations became an important institutional base for bringing green issues into the public sphere. Many organizations published newsletters and special reports in print or electronic form. Some produced television programs and published books. For example, between 22 April 1996 and March 2001 Global Village of Beijing produced 300 shows for its weekly television program, *Time for the Environment,* on CCTV-7. It also published books on environmental issues which could be ordered via its website. Titles included *Citizen's Environmental Guide, Children's Environmental Guide, Green Community Guide,* and *Environmental Song Book.*

Two environmental organizations in Beijing organized public forums on a regular basis. Tianxia Xi Education Institute, an educational and environmental NGO founded in 2003, organized forums on topics ranging from dam building to citizenship education and public health. Green Earth Volunteers, an unregistered NGO based in Beijing, began organizing monthly environmental salons for journalists in 1997. Featuring guest presentations on various environmental issues, these salons aimed to provoke broader discussions among Beijing's unofficial environmental circles, and helped journalists to write more and better environmental stories. They covered a wide range of topics. At a salon event in June 2002, a retired worker who introduced himself as an environmental volunteer made a slide presentation about the desertification of the grassland in China's western regions. Then the founder of Save the South China Tiger, a nonprofit organization registered in Great Britain, spoke about the protection of the tiger. Another event, held on 19 March 2003, featured three guest speakers who spoke on issues of animal protection and reproduction in nature reserves, the new challenges facing wildlife protection in the development of China's western regions, and the import and export of medicinal ingredients made from wild animals. Other topics for these salons included genetic modifications and ecological safety; sandstorms and their management; water-system design; the roles of government, NGOs, business, and media in environmental policymaking in the USA; and urban transportation.

Campaigns are an effective tactic used by NGOs to publicize environmental issues. They help to concentrate public attention on specific issues by creating media visibility and public discussions. Some have directly influenced policies. One of the first public campaigns was organized in 1995, to stop the felling of an old forest in Yunnan in order to protect the indigenous golden monkeys. Since then, not a year has passed without environmental campaigns. In 1997 a group of college students organized a campaign to promote recycling on university campuses in Beijing, while another group ran a campaign to boycott disposable chopsticks. We will return later in this chapter to the campaign launched in 2003 to stop dam building on the Nu River. The scope of these campaigns was wide-ranging. They included both moderate educational campaigns (e.g., promoting Earth Day activities) and more confrontational campaigns (boycotting commercial products and challenging industrial projects).

To understand why environmental NGOs could organize such a broad range of activities in China's constraining political context, it is essential to analyze why they developed in the first place. We underscore three conditions. First, the growth of environmental NGOs was part of a larger "associational revolution" in China (Wang and He, 2004). This has to do with many factors, including state decentralization and the government's recognition of a third sector. Second, the Chinese state was taking on shades of green, as reflected by the large body of environmental laws and regulations it has promulgated (Ho, 2001). This "greening of the state" provided a favorable setting for environmental NGOs. Third, the growth of environmental NGOs took place in a dynamic context of multiple social actors and complex social relationships. As the concept of "embedded environmentalism" indicates (Ho, 2007), environmental NGOs are embedded in social relations that enable their growth. One enabling type of social relations for example, is the growing ties between domestic NGOs and international organizations. Ranging from financial support to professional exchanges, these ties have promoted the visibility of Chinese NGOs both in China and in the international arena (Bentley, 2003; Morton, 2005).

THE MEDIA OF THE GREEN SPHERE: OFFICIAL, "ALTERNATIVE," AND THE INTERNET

We distinguish three types of media of China's green public sphere; namely, mass media, alternative media, and the Internet. Because they differ in their relationships to the state and in technological features, they are not equally accessible to Chinese environmentalists and they influence China's green sphere differently.

MASS MEDIA

Mass media—newspapers, television, and radio—had an enormous influence on the emergence of the green public sphere. Since the 1990s, mass media coverage of environmental issues in China has greatly increased. Surveys conducted by Friends of Nature found that the average number of articles on environmental issues published in national and regional newspapers was 125 in 1994; this number rose to 136 in 1995 and 630 in 1999.[5] The major environmental campaigns in recent years were all covered by the mass media. The bigger of these campaigns (such as those to protect the golden monkey and the Tibetan antelope and the one against dam-building on the Nu River) generated intense media publicity. Even a small campaign to guard two hatching wild geese in a public park in Beijing caught media attention.[6]

In principle, Chinese mass media are the organs of the state. How do we explain this growing media coverage of sometimes very contentious environmental issues? First, mass media had undergone de-ideologization, differentiation, and commercialization in the reform era (Akhavan-Majid, 2004; Zhao, 2000). Party organs such as the *People's Daily* and commercial papers, for example, are not subject to the same degree of state control. The increasing dependence on commercial revenues gave the mass media more latitude in covering issues of broad social interest. Growing environmental problems such as pollution became issues of great concern.

Second, the Chinese government supported media coverage of environmental issues by launching its own environmental media campaigns. The most ambitious project is the "China Environment Centennial Journey" (*Zhonghua huanbao shiji xing*). Funded by the government and led by a commission composed of high-level officials from various ministries, the campaign was inaugurated in 1993. Each subsequent year, the commission sponsored mass media institutions to send journalists out to the field to do investigative reporting on a selected environmental theme. The theme for 1993, for example, was "Fighting Environmental Pollution," and for 2005, "Clean Drinking Water." From 1993 to 2005, 50,000 journalists from across the country participated in the project and produced 150,000 reports on environmental issues (Wang, 2005: 114).

Third, Chinese environmentalists attach great importance to mobilizing the mass media and have been remarkably successful in this respect. Besides the structural changes in the media and a favorable political context, a major reason for their success is that many environmentalists and even leaders and founders of environmental NGOs are themselves media professionals. During our period of study, Green

Camp, Green Earth Volunteers, Green Plateau, Tianjin Friends of Green, and Panjin Black-Beaked Gull Protection Association were all led by journalists or former journalists. Friends of Nature, too, had some influential journalists in its membership. These environmentalist media professionals serve as direct links between the mass media and the environmentalists.

ALTERNATIVE MEDIA

Alternative media refers to the informal and "unofficial" publicity material produced and disseminated by NGOs, such as newsletters, special reports, brochures, flyers, and posters. During our period of study, they also included new media such as CD-ROMs and DVDs. These media materials are alternative in the sense that they are not controlled by the government, but are edited and produced by NGO staff or volunteers and distributed through informal channels. Some of these publications do not look very different from official publications, yet they are unofficial because they do not have official ISSN numbers. For example, Friends of Nature published a bimonthly newsletter from 1996. It looked just like a regular magazine in its professional appearance, yet it was not officially registered and had no ISSN number. The advantage of having no official registration is that environmental groups can largely publish what they want. The disadvantage is that without an official ISSN, the publications cannot be distributed publicly, which limits their reach (interview with FON staff, July 2002).

Almost all the NGOs we encountered in our research had publicity materials. The number and frequency of publications depended on their professional and financial resources. Production costs were usually covered with funding raised from corporate sponsors, foundations, or foreign donors. Staff and volunteers were responsible for editorial work and distribution. In membership-based organizations such as Friends of Nature, members automatically received the organization's official newsletter. Many organizations distributed their publications for free, while some sold them to defray production costs. These publications covered a broad range of environmental issues in a variety of genres. There were many personal stories and perspectives, indicating an emphasis on personal experience and individual viewpoints commensurate with an ethics of participation and respect.

The alternative publications of the green sphere have historical precedents. In production and distribution, they resemble the unofficial publications of the Democracy Wall movement in 1978 and 1979. Those earlier publications were also edited by self-organized groups and distributed through independent and informal channels (Goodman, 1981). They differ in contents and context, which partly explains why the Democracy Wall publications were quickly banned, while the alternative green media thrived. In contents, the unofficial publications during the Democracy Wall Movement were much more radical. They were full of direct denunciations of state policies and calls for democracy and political reform. At a time of great political uncertainty, the Democracy Wall movement helped to mobilize public support for Deng Xiaoping's political maneuvers against Mao's designated successor Hu Guofeng, but when its radicalism challenged Deng, it was suppressed. In contrast, alternative green media do not carry politically radical contents such as calls for democracy or political reform. Instead, they focus on environmental education and discussion. Rather than challenging state legitimacy, they operate largely within the parameters of state policies. This approach provides some degree of legitimacy and explains the survival of alternative green media.

THE INTERNET

Compared with print media, the Internet has the advantages of speed, broad reach, and interactivity. It favors open discussion, speedy communication, and wide dissemination. During our period of study, Chinese Internet users turned to the Internet for public expression and political activism even as the government was stepping up control (Barmé and Davies, 2004; Zhou, 2005; Yang, 2006b). How and why did Chinese environmentalists embrace the Internet? Environmentalists focused on three of the many different types of network functions. The first was websites. A survey of the web presence of environmental NGOs conducted in March 2004 found that of the 74 organizations surveyed, 40 (or 54 percent) had websites (Yang, 2006a). The second was mailing lists. Several environmental NGOs maintained active mailing lists that sent environmental information regularly to subscribers by email. The third was bulletin boards. Twenty-four (or 60 percent) of the 40 websites just mentioned had bulletin boards for public discussion (Yang, 2006a). In addition, some commercial portal sites and large online communities such as Tianya Club ran "green" web forums. Websites, mailing lists, and even bulletin boards continue to be used for public environment discussions as of 2012, but since 2010, microblogs have become the most popular online spaces for this purpose. Many environmental NGOs and activists regularly launch public debates through their microblog accounts.

The website of Friends of Nature offers a good example of the discourse produced in environmental websites. Friends of Nature went online in December 1998 and launched its first website in June 1999 (interview with FON staff, July 2002). Besides publishing activities and showcasing the organization's projects, the website supported an active bulletin board system and published the electronic version of the Friends of Nature print newsletter and an electronic digest. They featured interesting debates about such topics as the relationship between traditional culture and sustainable development, the meaning of "development," the environmental lessons of the SARS epidemic, and animal rights. The animal rights debate in the 4 March 2003 issue contains two lengthy articles. One explicates the importance of promoting animal rights from the perspective of environmental philosophy. The other argues that animal rights is a Western discourse with hidden imperialist pretensions, because in this discourse non-Western societies with different attitudes toward animals are portrayed as primitive and uncivilized.[7]

The editors were concerned less with who was right or wrong than with using web forums to foster discussion. As the editorial accompanying the two essays explains, "we want to provoke your thinking. We believe that the independent thinking of ordinary people is no less significant and no less valuable than that of the experts."[8] To facilitate discussion, at the end of each article a hot link was set up to Friends of Nature's bulletin board system, where dozens of messages were posted in response to the debate. One message said, "I haven't had time to read the articles ... but I'd like to state my views first. In my personal view, the rights of animals are the rights to existence and to free activity, which are endowed by Great Nature and shared by all creatures."[9] It is personal voices such as these that found channels of expression in the green websites.

THE GREEN SPHERE IN ACTION: THE CAMPAIGN TO STOP DAM BUILDING ON THE NU RIVER

We started this chapter with the campaign to stop dam building on the Nu River. We now return to the case to illustrate China's green public sphere in action. We highlight the interactive dynamics of civil society and different types of media in the campaign. The hydropower project on the Nu River was approved by the National Development and Reform Commission on 14 August 2003. Its core components were a series of 13 dams on the lower reaches of the river, which fall within Yunnan Province. According to the project design, the total installed capacity of the dams is 21 million kilowatts, exceeding even that of the ongoing project at the Three Gorges. The project aroused immediate controversy.[10] Supporters of the project claimed that it would accelerate the economic development of the river valley regions and help alleviate poverty. Environmentalists who campaigned against the project held that the ecological treasures of the Nu River—its breath-taking natural beauty and biodiversity—are unique in the world and belong not only to China; they are also a world heritage.[11] In framing the debate, they stressed that the Nu River is part of the Three Parallel Rivers of Yunnan Protected Areas, which had just been listed as a World Heritage Site by UNESCO on 3 July 2003. Damming the river, they argued, would threaten this world heritage site. They also argued that the project would benefit the developers more than the local residents, citing the potential problems of population resettlement and potential destructive effects on indigenous cultures. Finally, although they did not frame the issues in explicitly political terms, legal references were common. For example, a petition letter signed by 62 scientists, journalists, writers, artists, and environmentalists called for the enforcement of China's Environmental Impact Assessment Law.

The public views about the hydropower project on the Nu River thus fell into two opposing sides: the protection of natural heritage vs. economic development and poverty alleviation. The conflict was not surprising: this was only the most recent Chinese version of the tensions inherent in the global discourse of sustainable development.[12] What is remarkable in this case is that the arguments of both sides entered China's public sphere and influenced policy.

THE ROLE OF ENVIRONMENTAL NGOS IN PUBLICIZING THE CAMPAIGN

The campaign against dam building on the Nu River has some peculiar features. First, it enjoyed the support of the State Environmental Protection Administration officials. When the campaign started, China's first environmental impact assessment law had just gone into effect (1 September 2003). Perhaps to demonstrate its commitment to the new environmental law, the State Environmental Protection Administration organized forums to assess the environmental impact of the proposed project. The first forum took place on 3 September 2003. More than 30 scholars and researchers attended. The dominant view at the forum, voiced by a professor from the Asian International Rivers Center of Yunnan University, was harshly critical of the project.[13] National media, especially the *China Youth Daily*, covered the forum in reportage with dramatic titles such as "13 Dams to Be Built on the Last Ecological River, Experts Vehemently Oppose the Development of the Nu River" ("Zuihou de shengtai he shang yao xiu 13 dao ba, zhong zhuanjia banghe nujiang kaifa," *China Youth Daily*, 5 September 2003). From 14 to 19 October, the State Environmental Protection Administration led a group of experts on a study tour of the Nu River Valley and then held another forum on 20 and 21 October, this time in Kunming. This second forum invited representatives from relevant government agencies at the provincial and prefectural levels in Yunnan Province, as well as scientists and other scholars. At the forum, scholars from Beijing opposed the project, whereas the local Yunnan scholars and government officials defended it. The State Environmental Protection Administration officials were on the opposing side, but the controversy did not seem to be resolvable between the parties directly involved.

Second, environmental NGOs played a central role in tipping the balance in favor of the opponents of the project. They were instrumental in producing the "high level of social concern" cited by Wen Jiabao. Environmental NGOs launched a campaign as soon as they learned that the National Development and Reform Commission had approved the project. The China Environmental Culture Promotion Society organized one of the first influential public events. At its second membership congress on 25 October 2003, the organization issued a public petition to protect the Nu River. On 17 November 2003 the Tianxia Xi Education Institute organized a forum to educate the public about the Nu River. The forum featured a speaker from the Yunnan-based NGO Green Watershed.[14] In December 2003 an NGO in Chongqing City collected more than 15,000 petition signatures opposing the Nu River project (Yardley, 2004). On 8 and 9 January 2004 five research and environmental organizations, including Friends of Nature and Green Watershed, organized a forum in Beijing to discuss the economic, social, and ecological impact of hydropower projects, again focusing their criticisms on the Nu River project. From 16 to 24 February 2004 about 20 journalists, environmentalists, and researchers from Beijing and Yunnan conducted a study tour along the Nu River. They returned to Beijing to organize a photo exhibit. Indicating the international dimension of the campaign, they took the exhibit to the UNEP Fifth Global Civil Society Forum (GCSF) held in Jeju, South Korea, in March 2004 to mobilize international support. Together, these efforts created the momentum of a public campaign.

STRATEGIC USE OF THE MASS MEDIA AND THE INTERNET

Environmental NGOs made effective use of the mass media and the Internet to produce and disseminate opposition. The most active NGO in mobilizing the media was Green Earth Volunteers. Green Earth Volunteers organized monthly environmental salons for journalists. The two main organizers of the salons, Wang Yongchen and Zhang Kejia, were influential journalists and environmentalists: Wang, a senior journalist with China's Central People's Radio Station and a cofounder of Green Earth Volunteers, and Zhang, a journalist at *China Youth Daily* and a main force behind the newspaper's Green Net, an online section of the newspaper devoted to environmental issues. The environmental journalists' salon had already proved to be an important base for mobilizing the media. It played a crucial role, for example, in mobilizing media opposition to the Dujiangyan Dam incident in 2003.[15] In the Nu River case, both Wang and Zhang were signatories to the petition letter of 25 October 2003, the first major public action in the movement. Wang organized the study tour of the Nu River in February 2004 and the subsequent photo exhibit in

Beijing. Besides publishing many news reports about the debates surrounding the hydropower project, Zhang uses the Green Net of *China Youth Daily* to cover the debates.

In addition, websites were used to disseminate information and foster discussion. Integrating new media with the traditional print media, the Green Net of *China Youth Daily* presented a special column on the Nu River campaign and collected nearly 200 articles on the topic. The Institute for Environment and Development set up a campaign website which featured an online version of the aforementioned photo exhibit about the Nu River, beautiful scenic photographs of the river valley, and an archive of essays debating the issues. A campaign leader reported that after the website was set up, she received letters and telephone calls just about every day, and people would tell her how excitedly they were browsing the website and how they hoped that it could be updated more frequently.[16]

Debates about the case also appeared on the bulletin boards run by environmental NGOs and commercial websites. For example, a keyword search on 19 August 2004 for "Nu River" in the bulletin board system of the popular Tianyaclub.com yielded dozens of postings debating the Nu River project. The opinions in these postings were divided, and some were expressed in very angry tones. One posting laments: "Population and economic growth are the natural enemies of environmental protection!" Again, what matters here is not who was right and wrong, but that people were debating the issues in the public arena. All this shows that there was indeed a high level of "social concern" about the project, which prompted the central government to temporarily halt it in April 2004. It shows that China's fledging green public sphere was instrumental in producing this social concern.

CONCLUSION

We have shown that environmental NGOs were the primary discourse-producing publics of the fledging green public sphere in China. The mass media, alternative media, and the Internet provide the communicative spaces, but they were used in differential ways because of different institutional and political constraints associated with them. With the rise of a green public sphere, new ways of talking about the environment were introduced to the Chinese public. As much as the discourse itself was important, however, it is no less important to highlight the communicative spaces in which it appeared. These spaces are prerequisites for citizen involvement and political participation. They are essential for sustained and ongoing public discussion.

A green public sphere is not a homogeneous entity, but consists of multiple actors, multiple media, and multiple discourses. Nor is it completely autonomous or equal. Civil society actors must heed the political context; the different types of media are subject to varying degrees of political control. In a sense, the green sphere is a product of "embedded environmentalism" (Ho, 2007). It is embedded in politics, in civil society, and in communications technologies. Embeddedness can be both constraining and enabling.

A main concern in the scholarship on contemporary Chinese society is the relationship between the state and society. An influential perspective is state corporatism, which argues that the state permits the development of social organizations, such as NGOs, provided that they are licensed by the state and observe state controls on the selection of leaders and articulation of demands (Schmitter, 1974). In 1995 Unger and Chan made a strong case for a state corporatist perspective on Chinese society. Based on an analysis of more recent trends, however, Howell argues that the state corporatist perspective is no longer adequate for capturing some new directions in Chinese society, such as the emergence of new types of civil society organizations working on marginalized interests (2004). Our analysis lends support to Howell's conclusion. Although the Chinese government undoubtedly plays an important role in fostering the green public sphere, for example by sponsoring the "China Environment Centennial Journey" campaign, a state corporatist perspective does not give enough credit to the agency of non-state actors (see Akhavan-Majid, 2004). The evidence we present shows that the constitution of a Chinese green public sphere depended crucially on citizens and citizen organizations and on their creative uses of the Internet, alternative media, and the mass media. We need to develop a perspective that emphasizes the interpenetration and mutual shaping of state and society.

To some extent, the green public sphere is exemplary of the general development of the public sphere in China. There is an implicit politics to it quite beyond the environment, a politics of expanding general public discourse. This politics can also be discerned in other social arenas (such as rural poverty) where citizens and voluntary associations are similarly engaged in public discussion and in finding ways to engage policymakers. Public discussion in these other social arenas also depends on various types of media. A future research agenda, therefore, is to study the discourse, publics, and media in these other social arenas and explore the sources and consequences of potential synergies between different issue-specific public spheres.

ENDNOTES

1 This chapter first appeared in 2007 in *China Information*, 21(2), 211. Many thanks to the authors, editor Professor Tak-Wing Ngo, and the publisher, Sage, for the rights to reproduce an edited version here.

2 Environmentalists articulated these visions forcefully in a public debate in 2005 that has come to be known as the "respect/fear nature" (*jingwei ziran*) debate. See Bao and Bing (2006).

3 The original Chinese is *toucai shanhuo. Shanhuo,* here translated as "mountain stuff," refers to profitable products from the mountains such as expensive mushrooms and ginseng.

4 "Huanbao minjian zuzhi shuliang he renshu jiang yi 10 percent zhi 15 percent sududizeng" (Number and staff of environmental NGOs to increase by 10 to 15 percent) (*Xinhuanet,* 2006, 26 October). http://news.xinhuanet.com/environment/2006-10/28/content_ 5261309.htm (accessed 29 October 2006).

5 Friends of Nature, "Zhongguo baozhi de huanjing yishi" (Survey on environmental reporting in Chinese newspapers). (2000). Beijing: Friends of Nature publication.

6 "Cong hongda maque dao shouhu dayan" (From chasing and killing sparrows to guarding wild geese). (*Beijing Youth Daily,* 1999, 7 September). http://www.bjyouth.com. cn/Bqb/19990907/ GB/3998^0907B09918.htm (accessed 1 September 2004).

7 The two sides of the debates are represented in an issue of the electronic newsletter of Friends of Nature. See Friends of Nature (2003).

8 http://www.fon.org.cn/enl/view.php?id=49#pinglun (accessed 10 May 2005).

9 http://www.fon.org.cn/forum/showthread.php?threadid=2527 (accessed 10 May 2005).

10 The controversy analyzed here covered the period from August 2003 to April 2004.

11 The controversy was widely covered by the media. Transcripts of detailed arguments against the hydropower project made by scientists and environmentalists at a forum organized by the State Environmental Protection Administration are available on the Green Net web site of *China Youth Daily.* (Green Net, n.d.).

12 According to one study, the meaning of "sustainable development" is so ambiguous and so often contested that by 1992, five years after it was first introduced by the World Commission on Environment and Development, about 40 different definitions had appeared (see Torgerson, 1994).

13 See transcripts of the forum published on the Green Net web site of *China Youth Daily.* (Green Net, n.d.).

14 Announcement of Tianxia Xi's mailing list, 4 November 2003.

15 "Insiders' Reflections on the Fight to Protect the Dujiangyan Dam"—document circulated for discussion on a private environmental mailing list.

16 Discussions at an environmental salon held in Beijing on 17 May 2004. Source: Nu River mailing list, 31 May 2004.

Afterword

SENATOR CHRISTINE MILNE, LEADER OF THE AUSTRALIAN GREENS

This is an extract of an address to the Environmental Politics and Conflict in an Age of Digital Media symposium, University of Tasmania, 17–18 November 2011

My activist life in Australia has spanned the rise of the environment movement, beginning with the formation of the world's first Green party, the United Tasmania Group, in the early 1970s to protest against the flooding of Lake Pedder. In the early 1980s, I was arrested and jailed for my involvement in the Franklin River campaign, globally renowned as a major environmental campaign. I also led the Wesley Vale campaign, protesting a major pulp mill which was to be built in the North West of Tasmania on prime agricultural land.

On the back of the Wesley Vale campaign, I moved into activist politics as an elected member. I was elected to the Tasmanian parliament as an independent, then as a Green Independent, and then we formed the Tasmanian Greens and became a political party. Following that, we formed the Australian Greens. The founding of the Australian Greens led to the establishment of the Global Greens, and we had the first Global Greens conference in Canberra in 2001. There are now Green parties in seventy countries around the world, and we are about to have our third Global Greens Conference in Senegal in 2012.

The rise of the Greens began in Tasmania. Most Tasmanians would never talk about it because it is a fact they don't choose to embed into their consciousness. But it is something that's inspiring from the point of view of Green activists and Green parliamentarians. We have learnt over the years that our connection, our communications capacity, is critical because campaigns have not been won locally. We have recognized from the 1970s until now that it is the change in the enabling technology that has allowed environmental and political activism to get to where it is, and it is our history of needing to go around mainstream media that has allowed the Greens to be the leaders in using new technology in the political process in Australia.

When I first started organizing a campaign—the Wesley Vale campaign in 1987/88—I had a pen and paper, a kitchen table and a wall phone. No fax machines were readily available. One solicitor's office and the post office in Ulverstone in the North West of Tasmania had a fax machine, and if I wanted to be very daring, I could go down there and pay them to send a fax or receive a fax as part of the campaign. The rest of the time it was handwritten press releases, telephone calls to the newsroom and dictation over the phone to whoever happened to be on duty at the time.

We were without computers, without the Internet, without email, without mobile phones. Getting information on pulp mill technology around the world meant sourcing it from a public library, and then you could borrow the report or get it somehow and read it yourself. To get information to the only politicians in Tasmania who would use that information—that is, Bob Brown and Gerry Bates who were in the Tasmanian parliament at the time as Independents—I would send it down on the bus because that was faster than posting it. I would also obtain a report and send it on the bus to the university, where there was a group of scientists who knew how to assess and express the concerns about organochlorines, particularly in the marine environment, and about dioxins.

For the Greens, social media and all new media are central to us, and not only as an add-on to traditional media. This approach comes from our experience of being ignored by the old media. We have to use social media to bypass the traditional media; for the old parties in Australia, social media tends to be only an add-on. When the old parties look at social media, they're looking at it to see what they should react to and how they should react, whereas we use it to try and drive a different news agenda. So their approach is about shutting it down and our approach is about creating a different agenda.

I agree with Jared Cohen, who was in the U.S. administration before moving to Google Ideas, who has made the point that new media doesn't create cohesive, permanent, political movements, and new technology doesn't create democratic leaders and democratic institutions: what new media does do is enable them to emerge and solidify.

With new media, if you have a plan you can end up achieving a goal. Without a plan, a movement is easy to start but it is very difficult to finish because you have no idea where it is going to end up—the uprising in the Middle East (Arab Spring) and the Occupy Movement have demonstrated this. What new technology has done is enable people who want to organize to take action, so it has created a non-cohesive, impermanent, political movement. But it will be an opportunity lost if someone does not get in there and work out what, for example, the Occupy Movement wants to achieve globally, and where the plan is to take the activism into governments, into parliaments to change the institutional structure that will enable the changes that they are campaigning for.

The Greens see that we have an enabling tool in new media, but we also have a plan and organization to create a cohesive, permanent, political movement over time, which will support democratic leaders in the Greens, moving to democratize existing institutions. Our challenge is to be able to transform the online activism of movements like the Occupy Movement to offline activism when it comes to elections and political engagement, going from a protest movement on a set of issues which are often unclear even to the people engaged, to political action and electoral outcomes.

In terms of the Greens, our next challenge is how we maintain enough control and discipline in a plan that uses social media to make sure we don't run into all the risks and problems inherent in the form. On the other hand, we need to let enough control go to allow the movement to become spontaneous. This is a point made in Joe Trippi's book, *The Revolution Won't Be Televised*, looking at the Howard Dean campaign, which was the first time social media was really used effectively in political activism. It is a classic case of technology not creating a democratic leader. The technology turned Howard Dean into something he was not. He wasn't up to it and he recognized he wasn't up to it, and the campaign collapsed. New media can create a person in terms of a media profile, but if the person isn't a democratic leader then the whole thing will fail because they will not have the personal capacity to continue on. That is where social media assisted Barack Obama, because he does have the personal characteristics to be the leader that social media helped to make him.

The Greens currently use social media in several important ways. The days of sending reports on buses and phone trees have largely gone. What the Greens have instead is the ability to use the Internet, a party database, email, Twitter and Facebook contacts to mobilize locally and globally. For example, we are better able to use things like webinars. We had an Australian Greens meeting with Tim Jackson from the UK coming in to give a paper on new economics; we can access the intelligensia from around the world, if you like, without having to fly them here. We are getting better at organization, but the capacity of the Global Greens is constrained by the technological capacity of the membership in countries which haven't yet got the same technological outreach.

Through social media we are also able to quickly correct misinformation. This is a critical component because we have suffered with negative press every last week of every election campaign in Australia. The negative blitz will come three or four days out from the poll, and it has taken between 2 and 5 percent off the primary vote of the Greens at every election. Using social media now gives us the opportunity to correct the misinformation in television ads. In 2004, when I first stood for the Senate, a far-right party ran an advertisement that had a woman at the clothesline, a scenario of domestic bliss with kids playing in the background, stating "they're coming to give our children drugs" and at the end, "that's not Green, that's extreme." This advertisement was in the national media across Australia at the same time as the *Herald Sun* in Melbourne ran three front-page stories on all the "terrifying" policies of the Greens, and it took several percentage points off our vote. Now we have the opportunity to get out there straight away using all forms of social media to correct any misinformation, and we attempt to direct people back to our websites to have a look at what our policies actually are.

The Greens also use social media at critical times during a campaign. Midway through 2011, the Government was negotiating carbon prices, and vested interests were ramping up negatively in the mainstream media, running stories every day about how the sky is going to fall in if we get the carbon price, jobs will go, industries will close, will shut down. We ran an appeal through our party database saying, "we need some money to run a TV campaign to counter all of this," and people contributed straight away. We were able to run TV ads to counter the fact that we could see that the pro-climate action was losing and we were able to use the social media to mobilize donations.

In terms of Facebook, we are trying to build an online community that is supportive of one another, an exchange place for ideas and so on. One of the ways we try and use Twitter is not only to bypass mainstream media in terms of messaging, but to influence the topic of the day's discussion, because the press gallery feeds off and into the Twitter conversation on daily issues: Twitter enables us to monitor the events and discussions of the day, address misinformation before it gains momentum and rebuff incorrect assertions. We can also mobilize online activists to initiate action through Twitter and Facebook, asking people to write letters to the editor, comment on any high-profile news site that has some story up there, either pro- or anti- or whatever we want to use it for. Using social media provides a quick way of addressing specific queries people might have, and it enables us to make public statements that do not require a media release. It is also useful to humanize politicians because politicians find it difficult to be humanized in the media, especially when you are part of an issues-based party. The Greens do not appear on morning chat shows, on cooking shows, or get a media profile by singing and dancing like Kevin Rudd and Joe Hockey. One example of how we use Twitter was after Barack Obama's speech in the House of Representatives. Obama shook hands with everyone, and when he was introduced to the Greens, he said "keep up the good work." We thought that was an ideal thing to put out on Twitter—it was quick, easy and didn't warrant a media statement, but it was a good way of getting the message out there.

Social media enable us to share other content from across the web with our supporters. We can link to important articles that do not get into the mainstream media and spread them around. Social media also allow us to engage during formal sittings of the Parliament. For example, Greens senator Scott Ludlam uses Twitter during the Senate estimates process, where he'll put out a tweet to say, "this government department is about to take a seat and come before the senate estimates, do you have any particular questions," or "send me the question now because they're about to start." This is a way of engaging the community in that process. Senate committees are now streamed online, so for activist groups who might be working on a certain issue, they can sit on their computer, watch the Senate estimates process, email one of the Greens senators who is sitting at the front with a question, or when a bureaucrat gives an answer, someone can write in and say, "that is absolute rubbish, go back and ask him this" and so you engage in conversation with people who are interested in the issues.

Social media also enables the Greens to create content. Due to the cutbacks on investigative reporting and technical support for journalists in the mainstream or traditional media, stories about, for example, clear-fell logging in Tasmania are not appearing. The Greens can create their own content and put it up on YouTube and our website.

I'm not very good at Twitter and I'm only just learning how you might use it; some of our other members are much better than I am. With social media you have to have an authentic voice. I'm not a person who would put all my photos up somewhere for the world to see because it's not how I engage with my friends, so I find it quite hard to use Facebook in the way that other people might use it. My staff were at me, saying "you cannot be a 21st-century politician without an authentic social media presence. What are you going to do?" They suggested I write a blog about politics and I said "why would I want to do that, I'm in it every day, all day?" The only personal time I have that is not related to politics is my garden. That is my recreation and something I love. So I write about my garden. The blog has turned into this huge success which is extraordinary—it is all about growing food in an urban environment, and it gives me a chance to talk about food security issues, genetic engineering and all the issues pertaining to agriculture, but more particularly invasive species, attracting native birds, local habitat and the frustrations of trying to be an urban gardener. Writing about my garden built me a presence. On Twitter, at one point I mentioned that I was going to have a water feature in my garden. I also mentioned that I was growing vegetables in my garden. At the same time on Twitter, a New South Wales (NSW) MP wrote he'd been up at 4am and was going to cook some vegetables, and the mainstream media (the *Sydney Morning Herald*) wrote that they now knew where the NSW MP was getting his veges from; he was raiding Milne's vege patch in Hobart. *The Chaser* from ABC TV picked up on my Twitter conversation and ambushed me outside the ABC studios about my garden. This was during a critical time in the election campaign, and as *The Chaser* is one of the most popular TV programs during election campaigns, the upshot was that I got national TV coverage for the water feature. It was incredibly humanizing and very popular for the Greens, and I got another 150 odd followers overnight. It has built me a whole online community of people who are really concerned about all those issues of food security and food in urban gardening. It has led to a lot of speaking engagements at community fairs and farmers' markets, and has built me a whole constituency that was always onside and I was onside with, but I never had a particular engagement with. Although the blog is not a political hard sell, it does deal with issues surrounding food that people want to talk about, and it will be this community that will engage as we ratchet up campaigns on regulation and labelling of GE products. What started as my staff nagging me about having an authentic presence in social media has led to a complete, almost seamless transfer from one media form to another.

There are many downsides and risks in using social media. If you make a mistake it could go to thousands of people before you can stop it. You have to put up with trolling, people who deliberately misrepresent you, attempt to destroy you, provoke outbursts and try to discredit people. We have to put up with the influence of the third-party type component, which looks like a spontaneous level of activity but is fake and paid for. Also in the online world, opinion becomes fact, and opinion pieces are now starting to replace the news. Opinion is getting onto the front page.

Everything is measurable now with social media, and this can be useful. We know how many people clicked on a link, how many people took action, how many people gave a donation. On the other hand, it raises some big ethical questions, especially surrounding privacy and the use of new media. If you are going to be good at using social media, you have to put a lot of time into it. There are only a certain number of hours in a day to do the work you have to do, plus the media you have to do, plus the social media you have to do, and for every hour you put into social media, it's an hour you're not thinking about where you might go in a strategic sense.

One of the aims of social media is to try and get something to go viral. This is a fabulous outcome if you succeed in doing it well. I've only had one experience of trying to get something to go viral, and it was an utter nightmare. I had to stop it immediately because it made me realize that there is a whole new area that I don't know a great deal about, that I rapidly need to get across, namely intellectual property and copyright. Using my publications allowance, which is actually Australian Government money, I made an animated little cartoon on polar bears to go up on my website. All that was legal. The animated cartoon was fabulously successful. The cartoon showed two polar bears talking to one another on the ice about what causes climate change, and at the end the ice cracks, and it is obvious they are going to drown. It was taken up all over the world and the next thing I knew it was entered in the world's short film competition and it made the final 25. Suddenly, I was in this nightmare where US banks were ringing up saying, "how much would you want to charge for us to use that as part of our material?" Companies and PR firms wanted it, and I'm going "Oh, no, no, no, I don't want any of you to have it for greenwash. Only NGOs and not for profits with a genuine interest in climate change can have it." Then the company that actually made it said to me, "you actually don't own that, we own that, it's our copyright and we will decide who has it." So in the end I thought, this is a complete nightmare, I'm stopping it, it's not going anywhere. It was a real shame because it was a perfect formula for advertising on climate change. I actually got legal advice from the Government and they said, "actually you don't own it, we own it because it's Australian Government money that paid for it." Just a warning to everybody, if you're going to produce something that goes viral, make sure at the very beginning you actually have somebody to advise you carefully on who owns what and who owns the copyright.

In conclusion, the Greens are learning how to use new media well. But no matter how successful we are in using social media, it cannot replace an engagement with mainstream media, and it cannot replace democratic and institutional leadership and change through politics. It is an enabler, not an outcome in itself. Diversification of ownership and support for independent media are incredibly important, and they are just as important in new media as they have always been in terms of the traditional media.

References

Abbott, T. (2009, 27 July). A realist's approach to climate change. http://www.tonyabbott.com.au/News/tabid/94/articleType/ArticleView/articleId/7087/A-REALISTS-APPROACH-TO-CLIMATE-CHANGE.aspx (accessed 25 October 2012).

Abbotti, P., Wallace, C., & Beck, M. (2006). Chernobyl: Living with risk and uncertainty. *Health, Risk & Society*, 8(2), 105–121.

ABC. (2012a, 21 July). Pro-logging protester joins tree-sit activist. ABC News. http://www.abc.net.au/news/2012-07-20/pro-forestry-protester-joins-tree-sit-activist/4142618?section=tas (accessed 21 July 2012).

ABC. (2012b, 13 September). Forest activists' camp torched. ABC News. http://www.abc.net.au/news/2012-09-13/forest-activists27-camp-torched/4259124 (accessed 13 September 2012).

Aboud, L., & Museri, A. (2007). En caída libre: del diferendo al conflicto. In V. Palermo, & C. Reboratti (Eds.), *Del otro lado del río* (pp. 15–56). Buenos Aires: Editorial Edhasa.

Abrahamsson, K. V. (1999). Landscapes lost and gained: On changes in semiotic resources. *Human Ecology Review*, 6(2), 51–61.

Acevedo Rojas, J. (2010). Comunicación y conflictos socioambientales en el Perú: Radios educativas y comunitarias en la encrucijada. *Dialogos de la Comunicación*, 81.

Activists shut down Gunns Ltd Triabunna woodchip mill [Video file]. (2009, 25 February). http://www.youtube.com/watch?v=_xk8iYWjhvw (accessed 2 January 2012).

Adam, G. S. (1993). *Notes towards a definition of journalism: Understanding an old craft as an art form*. Poynter Papers 2. St. Petersburg: The Poynter Institute for Media Studies.

Adams, W. M. (2004). *Against extinction: The story of conservation*. London: Earthscan.

A. G. (Arthur Groom). (1929, May 11). A night in the gorge. *Brisbane Courier*, p. 27.

Agardy, T., Bridgewater, P., Crosby, M. P., Day, J., Dayton, P. K., Kenchington, R., Laffoley, D., McConney, P., Murray, P. A., Parks, J. E., & Peau, L. (2003). Dangerous targets? Unresolved issues and ideological clashes around marine protected areas. *Aquatic Conservation: Marine and Freshwater Ecosystems*, 13(4), 353–367.

Agence France-Presse. (2011, 14 March). Fewer Americans worry about climate change: Poll. http://www.physorg.com/news/2011-03-americans-climate-poll.html (accessed 28 December 2011).

Akhavan-Majid, R. (2004). Mass media reform in China: Toward a new analytical framework. *Gazette*, 66(6), 553–565.

Alexander, J. (2006a). Cultural pragmatics: Social performance between ritual and strategy. In J. Alexander, B. Giesen, & J. Mast (Eds.), *Social performance: Symbolic action, cultural pragmatics, and ritual* (pp. 29–90). Cambridge, UK: Cambridge University Press.

Alexander, J. (2006b). *The civil sphere*. Oxford: Oxford University Press.

Allan, S. (2002). *Media, risk and science*. Buckingham: Open University Press.

Allan, S., Adam, B., & Carter, C. (Eds.). (2000). *Environmental risks and the media*. London: Routledge.

Allan, S., & Thorsen, E. (Eds.). (2009). *Citizen journalism: Global perspectives*. London: Peter Lang.

Allen, D. E. (1994). *The naturalist in Britain: A social history* (2nd ed.). Princeton: Princeton University Press.

Allen, M. (2010). The experience of connectivity: Results from a survey of Australian Internet users. *Information, Communication & Society*, 13(3), 350–374.

Alliance of Small Island States. (2012). http://aosis.info/ (accessed 14 March 2012).

Amazing spotted owl mouse grab [Video file]. (2007, 10 May). http://www.youtube.com/watch?v=jqdAM95pBmU (accessed 2 January 2012).

Anderegg, W. R. L., Prall, J. W., Harold, J., & Schneider, S. H. (2010, 21 June). Expert credibility in climate change. *PNAS*. Epub ahead of print. doi: 10.1073/pnas.1003187107.

Anderson, A. (1997). *Media, culture and the environment*. London: UCL Press.

Anderson, A. (2003). Environmental activism and news media. In S. Cottle (Ed.), *News, public relations and power* (pp. 117–132). London: Sage.

Anderson, A. (2009). Media, politics and climate change: Towards a new research agenda. *Sociology Compass*, 3(2), 166–182.

Anderson, J. (2004). The ties that bind? Self- and place-identity in environmental direct action. *Ethics, Place & Environment*, 7(1/2), 45–57.

Andrews, J. C. (2011). Does the Fukushima accident significantly increase the nuclear footprint? *Bulletin of the Atomic Scientists*, 24(6), 36–39.

Antilla, L. (2005). Climate of scepticism: US newspaper coverage of the science of climate change. *Global Environmental Change*, 15(4), 338–352.

APN Australian Regional Media. (2012). QT—The Queensland Times. http://apnarm.com.au/newspapers/daily/4640.html (accessed 27 January 2012).

Arceneaux, N., & Schmitz Weiss, A. (2010). Seems stupid until you try it: Press coverage of Twitter, 2006–9. *New Media & Society*, 12(8), 1262–1279.

Arendt, H. (1983). *Men in dark times*. New York: Harcourt Brace & Company.

Attenborough, D. (2002, 19 December). Attenborough defends his views of life on earth. *The Guardian*. http://www.guardian.co.uk/theguardian/2002/dec/19/guardianletters2 (accessed 30 October 2012).

Attenborough, D. (2012, 3 February). Unpublished interview with Morgan Richards.

Austin, E. (1999). All expenses paid: Exploring the ethical swamp of travel writing. *Washington Monthly*, 31(7). http://www.washingtonmonthly.com/features/1999/9907.austin.expenses.html (accessed 26 January 2011).

Australian Academy of Science. (2010). *The science of climate change: Questions and answers.* Canberra: Australian Academy of Science.

Australian and New Zealand Environment and Conservation Council Task Force on Marine Protected Areas. (1998). *Guidelines for establishing the national representative system of marine protected areas.* Canberra: Environment Australia.

Australian Electoral Commission. (2010, 30 July). Election 2010: 1198 candidates to contest the 2010 federal election. http://www.aec.gov.au/about_aec/media_releases/e2010/30-07.htm (accessed 18 January 2012).

Australian Greens. (2010, 24 July). Greens lead the way in social media. http://www.greens.org.au/content/greens-lead-way-social-media (accessed 18 January 2012).

Australian Museum. (2011). 2011 winner: Science photography: Receding glacial cap with cryoalgae, Jason Edwards. http://www.eureka.australianmuseum.net.au/552B5840-7783-11E0-A87E005056B06558?DISPLAYENTRY=true (accessed 27 August 2011).

Australian, The. (2009a, 11 December). Met pushes climate unity. *The Australian,* p. 4.

Australian, The. (2009b, 11 December). Why can't we have an ETS just like Europe? *The Australian,* p. 13.

Australian, The. (2009c, 12 December). When world views collide. *The Australian,* p. 15.

Australian, The. (2009d, 17 December). The parallel universe with a life of its own. *The Australian,* p. 13.

Australian, The. (2009e, 19 December). How blue is your reef? *The Australian,* p. 13.

Bacon, W., & Nash, C. J. (2003). How the Australian media cover humanitarian issues. *Australian Journalism Review,* 25(2), 5–30.

Bacon, W., & Nash, C. (2012). Playing the media game. *Journalism Studies,* 13(2), 243–258.

Bailey, S. (2002, November 23). Architect of anti-forestry site hits out as identity revealed. *The Mercury,* p. 3.

Banks, S. A., & Skilleter, G. A. (2010). Implementing marine reserve networks: A comparison of approaches in New South Wales (Australia) and New Zealand. *Marine Policy,* 34, 197–207.

Bao, H., & Bing, Liu. (2006). Zhongsheng xuanhua: "Jingwei ziran" da taolun (The big debate regarding "respect or fear nature"). In L. Congjie (Ed.), *Huanjing lüpishu 2005 nian: Zhongguo de huanjing weiju yu tuwei* (Green book of environment 2005: Crisis and breakthrough of China's environment) (pp.119–130). Beijing: Shehui kexue wenxian chubanshe.

Baren, J. v., IJsselsteijn, W., Romero, N., Markopoulos, P., & de Ruyter, B. (2003, 6–8 October). Affective benefits in communication: The development and field-testing of a new questionnaire measure. Paper presented at PRESENCE 2003, Aalborg, Denmark. http://www.ijsselsteijn.nl/papers/p2003.pdf (accessed 12 November 2012).

Barmé, G. R., and Davies, G. (2004). Have we been noticed yet? Intellectual contestation and the Chinese web. In E. Gu & M. Goldman (Eds.), *Chinese intellectuals between state and market* (pp. 75–108). London and New York: RoutledgeCurzon.

Barnett, J. (2001). Adapting to climate change in Pacific island countries: The problem of uncertainty. *World Development,* 29, 977–993.

Barnett, J., & Campbell, J. (2010). *Climate change and small island states: Power, knowledge and the South Pacific.* London: Earthscan.

Barnett, L. (2007). Psychosocial effects of the Chernobyl nuclear disaster. *Medicine, Conflict and Survival,* 23(1), 46–57.

Barrett , N. S., Edgar, G. J., Buxton, C. D., & Haddon, M. (2007). Changes in fish assemblages following ten years of protection in Tasmanian marine protected areas. *Journal of Experimental Marine Biology and Ecology,* 345(2), 141–157.

Barthes, R. (1977). *Image, music, text.* London: Fontana.

Bauman, Z. (1987). *Legislators and interpreters.* Cambridge, UK: Polity Press.

Bay, I., & Oughton, D. H. (2005). Social and economic effects. In J. Smith & N. A. Beresford (Eds.), *Chernobyl, catastrophe and consequences* (pp. 239–262). Berlin: Springer-Verlag.

BBC Online. (2012, 27 January). Japan did not keep records of nuclear disaster meetings. *BBC Online.* http://www.bbc.co.uk/news/world-asia-16754891 (accessed 20 April 2012).

Bebbington, A. (2009). Latin America: Contesting extraction, producing geographies. *Singapore Journal of Tropical Geography,* 30(1), 7–12.

Bebbington, A., Hinojosa, L., Bebbington, D. H., Burneo, M. L., & Warnaars, X. (2008). Contention and ambiguity: Mining and the possibilities of development. *Development and Change,* 39(6), 887–914.

Beck, U. (1992). *Risk society: Towards a new modernity.* London: Sage Publications.

Beck, U. (1999). *World risk society.* Cambridge, UK: Polity Press.

Beck, U. (2006). *Cosmopolitan vision.* Cambridge, UK: Polity Press.

Beck, U. (2008). World at risk: The new task of critical theory. *Development and Society,* 37(1), 1–21.

Beck, U. (2009). *World at risk.* Cambridge, UK: Polity Press.

Becker, H. (1967). Whose side are we on? *Social Problems,* 14, 239–47.

Belyakov, O. V. (1999). Notes about the Chernobyl liquidators. http://www.ossh.com/chernobyl/www.progettohumus.it/include/chernobyl/nodimentica/liquidatori/docs/13%20.pdf (accessed 20 February 2013).

Benjamin, W. (1973). The work of art in the age of mechanical reproduction. In *Illuminations* (pp. 219–253). London: Fontana.

Bennett, J. (2010). *Vibrant matter: A political ecology of things.* Durham: Duke University Press.

Bennett, L. (1990). Toward a theory of press-state relations in the United States. *Journal of Communication,* 40(2), 103–127.

Bennett, L. (2003). New media power: The Internet and global activism. In N. Couldry & J. Curran (Eds.), *Contesting media power: Alternative media in a networked world* (pp. 17–38). Oxford: Rowman and Littlefield.

Bensen, J. (1998). *Souvenirs from high places: A history of mountaineering photography.* London: Reed Books.

Bentley, J. G. (2003). The role of international support for civil society organizations in China. *Harvard Asia Quarterly,* 7(1), 11–20.

Berglez, P. (2008). What is global journalism? Theoretical and empirical conceptualisations. *Journalism Studies,* 9(6), 845–858.

Berlant, L. (1997). *The queen of America goes to Washington City.* Durham: Duke University Press.

Berlant, L. (2008). *The female complaint: The unfinished business of sentimentality in American culture.* Durham: Duke University Press.

Berlowitz, V. (2012, 15 February). Unpublished interview with Morgan Richards.

Bertell, R. (2008). Chernobyl: An unbelievable failure to help. *International Journal of Health Services*, 38(3), 543–560.

Biggar, J. (2010). Crowdsourcing for the environment: The case of brighter planet. *PLATFORM: Journal of Media and Communication*, 2(2), 8–23.

Bloxham, A. (2011, 15 November). BBC drops Frozen Planet's climate change episode to sell show better abroad. *The Telegraph*. http://www.telegraph.co.uk/earth/earthnews/8889541/BBC-drops-Frozen-Planets-climate-change-episode-to-sell-show-better-abroad.html (accessed 28 October 2012).

Boelens, R., de Mesquita, M. B., Gaybo, A., & Pena, F. (2011). Threats to a sustainable future: Water accumulation and conflict in Latin America. *Sustainable Development Law & Policy*, 12(1), 41–45, 67–69.

Bogard, W. (1989). *The Bhopal tragedy: Language, logic, and politics in the production of a hazard.* Boulder, San Francisco: Westview Press.

Bolter, J. D., & Grusin, R. (1999). *Remediation: Understanding new media.* Cambridge, MA: MIT Press.

Bonyhady, T. (2000). *The colonial earth.* Melbourne: Miegunyah Press.

Bonyhady, T., & Griffiths, T. (2002). *Words for country: Landscape and language in Australia.* Sydney: UNSW Press.

Boorstin, D. J. (1992). *The image: A guide to pseudo-events in America.* New York: Vintage Books, Random House. (Original work published 1961).

Bormann, E. G. (1985). Symbolic convergence theory: A communication formulation. *Journal of Communication*, 35(4), 128–138.

Bos, S., & Lam, S. (2009). *Free as air and water.* New York: The School of Art at the Cooper Union.

Bourdieu, P., & Wacquant, L. (1992). *An invitation to reflexive sociology.* Chicago: University of Chicago Press.

Bousé, D. (2000). *Wildlife films.* Philadelphia: University of Pennsylvania Press.

Boyce, T., & Lewis, J. (Eds.). (2009). *Climate change and the media.* New York: Peter Lang.

Boykoff, M. T. (2012, January). Media coverage of climate change/global warming. http://science-policy.colorado.edu/media_coverage/us (accessed 17 February 2012).

Boykoff, M. T., & Boykoff, J. M. (2004). Balance as bias: Global warming and the US prestige press. *Global Environmental Change*, 14(2), 125–136.

Boykoff, M. T., & Boykoff, J. M. (2007). Climate change and journalistic norms: A case study of US mass-media coverage. *Geoforum*, 38, 1190–1204.

Boykoff, M. T., & Goodman, M. K. (2009). Conspicuous redemption? Reflections on the promises and perils of the "celebritization" of climate change. *Geoforum*, 40, 395–406.

Boykoff, M. T., & Timmons, R. (2007). *Media coverage of climate change: Current trends, strengths, weaknesses.* Human Development Report, Background Paper, United Nations Development Program.

Braasch, G. (2007). *Earth under fire: How global warming is changing the world.* Los Angeles: University of California Press.

Bratt Paulston, C., Kiesling, S. F., & Rangel, E. S. (Eds.). (2012). *The handbook of intercultural discourse and communication*. Hoboken: John Wiley & Sons.

Brebbia, C. A., Conti, M. E., & Tiezzi, E. (2007). *Management of natural resources, sustainable development and ecological hazards*. Southampton: WIT Press.

Brisbane Courier. (1921, 5 March). The national park: Tourist traffic increasing. *Brisbane Courier*, p. 12.

Brisbane Courier. (1929, 31 October). National parks: League formed. *Brisbane Courier*, p. 15.

Brockington, D. (2002). Fortress conservation: The preservation of the Mkomazi Game Reserve, Tanzania. Oxford: James Currey.

Brockington, D. (2008). Powerful environmentalisms: Conservation, celebrity and capitalism. *Media, Culture & Society*, 30(4), 551–568.

Brockington, D. (2009). *Celebrity and the environment: Fame, wealth and power in conservation*. London: Zed Books.

Brockington, D. (2011, 8 November). Analysis: Charities need to rethink celebrity. *Third Sector*. http://www.thirdsector.co.uk/news/1102612/analysis-charities-need-rethink-celebrity/ (accessed 26 October 2012).

Brockington, D., & Duffy, R. (2010). Capitalism and conservation: The production and reproduction of biodiversity conservation. *Antipode*, 42(3), 469–484.

Brockington, D., & Duffy, R. (2011). *Capitalism and conservation*. London: Wiley.

Brockington, D., Duffy, R., & Igoe, J. (2008). *Nature unbound: Conservation, capitalism and the future of protected areas*. London: Earthscan.

Brockington, D., & Scholfield, K. (2010). The conservationist mode of production and conservation NGOs in sub-Saharan Africa. *Antipode*, 42(3), 551–575.

Brown, B., & Singer, P. (1996). *The Greens*. Melbourne: The Text Publishing Company.

Brown, D., Lemons, J., & Tuana, N. (2006). The importance of expressly integrating ethical analyses into climate change policy formation. *Climate Policy*, 5, 549–552.

Brown, G., & Pickerill, J. (2009). Space for emotion in the spaces of activism. *Emotion, Space and Society*, 2(1), 24–35.

Brulle, R. J., Carmichael, J., & Jenkins, J. (2012). Shifting public opinion on climate change: An empirical assessment of factors influencing concern over climate change in the US, 2002–2010. *Climatic Change*, 114(2), 169–188.

Brumfiel, G., & Fuyuno, I. (2012, 8 March). Japan's nuclear crisis: Fukushima's legacy of fear. Nature 483(7388): 138-140.

Bruns, A. (2011). Towards distributed citizen participation: Lessons from Wikileaks and the Queensland floods. In P. Parycek, M. J. Kripp, & N. Edelmann (Eds.), *CeDEM11: Conference for E-Democracy and Open Government* (pp. 35–52). Krems: Edition Donau-Universität Krems.

Buchanan, M. (2003). *Small world: Uncovering nature's hidden networks*. London: Phoenix Press.

Buckingham, D. (2000). *The making of citizens: Young people, news, politics*. London: Routledge.

Burdick, J., Oxhorn, P., & Roberts, K. M. (Eds.). (2009). *Beyond neoliberalism in Latin America: Societies and politics at the crossroads*. New York: Palgrave Macmillan.

Burgess, H. J., & Green, J. (2009). *YouTube: Online video and participatory culture*. Malden: Polity Press.

Burgess, J., & Unwin, D. (1984). Exploring the Living Planet with David Attenborough. *Journal of Geography in Higher Education*, 8(2), 93–113.

Burke, K. (1965). *Permanence and change*. Indianapolis: The Bobbs-Merrill Company.

Burke, K. (1969). *A rhetoric of motives*. Berkeley: University of California Press.

Büscher, B., Sullivan, S., Neves, K., Igoe, J., & Brockington, D. (2012). Towards a synthesized critique of neoliberal biodiversity conservation. *Capitalism, Nature, Socialism*, 23(2), 4–30.

Calhoun, C. (1989). Tiananmen, television and the public sphere: Internationalization of culture and the Beijing Spring of 1989. *Public Culture*, 2(1), 54–71.

Calhoun, C. (1995). "New social movements" of the early nineteenth century. In M. Traugott (Ed.), *Repertoires and cycles of collective action* (pp. 173–215). Durham: Duke University Press.

Calhoun, C. (2004). Information technology and the international public sphere. In D. Schuler & P. Day (Eds.), *Shaping the network society: The new role of civil society in cyberspace* (pp. 229–51). Cambridge, MA: MIT Press.

Cameron, F. (2011). Saving the "disappearing islands": Climate change governance, Pacific island states and cosmopolitan dispositions. *Continuum*, 25(6), 873–886.

Cammaerts, B. (2012). Protest logics and the mediation opportunity structure. *European Journal of Communication*, 27(2), 117–134.

Carneiro, G. (2011). The Luiz Saldanha Marine Park: An overview of conflicting perceptions. *Conservation and Society*, 9, 325–333.

Carrier, J., & West, P. (2009). Introduction. In J. Carrier & P. West (Eds.), *Virtualism, governance and practice: Vision and execution in environmental conservation* (pp. 1–23). New York: Berghahn.

Carrier, J. G. (1998). Introduction. In J. G. Carrier & D. Miller (Eds.), *Virtualism: A new political economy* (pp. 1–24). Oxford: Berg.

Carruthers, D. (Ed.). (2008). *Environmental justice in Latin America: Problems, promise, and practice*. Cambridge, MA: MIT Press.

Carruthers, D., & Rodriguez, P. (2009). Mapuche protest, environmental conflict and social movement linkage in Chile. *Third World Quarterly*, 30(4), 743–760.

Carvalho, A. (2005). "Governmentality" of climate change and the public sphere. In M. E. Rodrigues & H. Machado (Eds.), *Scientific proofs and international justice: The future for scientific standards in global environmental protection and international trade* (pp. 1–24). Braga: Nucleo de Estudos em Sociologia, Universidade do Minho.

Carvalho, A. (2007). Ideological cultures and media discourse on scientific knowledge: Re-reading news on climate change. *Public Understanding of Science*, 16, 223–243.

Castells, M. (1996). *The rise of the network society*. Oxford: Blackwell.

Castells, M. (Ed.) (2004). *The network society: A cross-cultural perspective*. Cheltenham: Edward Elgar.

Castells, M. (2007). Communication, power and counter-power in the network society. *International Journal of Communication*, 1, 238–266.

Castells, M. (2009). *Communication power*. Oxford: Oxford University Press.

Castree, N. (2007a). Neoliberalizing nature: Processes, effects and evaluations. *Environment and Planning*, 40(1), 153–173.

Castree, N. (2007b). Neoliberalizing nature: The logics of deregulation and reregulation. *Environment and Planning*, 40(1), 131–152.

Chakrabarty, D. (2009). The climate of history: Four theses. *Critical Inquiry*, 35, 197–222.

Chambers, K., & Chambers, A. (2001). *Unity of heart: Culture and change in a Polynesian society.* Prospect Heights: Waveland Press.

Chandler, L., & Baldwin, C. (2010). Reflections from the water's edge: Collaborative photographic narratives addressing climate change. *Social Alternatives*, 29(3), 30–36.

Chang, H. S. (2011). The implications of Fukushima: The South Korean perspective. *Bulletin of the Atomic Scientists*, 67(4), 18–22.

Chapman, G., Keval, K., Fraser, C., & Gaber, I. (1997). *Environmentalism and the mass media: The north–south divide.* New York: Routledge.

Charlton, A. (2011). Man-made world: Choosing between progress and planet. *Quarterly Essay*, 44, 1–72.

Chen, P. (2010). Adoption and use of digital media in election campaigns: Australia, Canada and New Zealand. *Public Communication Review*, 1(1), 3–26.

Chess, C., Salomone, K., & Hance, B. (1995). Improving risk communication in government: Research priorities. *Risk Analysis*, 15, 127–135.

Chris, C. (2006). *Watching wildlife.* Minneapolis: University of Minnesota Press.

Chubb, P., & Bacon, W. (2010). Australia: Fiery politics and extreme events. In E. Eide, R. Kunelius, & V. Kumpu (Eds.), *Global climate—local journalisms: A transnational study of how media make sense of climate summits* (pp. 51–66). Global Journalism Research Series vol. 3. Freiburg: Projekt Verlag.

Clark, J., & Slyke, T. V. (2011). How journalists must operate in a new networked media environment. In R. W. McChesney & V. Pickard (Eds.), *Will the last reporter please turn out the lights: The collapse of journalism and what can be done to fix it* (pp. 238–248). New York: New Press.

ClickGreen. (2011, 28 January). Green-fatigue grows as number of climate change sceptics doubles. http://www.clickgreen.org.uk/news/national-news/121834-green-fatigue-grows-as-number-of-climate-change-sceptics-doubles.html (accessed 13 January 2012).

CNN World. (2010, 27 January). Americans cooling on climate change, survey says. *CNN World*. http://articles.cnn.com/2010-01-27/world/climate.report.america.trust_1_climate-change-climate-skeptics-climate-leaders?_s=PM:WORLD (accessed 28 December 2011).

Coghlan, A. (2011, 30 September). Fukushima's radioactive sea contamination lingers. *New Scientist*. http://www.newscientist.com/article/dn20990-fukushimas-radioactive-sea-contamination-lingers.html (accessed 28 October 2012).

Cohen, J., & Agiesta, J. (2009, 18 December). On environment, Obama and scientists take hit in poll. *The Washington Post*. http://www.washingtonpost.com/wp-dyn/content/article/2009/12/18/AR2009121800002.html (accessed 17 December 2011).

Cole, N., & Watrous, S. (2007). Across the great divide: Supporting scientists as effective messengers in the public sphere. In S. C. Moser & L. Dilling (Eds.), *Creating a climate for change: Communicating climate change and facilitating social change* (pp. 180–199). Cambridge, UK: Cambridge University Press.

Come to Camp Weld [Video file]. (2006, 30 October). http://www.youtube.com/watch?v=oLGXf4UwAoc (accessed 2 January 2012).

Cong hongda maque dao shouhu dayan (From chasing and killing sparrows to guarding wild geese). (1999, 7 September). *Beijing Youth Daily.* http://www.bjyouth.com.cn/Bqb/19990907/GB/3998^0907B09918.htm (accessed 1 September 2004). This webpage is no longer available. Printed hard copy of webpage available from author.

Connell, J. (1993). Climatic change: A new security challenge for the atoll states of the southwest Pacific. *Journal of Commonwealth and Comparative Politics,* 31, 173–192.

Connell, J. (2003). Losing ground? Tuvalu, the greenhouse effect and the garbage can. *Asia Pacific Viewpoint,* 44(2), 89–107

Connell, J., & Lea, J. (1992). My country will not be there: Global warming, development and the planning response in small island states. *Cities,* 9, 295–309.

Connor, J., & Stefanova, K. (2012). Climate of the nation 2012: Australian attitudes on climate change. http://www.climateinstitute.org.au/verve/_resources/TheClimateOfTheNation2012_Final.pdf (accessed 26 October 2012).

Coombs, W. T. (2006). The protective powers of crisis response strategies: Managing reputational assets during a crisis. *Journal of Promotion Management,* 12(3), 241–260.

Cooper, A. F. (2008). *Celebrity diplomacy.* Boulder: Paradigm.

Cooper, M. (2011). The implications of Fukushima: The US perspective. *Bulletin of the Atomic Scientists,* 67(4), 8–13.

Coorey, P. (2011, May 27). Liberals split by brawling. *The Sydney Morning Herald.* http://www.smh.com.au/national/liberals-split-by-brawling-20110526-1f6hr.html (accessed 28 October 2012).

Corner, J., Richardson, K., & Fenton, N. (1990). *Nuclear reactions: Form and response in public issue television.* London: John Libbey.

Corson, C. (2010). Shifting environmental governance in a neoliberal world: US AID for conservation. *Antipode,* 42(3), 576–602.

Cottle, S. (2000a). Rethinking news access. *Journalism Studies,* 1, 427–448.

Cottle, S. (2000b). TV news, lay voices and the visualisation of environmental risks. In S. Allan, B. Adams, & C. Carter (Eds.), *Environmental risks and the media* (pp. 29–44). London: Routledge.

Cottle, S. (2004). Producing nature(s): On the changing production ecology of natural history TV. *Media, Culture & Society,* 26(1), 81–101.

Cottle, S. (2006). *Mediatized conflict: Developments in media and conflict studies.* Maidenhead: Open University Press.

Cottle, S. (2009a). Global crisis in the news: Staging new wars, disasters, and climate change. *International Journal of Communication,* 3, 494–516. http://ijoc.org/ojs/index.php/ijoc/article/view/473 (accessed 18 November 2012).

Cottle, S. (2009b). *Global crisis reporting: Journalism in the global age.* Maidenhead: Open University Press.

Cottle, S. (2011a). Cell phones, camels and the global call for democracy. In J. Mair & R. Keeble (Eds.), *Mirage in the desert? Reporting the Arab Spring* (pp. 196–210). Bury St Edmunds: Arima Publishing.

Cottle, S. (2011b). Media and the Arab uprisings 2011: Research notes. *Journalism: Theory, Practice & Criticism*, 12(5), 647–659.

Cottle, S. (2011c). Taking global crises in the news seriously: Notes from the dark side of globalization. *Global Media and Communication*, 7(2), 77–95.

Cottle, S., & Lester, L. (Eds.). (2011). *Transnational protest and the media*. New York: Peter Lang.

Cottle, S., & Rai, M. (2006). Between display and deliberation: Analyzing TV news as communicative architecture. *Media, Culture & Society*, 28(2), 163–189.

Couldry, N. (2000). *The place of media power: Pilgrims and witnesses of the media age*. London: Routledge.

Couldry, N. (2010). *Why voice matters: Culture and politics after neoliberalism*. London: Sage Publications.

Couldry, N. (2012). *Media, society, world: Social theory and digital media practice*. Cambridge, UK: Polity Press.

Cox, R. (2010). *Environmental communication and the public sphere*. 2nd ed. Thousand Oaks: Sage Publications.

Crikey. (2004, March 4). Mad Monk vs the porn-pusher. *Crikey*. http://www.crikey.com.au/2004/03/04/mad-monk-vs-the-porn-pusher/ (accessed 21 June 2010).

Crouch, D., & Damjanov, K. (2011). Piracy up-linked: Sea Shepherd and the spectacle of protest on the high seas. In S. Cottle & L. Lester (Eds.), *Transnational protests and the media* (pp. 185–196). New York: Peter Lang.

Cubitt, S. (2005). *Ecomedia*. Amsterdam: Rodopi.

Curran, J. (2002). *Media and power*. London: Routledge.

Curran, J., Fenton, N., & Freedman, D. (2012). *Misunderstanding the Internet*. London: Routledge.

Cushman, J. H. (1998, 26 April). Industrial group plans to battle climate treaty. *The New York Times*, A1.

Cvetkovich, A. (2003). *An archive of feeling: Trauma, sexuality and lesbian public cultures*. Durham: Duke University Press.

Cvetkovich, A., & Pellegrini, A. (2003). Introduction. *The Scholar and Feminist Online*, 2(1), 1–12.

Cyranoski, D. (2012, June 7). Japan considers nuclear-free future. *Nature*, 486(7401), 13.

Dahlberg, L., and Siapera, E. (2007). *Radical democracy and the Internet*. Basingstoke: Palgrave.

Dahlgren, P. (2009). *Media and political engagement: Citizens, communication and democracy*. Cambridge, UK: Cambridge University Press.

Daley, J. (2010, 7 January). Why the decline and rebirth of environmental journalism matters. http://www.yaleclimatemediaforum.org/2010/01/why-decline-rebirth-of-environmental-journalism-matters (accessed 12 July 2010).

Darley, A. (2004). Simulating natural history: Walking with dinosaurs as hyper-real edutainment. *Science as Culture*, 12(2), 227–256.

Davis, A. (2003). Public relations and news sources. In S. Cottle (Ed.), *News, public relations and power* (pp. 27–42). London: Sage Publications.

Dean, J. (2010). *Blog theory*. London: Polity Press.

Debord, G. (1995). *Society of the spectacle*. New York: Zone Books. (Original work published 1967).

Delingpole, J. (2009, 29 November). Climategate: How the "greatest scientific scandal of our generation" got its name. *The Telegraph*. http://blogs.telegraph.co.uk/news/jamesdelingpole/100018246/climategate-how-the-greatest-scientific-scandal-of-our-generation-got-its-name/ (accessed 3 January 2010).

della Porta, D., & Tarrow, S. (Eds.). (2005). *Transnational protest and global activism*. Oxford: Rowman and Littlefield.

DeLuca, K. M. (1999). *Image politics: The new rhetoric of environmental activism*. London: Guilford Press.

DeLuca, K. M., & Demo, A. T. (2000). Imaging nature: Watkins, Yosemite and the birth of environmentalism. *Critical Studies in Media Communication*, 17(3), 241–250.

Department for Environment, Food and Rural Affairs. (2011a, 17 February). The future of forestry in England. http://www.defra.gov.uk/news/2011/02/17/futureforestry (accessed 28 January 2012).

Department for Environment, Food and Rural Affairs. (2011b, 27 January). The future of the public forest estate (PFE) in England: A public consultation. http://www.forestry.gov.uk/fr/INFD-8HUB8N (accessed 29 January 2012).

Department of Economic Development, Tourism and the Arts. (n.d.). *Economic development plan. Goal two: Sector development*. http://www.development.tas.gov.au/economic/economic_development_plan/achieving_our_vision/goal_two (accessed 11 March 2012).

Department of Premier and Cabinet. (2006). *Tasmanian Brand Guide*. Hobart: Department of Premier and Cabinet.

Department of Sustainability, Environment, Water, Population and Communities. (2011). Commonwealth Marine Reserves. http://www.environment.gov.au/marinereserves/index.html (accessed 26 October 2012).

Deuze, M. (2003). The Web and its journalisms: Considering the consequences of different types of news media online. *New Media & Society*, 5(2), 203–226.

Deuze, M. (2011). Media life. *Media, Culture & Society*, 33(1), 137–148.

Diamond, J. (2005). *Collapse: How societies choose to fail or survive*. Melbourne: Penguin.

Donaldson, S. (2011, August 11). *Study: Drudge Report one of the Web's biggest traffic drivers*. http://technology.inc.com/2011/08/11/study-drudge-report-one-of-the-webs-biggest-traffic-drivers/ (accessed 26 October 2012).

Donner, S. D. (2007). Domain of the gods: An editorial essay. *Climatic Change*, 85, 231–236.

Doran, P. D., & Zimmerman, M. K. (2009). Examining the scientific consensus on climate change. *EOS*, 90(3), 21–22.

Dougherty, M. L. (2011). The global gold mining industry, junior firms, and civil society resistance in Guatemala. *Bulletin of Latin American Research*, 30(4), 403–418.

Douglas, M., & Wildavsky, A. (1982). *Risk and culture: An essay on the selection of technical and environmental dangers*. Berkeley: University of California Press.

Doyle, J. (2007a). Picturing the clima(c)tic: Greenpeace and the representational politics of climate change communication. *Science as Culture*, 16(2), 129–150.

Doyle, J. (2007b). Seeing the climate? The problematic status of visual evidence in climate change campaigning. In S. Dobrin & S. Morey (Eds.), *Ecosee: Image, rhetoric, and nature* (pp. 279–298). Albany: State University of New York Press.

Doyle, J. (2011). *Mediating climate change*. Farnham: Ashgate.

Doyle, T. (2010). Surviving the gang bang theory of nature: The environment movement during the Howard years. *Social Movement Studies, 9*, 155–169.

Draper, M. (1998). Zen and the art of garden province maintenance: The soft intimacy of hard men in the wilderness of KwaZulu-Natal, South Africa, 1952–1997. *Journal of Southern African Studies, 24*(4), 801–828.

Draper, M., & Maré, G. (2003). Going in: The garden of England's gaming zookeeper and Zululand. *Journal of Southern African Studies, 29*(2), 551–569.

Dunlap, T. R. (2004). Faith in nature: Environmentalism as religious quest. Seattle: University of Washington Press.

Dyer, R. (1979). *Stars*. London: British Film Institute.

Dyer, R. (1993). *The matter of images: Essays on representation*. London: Routledge.

Earl, J. (2010). The dynamics of protest-related diffusion on the web. *Information, Communication & Society, 13*(2), 209–225.

Earth Tribe TV.org—old growth forest [Video file]. (2006, 4 December). http://www.youtube.com/watch?v=WkPz51Ei7nA (accessed 2 January 2012).

Eaton, M. (2010). Manufacturing community in an online activist organization: The rhetoric of MoveOn.org's e-mails. *Information, Communication & Society, 13*(2), 174–192.

Economist, The. (2011, 9 July). The people formerly known as the audience: Special report. *The Economist, 400* (8741), 9–12.

Eide, E., Kunelius, R., & Kumpu, V. (Eds.). (2010). *Global climate—local journalisms: A transnational study of how media make sense of climate summits*. Global Journalism Research Series vol. 3. Freiburg: Projekt Verlag.

Ellis, R. J. (1990). A geography of vertical margins: Twentieth-century mountaineering narratives and the landscapes of neo-imperialism. Doctoral dissertation, University of Colorado.

"Endangered" Tasmania's wild places [Video file]. (2008, 27 June). http://www.youtube.com/watch?v=PDB5a_36kbQ (accessed 2 January 2012).

Entman, R. M. (1993). Framing: Toward clarification of a fractured paradigm. *Journal of Communication, 43*(4), 51–58.

Ericson, R., Baranek, P., & Chan, J. (1989). *Negotiating control: A study of news sources*. Milton Keynes: Open University Press.

Ericsson, M., & Larsson, V. (2012). E&MJ's annual survey of global mining investment. *Engineering and Mining Journal*. http://www.e-mj.com/index.php/features/1610-eamjs-annual-survey-of-global-mining-investment.html (accessed 27 October 2012).

Ethical Traveler. (2012, May). Future remains uncertain for Tasmania's old-growth forests. http://www.ethicaltraveler.org/act/future-remains-uncertain-for-tasmanias-old-growth-forests/ (accessed 27 October 2012).

Etkin, D., & Ho, E. (2007). Climate change: Perceptions and discourses of risk. *Journal of Risk Research, 10*(5), 623–641.

Evans, M. (2011). Climate change science and the politics of spin. *Advocate, 18*(2), 21. http://issuu.com/nteu/docs/advocate_18_02 (accessed 15 May 2012).

Fackler, M. (2011, 6 April). Japanese city's cry resonates around the world. *The New York Times.* http://www.nytimes.com/2011/04/07/world/asia/07plea.html?pagewanted=all (accessed 27 October 2012).

Fairclough, N. (2001). The dialectics of discourse. *Textus,* 14(2), 3–10.

Farbotko, C. (2005). Tuvalu and climate change: Constructions of environmental displacement in the *Sydney Morning Herald. Geografiska Annaler. Series B, Human Geography,* Special issue: *Islands: Objects of representation,* 87(4), 279–293.

Farbotko, C. (2010a). The global warming clock is ticking so see these places while you can: Voyeuristic tourism and model environmental citizens on Tuvalu's disappearing islands. *Singapore Journal of Tropical Geography,* 31, 224–238.

Farbotko, C. (2010b). Wishful sinking: Disappearing islands, climate refugees and cosmopolitan experimentation. *Asia Pacific Viewpoint,* 51(1), 47–60.

Farbotko, C., & McGregor, H. (2010). Copenhagen, climate science and the emotional geographies of climate change. *Australian Geographer,* 41(2), 159–166.

Fearn-Banks, K. (1996). *Crisis communications: A case book approach.* Mahwah: Lawrence Erlbaum Associates.

Ferguson, S. D. (1999). *Communication planning: An integrated approach.* Thousand Oaks: Sage Publications.

Fink, S. (1986). Crisis management: Planning for the inevitable. New York: AMACOM.

Fischhoff, B. (1995). Risk perception a communication unplugged: Twenty years of process. R*isk Analysis,* 15, 137–145.

Fiske, S. J. (1992). Sociocultural aspects of establishing marine protected areas. *Ocean & Coastal Management,* 17, 25–46.

Flam, H., & King, D. (2005). *Emotions and social movements.* London: Routledge.

Flanagan, R. (2010). The outsiders: Olegas Truchanas and Peter Dombrovskis. *Art and Australia,* 48(1), 124–127.

Flew, T. (2009). Beyond globalisation: Rethinking the scalar and the relational in global media studies. *Global Media Journal – Australian edition,* 3(1). http://www.commarts.uws.edu.au/gmjau/v3_2009_1/3vi1_terry_flew.html

Flyvbjerg, B. (2001). *Making social science matter: Why social enquiry fails and how it can succeed Again.* Cambridge, UK: Cambridge University Press.

Foley, P. (2010a, 6 July). Don't worry says minister. *The Queensland Times,* p. 7.

Foley, P. (2010b, 20 August). Preference grab labeled "gutless". *The Queensland Times,* p. 14.

Foley, P. (2010c, 20 July). This battle is not over. *The Queensland Times,* p. 1.

Forestry Tasmania (n.d.). Our vision. http://www.forestrytas.com.au/about-us/our-vision (accessed 13 March 2012).

Fothergill, A. (2012, 15 February). Unpublished interview with Morgan Richards.

Foucault, M. (1972). The discourse on language. In M. Foucault, *The archaeology of knowledge: And the discourse on language* (pp. 215–237). New York: Pantheon Books.

Foucault, M. (1991). Governmentality. In G. Burchell, C. Gordon, & P. Miller (Eds.), *The Foucault effect: Studies in governmentality* (pp. 87–104). Chicago: University of Chicago Press.

Fox Nation. (2011, 4 August). Poll: Majority thinks global warming "scientists" lie. Fox Nation. http://nation.foxnews.com/global-warming/2011/08/04/poll-majority-thinks-global-warming-scientists-lie (accessed 28 December 2011).

FoxNews.com. (2010, 11 January). 30 years of global cooling are coming, leading scientist says. FoxNews.com. http://www.foxnews.com/scitech/2010/01/11/years-global-cooling-coming-leading-scientist-says (accessed 26 October 2010).

Franklin, A. S. (2006). The humanity of wilderness photography? *Australian Humanities Review*, 38, 1–16.

Fraser, N. (1992). Rethinking the public sphere—A contribution to the critique of actually existing democracy. In C. Calhoun (Ed.), *Habermas and the Public Sphere* (pp. 109–141). Cambridge, MA: MIT Press.

Fraser, N. (2007). Transnationalizing the public sphere: On the legitimacy and efficacy of public opinion in a post-Westphalian world. *Theory, Culture and Society*, 24(4), 7–30.

Freelon, D. (2010). Analyzing online political discussion using three models of democratic communication. *New Media & Society*, 12(7), 1172–1190.

Friends of Nature. (2000). Zhongguo baozhi de huanjing yishi (Survey on environmental reporting in Chinese newspapers). Beijing: Friends of Nature.

Friends of Nature. (2003, 4 March). Electronic newsletter 5. http://www.fon.org.cn/channal.php?cid=552 (accessed 7 December 2005).

Frost, W. (2004). Tourism, rainforests and worthless lands: The origins of national parks in Queensland. *Tourism Geographies*, 6(4), 493–507.

FRYING DUTCHMAN "humanERROR" [Video file]. (2012, 16 January). http://www.youtube.com/watch?v=Q5p283KZGa8&feature=relmfu (accessed 19 November 2012).

Fukuwara, A. (2007). *Village life in modern Japan: An environmental perspective*. Melbourne: Trans Pacific Press.

Fürsich, E., & Kavoori, A. P. (2001). Mapping a critical framework for the study of travel journalism. *International Journal of Cultural Studies*, 4(2), 149–171.

Gaber, I. (2000). The greening of the public, politics, and the press, 1985–1999. In J. Smith (Ed.), *The daily globe: Environmental change, the public, and the media* (pp. 115–127). London: Earthscan.

Gall, W. T. (1924, 5 April). Meston: The man and his work. *Daily Mail*, p. 14.

Gamson, J. (1994). *Claims to fame: Celebrity in contemporary America*. Berkeley: University of California Press.

Gamson, W. A., & Meyer, D. S. (1996). Framing political opportunity. In D. McAdam, J. D. McCarthy, & M. N. Zald (Eds.), *Comparative perspectives on social movements: Political opportunities, mobilizing structures, and cultural framings* (pp. 275–290). Cambridge, UK: Cambridge University Press.

Gamson, W. A., & Modigliani, A. (1989). Media discourse and public opinion on nuclear power: A constructionist approach. *American Journal of Sociology*, 95, 1–37.

Gamson, W. A, & Wolfseld, G. (1993). Movements and media as interacting systems. *The Annals of the American Academy of Political and Social Science*, 528 (1), 114–125.

Gandy, O. H. (1980). *Beyond agenda setting: Information subsidies and public policy*. Norwood: Ablex Publishing.

Garcia, D., & Lovink, G. (1997, 16 May). The ABC of tactical media. http://www.nettime.org/Lists-Archives/nettime-l-9705/msg00096.html (accessed 1 November 2012).

Gardiner, S. M. (2004). Ethics and global climate change. *Ethics, 114*, 555–600.

Gardiner, S. M. (2006). A perfect moral storm: Climate change, intergenerational ethics and the problem of moral corruption. *Environmental Values, 15*, 397–413.

Garnaut, R. (2008). *The Garnaut Climate Change Review: Final report*. Melbourne: Cambridge University Press.

Garnaut, R. (2011). *The Garnaut Review 2011: Australia in the global response to climate change*. Melbourne: Cambridge University Press.

Garry, C. (2010a, 5 May). Abbott rival chases seat with LNP. *The Queensland Times*. http://www.qt.com.au/story/2010/05/05/abbott-tony-lnp-blair-petersen-patricia/ (accessed 12 January 2012).

Garry, C. (2010b, 4 August). Green says no to deal. *The Queensland Times*, p. 6.

Garry, C. (2010c, 17 March). Political campaign cashes in. *The Queensland Times*. http://www.qt.com.au/story/2010/03/17/candidate-offers-1000-prize-to-back-campaign/ (accessed 17 June 2010).

Gartner, W. C. (1993). Image formation process. In M. Uysal & D. R. Fesenmaier (Eds.), *Communication and channel systems in tourism marketing* (pp. 191–215). Binghamton: Haworth.

Gay, D. (2009, 24 September). Marine parks moratorium—stop Labor's lockout [Video file]. http://www.youtube.com/watch?v=x3RclY_6qGw (accessed 27 October 2012).

Gelbspan, R. (2004). *Boiling point: How politicians, big oil and coal, journalists and activists are fuelling the climate crisis and what we can do to avert disaster*. New York: Basic Books.

Geman, B. (2011, 16 September). Polls show growing belief that global warming is occurring. *The Hill*. http://thehill.com/blogs/e2-wire/e2-wire/182007-poll-shows-growing-belief-that-global-warming-is-occurring (accessed 8 January 2012).

Giarracca, N. (2007). The tragedy of development: Disputes over natural resources in Argentina. *Sociedad (Buenos Aires)*, 3(26), 26.

Gibson, M. (2012a, 16 April). Unpublished interview with Brett Hutchins.

Gibson, M. (2012b, 15 September). Miranda's daily blog: Day 275. *The Observer Tree*. http://observertree.org/2012/09/15/mirandas-daily-blog-day-275/ (accessed 16 September 2012).

Gibson, R., & Cantijoch, M. (2011). Comparing online elections in Australia and the UK. *Communication, Politics & Culture*, 44(2), 4–17.

Giddens, A. (1990). *The consequences of modernity*. Cambridge, UK: Polity Press.

Giddens, A. (1991). *Modernity and self-identity: Self and society in late modern age*. Cambridge, UK: Polity Press.

Giddens, A. (2009). *The politics of climate change*. Cambridge, UK: Polity Press.

Gillis, J. (2011, 24 December). Harsh political reality slows climate studies despite extreme year. *The New York Times*. http://www.nytimes.com/2011/12/25/science/earth/climate-scientists-hampered-in-study-of-2011-extremes.html?pagewanted=all (accessed 27 December 2011).

Gillis, J., & Kaufman, L. (2012, 16 February). In documents, a plan to discredit climate teaching. *The New York Times*, p. A21.

Gitlin, Todd. (1980). *The whole world is watching: Mass media in the making and unmaking of the new left*. Berkeley: University of California Press.

Gladstone, W. (2007). Requirements for marine protected areas to conserve the biodiversity of rocky reef fishes. *Aquatic Conservation: Marine and Freshwater Ecosystems, 17,* 71–87.

Glaser, A. (2011). After Fukushima: Preparing for a more uncertain future of nuclear power. *The Electricity Journal,* 24(6), 27–35.

Glover, D. (2007). Speechwriters and political speech: Pitting the good angels against the bleak. In S. Young (Ed.), *Government communication in Australia* (pp. 144–157). Melbourne: Cambridge University Press.

Goffman, E. (1974). *Frame analysis: An essay on the organization of experience.* Cambridge, MA: Harvard University Press.

Goggin, G. (2011). *Global mobile media.* London: Routledge.

Golan, K. (2003). Surviving a public health crisis: Tips for communicators. *Journal of Health Communication,* 8, 126–127.

Goldman, J., Shilton, K., Burke, J., Estrin, D., Hansen, M., Ramanathan, N., Reddy, S., Samanta, V., Srivastava, & M., West, R. (2009). *Participatory sensing: A citizen-powered approach to illuminating the patterns that shape our world.* Los Angeles: Woodrow Wilson Center for International Scholars. http://wilsoncenter.org/topics/docs/participatory_sensing.pdf (accessed 27 October 2012).

Gonzales-Herrero, A., & Pratt, C. B. (1998). Marketing crises in tourism: Strategies in the United States and Spain. *Public Relations Review,* 24(1), 83–97.

Goodman, D. S. G. (1981). *Beijing street voices: The poetry and politics of China's democracy movement.* London: Marion Boyars.

Goodman, M. K. (2010). The mirror of consumption: Celebritization, developmental consumption and the shifting cultural politics of fair trade. *Geoforum,* 41(1), 104–116.

Gramsci, A. (1988). *A Gramsci reader: Selected writings 1916–1935* (Trans. & Ed. D. Forgacs). London: Lawrence and Wishart.

Granot, H. (1998). The dark side of growth and industrial disasters since the Second World War. *Disaster Prevention and Management,* 7(3), 195–204.

Grassroots Mapping. (2010). Gulf oil spill mapping. http://www.grassrootsmapping.org/gulf-oil-spill (accessed 16 September 2010).

Gray, P. (2002). *The human consequences of the Chernobyl nuclear accident: A strategy for recovery.* Report commissioned by UNCP and UNICEF with the support of UN-OCHA and WHO. http://chernobyl.undp.org/english/docs/strategy_for_recovery.pdf (accessed 15 July 2011).

Greaves, A. (2011, 28 February). Statement to the House of Lords (column 824). http://www.publications.parliament.uk/pa/ld201011/ldhansrd/text/110228-0001.htm (accessed 12 January 2012).

Greenfield, C., & Williams, P. (2001). "Howardism" and the media rhetoric of "battlers" vs. "elites". *Southern Review,* 34(1), 32–44.

Green Net. (n.d.). Web site of the *China Youth Daily.* http://202.99.23.201/cydgn/gb/cydgn/content.759348.htm (accessed 1 May 2004). This webpage is no longer available. Printed hard copy of webpage available from author.

Greenpeace International. (2011a). *The advanced energy [r]evolution. A sustainable energy outlook for Japan.* Amsterdam: Greenpeace International.

Greenpeace International. (2011b). *Chernobyl 25 years factsheet*. Amsterdam: Greenpeace International.

Greenpeace Japan. (2012). *Toxic assets: Nuclear reactors in the 21st century*. Amsterdam: Greenpeace International.

Greenslade, R. (2011, 16 March). How the papers are covering Japan's catastrophe. Greenslade blog. *The Guardian*. http://www.guardian.co.uk/media/greenslade/2011/mar/16/japan-earthquake-and-tsunami-national-newspapers (accessed 16 January 2013).

Greenwald, J. (2008, June). Sympathy for the devil. *Islands*. http://www.islands.com/article/Sympathy-for-the-Devil (accessed 31 December 2011).

Greenwald, J. (2010, March/April). Bedeviled island. *Afar*, 64–73.

Greenwald, J. (n.d.) Jeff Greenwald blog. http://www.jeffgreenwald.com/category/blog (accessed 8 January 2012).

Greer, J., & Bruno, K. (1996). *Greenwash: The reality behind corporate environmentalism*. Penang: Third World Network.

Guha, R. (1997). Radical American environmentalism and wilderness preservation: A Third World critique. In R. Guha & J. Martinez-Alier (Eds.), *Varieties of environmentalism: Essays north and south*. London: Earthscan.

Guha, R. (2003). The authoritarian biologist and the arrogance of anti-humanism: Wildlife conservation in the Third World. In V. Saberwal & M. Rangarajan (Eds.), *Battles over nature: Science and the politics of conservation*. Delhi: Permanent Black.

Guha, R., & Martinez-Alier, J. (1997). *Varieties of environmentalism: Essays north and south*. London: Earthscan.

Gulledge, J. (2011, March 1). Sixth independent investigation clears "climategate" scientists. Center for Climate and Energy Solutions. http://www.pewclimate.org/blog/gulledgej/sixth-independent-investigation-clears-climategate-scientists (accessed 13 March 2011).

Guskin, E., Rosenstiel, T., & Moore, P. (2011). Network: By the numbers. The State of the News Media 2011, Pew Research Center's Project for Excellence in Journalism. http://stateofthemedia.org/2011/network-essay/data-page-5/ (accessed 26 October 2012).

Habermas, J. (1989a). The public sphere. In S. Seidman (Ed.), *Jurgen Habermas on society and politics: A reader* (pp. 231–236). Boston: Beacon Press.

Habermas, J. (1989b). *The structural transformation of the public sphere*. Cambridge, MA: MIT Press.

Habermas, J. (1992). Further reflections on the public sphere. In C. Calhoun (Ed.), *Habermas and the public sphere* (pp. 421–461). Cambridge, MA: MIT Press.

Hall, S., Critcher, C., Jefferson, T., Clarke, J., & Roberts, B. (1978). *Policing the crisis: Mugging, the state and law and order*. London: Macmillan.

Hallin, D. (2000). Commercialism and professionalism in the American news media. In J. Curran & M. Gurevitch (Eds.), *Mass media and society* (3rd ed.) (pp. 218–237). London: Edward Arnold.

Hamilton, C. (2009, 7 July). *The rebirth of nature and the climate crisis*. A Sydney Ideas lecture. http://sydney.edu.au/sydney_ideas/lectures/2009/rebirth_of_nature.shtml (accessed 27 October 2012).

Hamilton, C. (2010). *Requiem for a species: Why we resist the truth about climate change*. Sydney: Allen & Unwin.

Hamilton-Smith, E. (1998). From cultural awakening to post-industrialism: The history of leisure, recreation and tourism in Australia. In H. Perkins & G. Cushman (Eds.). *Time out?* (pp. 34–50). Auckland: Addison Wesley Longman.

Hannigan, A. (2006). *Environmental sociology*. London: Routledge.

Hansen, A. (1991). The media and the social construction of the environment. *Media, Culture & Society*, 13, 443–458.

Hansen, A. (Ed.). (1993). *The mass media and environmental issues*. Leicester: Leicester University Press.

Hansen, A. (2010). *Environment, media and communication*. London: Routledge.

Hansen, A., & Machin, D. (2008). Visually branding the environment: Climate change as a marketing opportunity. *Discourse Studies*, 10(6), 777–794.

Hänska-Ahy, M., & Shapour, R. (2012). Who's reporting the protests? Converging practices of citizen journalists and two BBC World Service newsrooms, from Iran's election protests to the Arab uprisings. *Journalism Studies*, 7(2), 238–254.

Hanson, F. (2010). The Lowy Institute poll 2010: Australia and the world: Public opinion and foreign policy. Sydney: Lowy Institute for International Policy. http://lowyinstitute.cachefly.net/files/pubfiles/LowyPoll_2010_LR_Final.pdf (accessed 30 October 2012).

Hanson, F. (2011). The Lowy Institute poll 2011: Australia and the world: Public opinion and foreign policy. Sydney: Lowy Institute for International Policy. http://lowyinstitute.cachefly.net/files/pubfiles/Lowy_Poll_2011_WEB.pdf (accessed 30 October 2012).

Hanusch, F. (2012). A profile of Australian travel journalists' views and ethical standards. *Journalism: Theory, Practice & Criticism*, 13(5), 668–686.

Haraldson, D. (2010, 22 October). Pollies learn new tricks too little too late. *Now UC*. http://www.nowuc.com.au/2010/10/22/pollies-learn-new-tricks-too-little-too-late (accessed 18 January 2012).

Harre, R., Brockmeier, J., & Muhlhausler, P. (1999). *Greenspeak: A study of environmental discourse*. Thousand Oaks: Sage Publications.

Harris Poll, The. (2011, 7 July). Most Americans think devastating natural disasters are increasing. *PR Newswire*. http://www.prnewswire.com/news-releases/most-americans-think-devastating-natural-disasters-are-increasing-125131439.html (accessed 14 December 2011).

Harrison, J. (2000). *Terrestrial TV news in Britain*. Manchester: Manchester University Press.

Harvey, D. (2006). *Spaces of global capitalism*. London: Verso.

Harvey, F. (2011, 17 February). Forest sell-off: Social media celebrates victory. *The Guardian*. http://www.guardian.co.uk/environment/blog/2011/feb/17/forest-sell-off-social-media (accessed 25 January 2012).

Hauerwas, S. (1981). *A community of character*. Notre Dame: University of Notre Dame Press.

Hawkins, D. (1962, 18 September). BBC Natural History Unit: Report by Head of West Regional Programmes. Courtesy of the BBC Written Archives Centre.

Heath, R. L., Palenchar, M. J., Proutheau, S., & Hocke, T. M. (2007). Response to Cox: Nature, crisis, risk, science, and society: What is our ethical responsibility? *Environmental Communication: A Journal of Nature and Culture*, 1(1), 34–48.

Hegarty, J. (2010). Out of the consulting room and into the woods? Experiences of nature-connectedness and self-healing. *European Journal of Ecopsychology*, 1(1), 64–84.

Heinich, N. (2011). La culture de la celebrite en France et dans les pays anglophones. *Revue française de sociologie, 52*(2), 353–372.

HELLog [Video file]. (2007, 19 January). http://www.youtube.com/watch?v=dgEfTgokkpQ (accessed 2 January 2012).

HELLog 2: No log gets left behind [Video file]. (2007, 21 January). http://www.youtube.com/watch?v=oH3041MKqBA (accessed 2 January 2012).

Hermida, A. (2010). Twittering the news. *Journalism Practice, 4*(3), 297–308.

Herrera Huerfano, E. (Ed.). (2011). Experiencia de comunicación y desarrollo sobre medio ambiente: Estudios de caso e historias de vida en la región andina de Colombia. *Alianza interinstitucional, primera edición*. Bogota: UNAD-UNIMINUTO-Universidad Santo Tomas.

Herrick, C. (2010). Federal election 2010: The Australian Greens' social networking strategy. http://www.computerworld.com.au/article/355348/federal_election_2010_australian_greens_social_networking_strategy/ (accessed 14 October 2011).

Heynen, N., McCarthy, J., Prudham, W. S., & Robbins, P. (2007). *Neoliberal environments: False promises and unnatural consequences*. London: Routledge.

Hickman, L. (2010, July 5). US climate scientists receive hate mail barrage in wake of UEA scandal. *The Guardian*. http://www.guardian.co.uk/environment/2010/jul/05/hate-mail-climategate (accessed 18 November 2012).

Hilgartner, S., & Bosk, C. L. (1988). The rise and fall of social problems: A public arenas model. *American Journal of Sociology, 94*(1), 53–78.

Hill, J. (2001). *The legacy of Luna*. St. Louis: Turtleback Books.

Hinds, J., & Sparks, P. (2008). Engaging with the natural environment: The role of affective connection and identity. *Journal of Environmental Psychology, 28*(2), 109–120.

Hirata, K. (2004). Civil society and Japan's dysfunctional democracy. *Journal of Developing Societies, 20*(1–2), 107–124.

Hirokawa, K. (2012). *Verifying nuclear disaster report: What was told then*. Tokyo: Days Japan.

Ho, F., & Hallahan, K. (2004). Post-earthquake crisis communications in Taiwan: An examination of corporate advertising and strategy motives. *Journal of Communication Management, 8*(3), 291–306.

Ho, P. (2001). Greening without conflict? Environmentalism, NGOs and civil society in China. *Development and Change, 32*(5), 893–921.

Ho, P. (2007). Embedded activism and political change in a semiauthoritarian context. *China Information, 21*(2), 187–210.

Hodgson, J. D., Sano, Y., & Graham, J. L. (Eds.). (2000). *Doing business with the new Japan*. Lanham: Rowman & Littlefield Publishers.

Hoggett, C. (2011a, 15 December). Hi-tech forest battle. *The Mercury*. http://www.themercury.com.au/article/2011/12/15/284721_tasmania-news.html (accessed 17 September 2012).

Hoggett, C. (2011b, 24 December). Santa lands in tree-tops. *The Mercury*. http://www.themercury.com.au/article/2011/12/24/287011_tasmania-news.html (accessed 17 September 2012).

Holmes, G. (2011). Conservation's friends in high places: Neoliberalism, networks, and the transnational conservation elite. *Global Environmental Politics, 11*(4), 1–21.

Hopke, J. E. (2012). Water gives life: Framing an environmental justice movement in the mainstream and alternative Salvadoran press. *Environmental Communication: A Journal of Nature and Culture*, 6(3), 365–382.

Horne, J. (2005). *The pursuit of wonder*. Carlton: Miegunyah Press.

Hough, A. (2011, 7 December). 'Frozen Planet': Controversial BBC climate change episode to air in America. *The Telegraph*. http://www.telegraph.co.uk/earth/earthnews/8939592/Frozen-Planet-controversial-BBC-climate-change-episode-to-air-in-America.html (accessed 28 October 2012).

Howard-Williams, R. (2009). Ideological construction of climate change in Australian and New Zealand newspapers. In T. Boyce & J. Lewis (Eds.), *Climate change and the media* (pp. 28–40). New York: Peter Lang.

Howell, J. (2004). New directions in civil society: Organizing around marginalized interests. In J. Howell (Ed.), *Governance in China* (pp. 143–71). Lanham: Rowman & Littlefield.

Huanbao minjian zuzhi shuliang he renshu jiang yi 10% zhi 15% sudu dizeng (Number and staff of environmental NGOs to increase by 10 to 15%). (2006, 28 October). *Xinhuanet*. http://news.xinhuanet.com/environment/2006-10/28/content_5261309.htm (accessed 29 October 2006). This webpage is no longer available. Printed hard copy of webpage available from author.

Hulme, M. (2009). *Why we disagree about climate change: Understanding controversy, inaction and opportunity*. Cambridge, UK: Cambridge University Press.

Hutchins, B., & Lester, L. (2006). Environmental protest and tap-dancing with the media in the information age. *Media, Culture & Society*, 28(3), 433–451.

Hutchins, B., & Lester, L. (2011). Politics, power and online protests in an age of environmental conflict. In S. Cottle & L. Lester (Eds.), *Transnational protests and the media* (pp. 159–171). New York: Peter Lang.

Igoe, J. (2010). The spectacle of nature in the global economy of appearances: Anthropological engagements with the spectacular mediations of transnational conservation. *Critique of Anthropology*, 30(4), 375–397.

Igoe, J., Neves, K., & Brockington, D. (2010). A spectacular eco-tour around the historic bloc: Theorizing the convergence of biodiversity conservation and capitalist expansion. *Antipode*, 42(3), 486–512.

Innis, H. (2007). *Empire and communications*. Lanham: Rowman & Littlefield. (Original work published 1950).

International Atomic Energy Agency. (2005). *Chernobyl's legacy: Health, environmental and socio-economic impacts*. The Chernobyl Forum. Vienna: International Atomic Energy Agency.

Intergovernmental Panel on Climate Change. (2001). Working Group II Report "impacts, adaptation and vulnerability" (IPCC third assessment report). http://www.ipcc-wg2.gov/publications/Reports/index.html#AR (accessed 14 March 2012).

Intergovernmental Panel on Climate Change. (2007a). Summary for policymakers. In S. Solomon, D. Qin, M. Manning, Z. Chen, M. Marquis, K. B. Averyt, M. Tignor, & H. L. Miller (Eds.), *Climate change 2007: The physical science basis: Contribution of Working Group I to the Fourth Assessment Report of the Intergovernmental Panel on Climate Change*. Cambridge, UK: Cambridge University Press. http://www.ipcc.ch/pdf/assessment-report/ar4/wg1/ar4-wg1-spm.pdf (accessed 28 October 2012).

Intergovernmental Panel on Climate Change. (2007b). Working Group II Report "impacts, adaptation and vulnerability" (IPCC fourth assessment report). http://www.ipcc-wg2.gov/publications/Reports/index.html#AR (accessed 14 March 2012).

Ipsos Global @dvisor. (2011, April 18). Energy security is a top concern for Brits. http://www.ipsos-mori.com/researchpublications/researcharchive/2771/Energy-security-is-a-top-concern-for-Brits.aspx (accessed 18 November 2012).

Ipswich. (2010, 30 July). Discuss: I'm not a criminal: Petersen [Facebook post].http://www.facebook.com/pages/Ipswich/61255807179 (accessed 1 August 2010). This post is no longer available on Facebook. Printed hard copy of webpage available from author.

Ipswich-City.com. forums (2010). De-friended & blocked by Patricia Petersen! (accessed 17 June 2010). This post is no longer available. Printed hard copy of webpage available from author.

Ito, M. (2012). *How did the TV broadcast nuclear accident?* Tokyo: Heibon-sha.

Iversen, M. (2007). Following pieces on performative photography. In J. Elkins (Ed.), *Photography theory* (pp. 91–108). New York: Routledge.

Iyengar, S. (1991). *Is anyone responsible? How television frames political issues.* Chicago: University of Chicago Press.

Jackson, Z. (2010, 10 July). Candidate appalled by a very dirty trick. *The Queensland Times*, p. 3.

Jacques, P. (2008). Ecology, distribution and identity in the world politics of environmental skepticism. *Capitalism Nature Socialism*, 19(3), 8–28.

Jacques, P. J., Dunlap, R. E., & Freedman, M. (2008). The organization of denial: Conservative think tanks and environmental skepticism. *Environmental Politics*, 17, 349–385.

Jamieson, D. (1992). Ethics, public policy, and global warming. *Science, Technology & Human Values*, 17, 139–153.

Japan Meteorological Agency. (2011, 11 March). http://www.jma.go.jp/jp/tsunami/observation_04_20110312193944.html (accessed 19 November 2012).

Jasper, J. (2011). Emotions and social movements: Twenty years of theory and research. *Annual Review of Sociology*, 37, 285–303.

Java, A., Song, X., Finin, T., & Tseng. B. (2007). Why we Twitter: Understanding microblogging usage and communities. *Proceedings of the 9th WebKDD and 1st SNA-KDD 2007 workshop on Web mining and social network analysis*. http://aisl.umbc.edu/resources/369.pdf (accessed 4 May 2012).

Jeffries, M. (2003). BBC natural history versus science paradigms. *Science as Culture*, 12(4), 527–545.

Johnsen, S. A. K. (2011). The use of nature for emotion regulation: Toward a conceptual framework. *Ecopsychology*, 3(3), 175–185.

Jones, B. (2012). New media, political infantilisation and the creativity paradox. In H. Sykes (Ed.), *More or less: Democracy and new media* (pp. 5–23). Sydney: Future Leaders.

Jones, J. M. (2011, March 14). In US, concerns about global warming stable at lower levels. *Gallup Politics*. http://www.gallup.com/poll/146606/concerns-global-warming-stable-lower-levels.aspx (accessed 8 February 2011).

Jones, O., & Cloke, P. (2002). *Tree cultures*. Oxford: Berg.

Jong, W., Stammers, N., & Shaw, M. (Eds.). (2005). *Global activism, global media*. London: Pluto Press.

Jowett, G., & O'Donnell, V. (2012). *Propaganda & persuasion* (5th ed.). Los Angeles: Sage Publications.

Kaldor, M. (2006). *New and old wars: Organized violence in a global era.* Cambridge, UK: Polity Press.

Kasperson, R. E., Renn, O., Slovic, P., Brown, H. S., Emel, J., Goble, R., Kasperson, J. X., & Ratick, S. (1988). The social amplification of risk: A conceptual framework. *Risk Analysis,* 8, 177–187.

Katz, E., & Liebes, T. (2007). "No more peace!" How disaster, terror and war have upstaged media events. *International Journal of Communication,* 1, 157–166.

Kavada, A. (2005). Civil society organizations and the Internet: The case of Amnesty International, Oxfam and the World Development Movement. In W. de Jong, M. Shaw, & N. Stammers (Eds.), *Global activism, global media* (pp. 208–222). Ann Arbor: Pluto Press.

Kawamura, M. (2011). *Nuclear power and nuclear bomb: Conceptual history of nuclear (genpatsu to genbaku).* Tokyo: Kawaide Shobo.

Keane, B. (2011, 9 March). Climate change cage match: Abbott debates Abbott. *Crikey.* http://www.crikey.com.au/2011/03/09/climate-change-cage-match-abbott-debates-abbott/ (accessed 11 November 2011).

Kearney, B. (2007a, November). The great Batemans MPA swindle: Not science, a sham! *Ausmarine,* 16–17.

Kearney, B. (2007b). The pros and cons of marine protected areas in New South Wales: Who's being hoodwinked? Address to the Australian Society for Fish Biology, Canberra. http://aerg.canberra.edu.au/reprints/2007_Kearney_pros_cons_marine_protected_areas_NSW.pdf (accessed 30 October 2012).

Kearney, B. (2009). *Response to ACORF on "The Torn Blue Fringe: Marine Conservation in NSW" (Winn 2008).* http://www.sportsfish.com.au/downloads/ACORF_Final_Report_19-03-09.pdf (accessed 30 October 2012).

Kelly, G., & Hosking, K. (2008). Nonpermanent residents, place attachment and "sea change" communities. *Environment and Behavior,* 40(4), 575–594.

Kelly, P., & Shanahan, D. (2010, August 20). Julia Gillard's carbon price promise. *The Australian.* http://www.theaustralian.com.au/national-affairs/julia-gillards-carbon-price-promise/story-fn59niix-1225907522983 (accessed 28 October 2012).

Kenterelidou, C. (2009). *Broadcast television network evening news (TV news) and government: The implosion of information and propaganda in the news.* Thessaloniki: Aristotle University of Thessaloniki.

Kenterelidou, C., & Panagiotou, N. (2006). Public communication in time of crisis: The mobilization of the Greek public opinion during the Asia tsunami crisis. In M. Zlateva & T. Petev (Eds.), *Public communication, globalization and democracy* (pp. 235–248). Sofia: FJMC Sofia University.

Kerr, C. (2009, 9 December). Climate change sceptic blasts "one-sided" media. *The Australian,* p. 7.

Kim. (2009, February 25). Queensland election 2009—day three. *Larvatus Prodeo.* http://larvatusprodeo.net/2009/02/25/queensland-election-2009-day-three/ (accessed 15 November 2011).

Kirchhoff, S. M. (2009, 8 July). *The U.S. newspaper industry in transition.* Congressional Research Service report for Congress. http://fpc.state.gov/documents/organization/126872.pdf (accessed 17 December 2011).

Kloor, K. (2011, 17 October). *Is climate fatigue setting in?* The Yale Forum on Climate Change & the Media. http://www.yaleclimatemediaforum.org/2011/10/is-climate-fatigue-setting-in/ (accessed 7 February 2013).

Korauaba, T. (2012). Media and the politics of climate change in Kiribati. Unpublished masters thesis, Auckland University of Technology.

Korner, A. (2010, 5 July). Done deal: Hundreds rally against "toxic dump" proposal. *The Queensland Times*, pp. 1–2.

Kovel, J. (2008). Ecosocialism, global justice, and climate change. *Capitalism Nature Socialism*, 19(2), 4–14.

Krugman, P. (2009, September 27). Cassandras of climate. *The New York Times*, A21.

Kurzman, C., Anderson, C., Key, C., Lee, Y. O., Moloney, M., Silver, A., & Van Ryn, M. W. (2007). Celebrity status. *Sociological Theory*, 25(4), 347–367.

Lahey, R. (1911). "To the head of the Coomera", Sketcher, *Queenslander*, 8 September, p. 8.

Lakoff, G. (2008). *The political mind: Why you can't understand 21st-century politics with an 18th-century brain*. New York: Viking Adult.

La Nación. (2006a, September 24). Movilización en Entre Ríos. http://www.lanacion.com.ar/843267-movilizacion-en-entre-rios (accessed 12 November 2012).

La Nación. (2006b, July 28). Taiana: "Causarán un daño irreparable". http://www.lanacion.com.ar/826832-taiana-causaran-un-dano-irreparable (accessed 12 November 2012).

Lange, J. I. (1993). The logic of competing information campaigns: Conflict over old growth and the spotted owl. *Communication Monographs*, 60(3), 239–257.

Lash, S., & Urry, J. (1994). *Economies of sign and space*. London: Sage.

Latour, B. (1993). *We have never been modern*. Cambridge, MA: Harvard University Press.

Latta, A., & Wittman, H. (Eds.). (2012). *Environment and citizenship in Latin America: Natures, subjects and struggles*. New York: Berghahn Books.

Lean, E. (2004). The making of a public: Emotions and media sensation in 1930s China. *Twentieth-Century China*, 29(2), 39–62.

Leiserowitz, A., Maibach, E., Roser-Renouf, C., & Smith, N. (2011). *Global warming's six Americas in May 2011*. Yale University and George Mason University. New Haven: Yale Project on Climate Change Communication. http://environment.yale.edu/climate/files/SixAmericasMay2011.pdf (accessed 28 October 2012).

Leiserowitz, A., Maibach, E., Roser-Renouf, C., Smith, N., & Dawson, E. (2010, 2 July). Climategate, public opinion, and the loss of trust. Working paper. New Haven: Yale Project on Climate Change Communication. http://environment.yale.edu/climate/files/Climategate_Opinion_and_Loss_of_Trust_1.pdf (accessed 4 January 2012).

Leiserowitz, A., Maibach, E., Roser-Renouf, C., Smith, N., & Hmielowski, J. D. (2011). *Climate change in the American mind: Americans' global warming beliefs and attitudes in November 2011*. New Haven: Yale Project on Climate Change Communication. http://environment.yale.edu/climate/files/ClimateBeliefsNovember2011.pdf (accessed 4 January 2012).

Leiserowitz, A., Smith, N., & Marlon, J. R. (2010). *Americans' knowledge of climate change*. New Haven: Yale Project on Climate Change Communication. http://environment.yale.edu/climate/files/ClimateChangeKnowledge2010.pdf (accessed 3 January 2012).

Leiss, W. (1996). Three phases in the evolution of risk communication practice. *Annals of the American Academy of Political and Social Science*, 545, 85–94.

Leone, S. (2012, 8 May). Hollande victory signals shift in France's renewable energy policy. *RenewableEnergyWorld.com*. http://www.renewableenergyworld.com/rea/news/article/2012/05/hollande-victory-signals-shift-in-frances-renewable-energy-policy (accessed 19 November 2012).

Lester, L. (2006). Lost in the wilderness? Celebrity, protest and the news. *Journalism Studies*, 7(6), 907–921.

Lester, L. (2007). *Giving ground: Media and environmental conflict in Tasmania*. Hobart: Quintus.

Lester, L. (2010a). Big tree, small news: Media access, symbolic power and strategic intervention. *Journalism*, 11(5), 589–606.

Lester, L. (2010b). *Media and environment: Conflict, politics and the news*. Cambridge, UK: Polity Press.

Lester, L., & Cottle, S. (2009). Visualizing climate change: Television news and ecological citizenship. *International Journal of Communication*, 3, 920–936. http://ijoc.org/ojs/index.php/ijoc/article/view/509/371 (accessed 30 October 2012).

Lester, L., & Hutchins, B. (2009). Power games: Environmental protest, news media and the Internet. *Media, Culture & Society*, 31(4), 579–595.

Lester, L., & Hutchins, B. (2012a). The power of the unseen: Environmental conflict, the media and invisibility. *Media, Culture & Society*, 34(7), 847–863.

Lester, L., & Hutchins, B. (2012b). Soft journalism, politics and environmental risk: An Australian story. *Journalism*, 13(5), 654–667.

Levinson, P. (1999). *Digital McLuhan: A guide to the information millennium*. Routledge: London.

Leviston, Z., Leitch, A., Greenhill, M., Leonard, R., & Walker, I. (2011). *Australians' views of climate change*. CSIRO report, Canberra. http://www.garnautreview.org.au/update-2011/commissioned-work/australians-view-of-climate-change.pdf (accessed 28 October 2012).

Liberal Party of Australia. (2010). *The coalition's plan for real action on the environment, climate change and heritage*. http://greghunt.com.au/portals/0/ArticleAttachments/Real-Action-on-the-Environment-Climate-Change-and-Heritage.pdf (accessed 28 October 2012).

Liebes, T. (1998). Television's disaster marathons: A danger for democratic processes? In T. Liebes & J. Curran (Eds.), *Media, ritual and identity* (pp. 71–84). London: Routledge.

Lievrouw, L. (2011). *Alternative and activist new media*. Cambridge, UK: Polity Press.

Lockwood, A. (2010). Seeding doubt: How sceptics have used new media to delay action on climate change. *Geopolitics, History and International Relations*, 2(2), 136–164.

Lockwood, A. (2012). The shore is not a beach. In C. Berberich, N. Campbell, & R. C. Hudson (Eds.), *Land and identity* (pp. 259-281). Amsterdam: Rodopi.

Lomborg, B. (2007). *Cool it: The skeptical environmentalist's guide to global warming*. New York: Knopf.

López Echagüe, H. (2006). *Crónica del ocaso: Apuntes sobre las papeleras y la devastación del litoral argentino y uruguayo*. Buenos Aires and Miami: Grupo Editorial Norma.

Lorenzoni, I., Leiserowitz, A., Doria, M., Portinga, W., & Pidgeon, N. (2006). Cross-national comparisons of image associations with "global warming" and "climate change" among lay people in the United States of America and Great Britain. *Journal of Risk Research*, 9(3), 265–281.

Lovelock, J. (2006). The revenge of Gaia: Earth's climate crisis and the fate of humanity. New York: Basic Books.

Lovink, G. (2002). *Dark fibre: Tracking critical Internet culture*. Cambridge, MA: MIT Press.

Lovink, G. (2005). Tactical media, the second decade. (Preface to the Brazilian Submidialogia publication). http://geertlovink.org/texts/tactical-media-the-second-decade/ (accessed 18 February 2013).

Lowe, I. (2007). Reaction time: Climate change and the nuclear option. *Quarterly Essay, 27.*

Lowe, P., & Morrison, D. (1984). Bad news or good news: Environmental politics and the mass media. *Sociological Review,* 32(1), 75–90.

Lower Weld Valley threatened forests from the air [Video file]. (2007, 6 February). http://www.youtube.com/watch?v=pH-XEZED7pk (accessed 2 January 2012).

Lu, Hongyan. (2003). Bamboo sprouts after the rain: The history of university student environmental associations in China. *China Environment Series,* 6, 55–65.

Luke, T. W. (1999). *Capitalism, democracy and ecology: Departing from Marx.* Urbana: University of Illinois Press.

Lynch, J., & McGoldrik, A. (2005). *Peace journalism.* Stroud: Hawthorn Press.

Lynch, R., & Veal, A. J. (1997). *Australian leisure.* Melbourne: Addison, Wesley and Longman Australia.

Ma, S. (1994). The Chinese discourse on civil society. *China Quarterly,* 137, 180–193.

MacDonald, K. (2010). The devil is in the (bio)diversity: Private sector "engagement" and the restructuring of biodiversity conservation. *Antipode,* 42(3), 513–550.

MacFarlane, R. (2003). *Mountains of the mind: A history of a fascination.* London: Granta Books.

MacKenzie, D. (2011, March 24). Fukushima radioactive fallout nears Chernobyl levels. *New Scientist.* http://www.newscientist.com/article/dn20285-fukushima-radioactive-fallout-nears-chernobyl-levels.html (accessed 28 October 2012).

Macnaghten, D., & Urry, J. (1998). *Contested natures.* London: Sage Publications.

Macnamara, J. (2008, July). E-electioneering: Use of new media in the 2007 Australian federal election. Paper presented at the Australian and New Zealand Communication Association Conference, Wellington, New Zealand.

Macnamara, J. (2011). Pre- and post-election 2010 online: What happened to the conversation? *Communication, Politics & Culture,* 44(2), 18–36.

Maibach, E., Wilson, K., & Witte, J. (2010). *A national survey of news directors about climate change: Preliminary findings.* Fairfax: George Mason University Center for Climate Change Communication. http://www.climatechangecommunication.org/images/files/TV_News_Directors_&_Climate%20Change%281%29.PDF (accessed 6 August 2011).

Mainichi Shimbun, The. (2012, 22 March). Fukushima Pref. deleted 5 days of radiation dispersion data just after meltdowns. *The Mainichi Shimbun.* http://mainichi.jp/english/english/mdn-news/news/20120322p2a00m0na012000c.html (accessed 21 April 2012).

Manne, R. (2011). Bad news, Murdoch's *Australian* and the shaping of the nation, *Quarterly Essay,* 43.

Marine Parks Authority. (2001). *Developing a representative system of marine protected areas in NSW—an overview.* Canberra: Heartland Publishing.

Marine Parks Authority. (2006a). *Summary of submissions on the Batemans Marine Park draft zoning plan.* Narooma, New South Wales: NSW Marine Parks Authority.

Marine Parks Authority. (2006b). *Summary of submissions on the Port Stephens–Great Lakes Marine Park draft zoning plan.* http://www.mpa.nsw.gov.au/pdf/PSGLMP-summary-submissions.pdf (accessed 26 October 2012).

Marr, D. (2010). Power trip: The political journey of Kevin Rudd. *Quarterly Essay, 38*, 1–91.

Marshall, D. P. (1997). *Celebrity and power. Fame in contemporary culture*. Minneapolis: University of Minnesota Press.

Martens, S. (2006). Public participation with Chinese characteristics: Citizen consumers in China's environmental management. *Environmental Politics, 15*(21), 211–230.

McAfee, K. (1999). Selling nature to save it? Biodiversity and green developmentalism. *Environment and Planning D: Society and Space, 17*, 133–154.

McColl, M. (2005). *Ethical Traveler*: New hope for Australia's old-growth forests. *Earth Island Journal, 20*(1). http://www.earthisland.org/journal/index.php/eij/article/ethical_traveler1/ (accessed 8 May 2012).

McCoombs, M., & Shaw, D. (1972). The agenda-setting function of the mass media. *Public Opinion Quarterly, 35*, 176–187.

McCright, A. M., & Dunlap, R. E. (2003). Defeating Kyoto: The conservative movement's impact on US climate change policy. *Social Problems, 50*(3), 348–373.

McGaurr, L. (2010). Travel journalism and environmental conflict: A cosmopolitan perspective. *Journalism Studies, 11*(1), 50–67.

McGaurr, L. (2012). The devil may care: Travel journalism, cosmopolitan concern, politics and the brand. *Journalism Practice, 6*(1), 42–58.

McKay, J. (1998). A good show: Colonial Queensland at international exhibitions: Memoirs of the Queensland Museum. *Cultural Heritage Series, 1*(2), 239–319.

McKee, Y. (2007). Art and the ends of environmentalism: From biosphere to the right to survival. In M. Feher (Ed.), *Nongovernmental politics* (pp. 539–583). New York: Zone Books.

McKnight, D. (2008, 12–13 February). Telling the story of global warming: News and opinion at News Corporation. Paper presented at the Politics/Media Conference, University of Melbourne.

McKnight ,D. (2012). *Rupert Murdoch: An investigation of political power.* Sydney: Allen and Unwin.

McLuhan, M., & Fiore, Q. (1967). *The medium is the massage*. Ringwood: Penguin Books.

Meadows, M. (2001). The changing role of Queensland newspapers in imagining leisure and recreation. *Ejournalist, 1*(2). http://ejournalist.com.au/v1n2/MEADOWS.pdf (accessed 15 March 2012).

Meadows, M. (2012). Putting the citizen back into journalism. *Journalism*. Epub ahead of print April 18. doi: 10.1177/1464884912442293

Media Matters. (2011). Black and white and re(a)d all over: The conservative advantage in syndicated op-ed columns. http://mediamatters.org/reports/oped/ (accessed 21 December 2011).

Meikle, G. (2002). *Future active: Media activism and the Internet*. Annandale: Pluto Press.

Mercer, C. (1989). Antonio Gramsci: Elaborare, or the work and government of culture. Paper presented at the Australian Sociological Association Conference, La Trobe University, Melbourne.

Merchant, C. (1993). *The death of nature: Women, ecology, and the scientific revolution*. San Francisco: Harper.

Meston, A. (1889a, May 18). Exploring the Bellenden Ker. *Brisbane Courier*, p. 7.

Meston, A. (1889b, May 25). Exploring the Bellenden Ker II. *Brisbane Courier*, p. 7.

Meyer, D. S., & Gamson, J. (1995). The challenge of cultural elites: Celebrities and social movements. *Sociological Inquiry*, 65(2), 181–206.

Michaels, D., & Monforton, C. (2005). Manufacturing uncertainty: Contested science and the protection of the public's health and environment. *Public Health Matters*, 95(1), 39–48.

Miles, M. (2010). Representing nature: Art and climate change. *Cultural Geographies*, 17(1), 19–35.

Miles, P. (2008a, 19–20 April). Remote possibilities. *Financial Times*, p. 11 (Life and Arts).

Miles, P. (2008b, March). Tasmania's forest under threat. *Condé Nast Traveller*, p. 34.

Miles, P. (n.d.) Paul Miles: Writer & photographer. http://www1.clikpic.com/paulmiles/ (accessed 8 January 2012).

Miller, D. (1999). Risk, science and policy: Definitional struggles, information management, the media and BSE. *Social Science Medicine*, 49, 1239–1255.

Miller, M., & Riechert, B. (2000). Interest group strategies and journalistic norms: News media framing of environmental issues. In S. Allan, B. Adam, & C. Carter (Eds.), *Environmental risks and the media* (pp. 45–54). London: Routledge.

Milliken, R. (2011). Lucky for some. *The Economist: The World in 2012*, 400(8741), 68.

Mills, S. (1997, 21 February). Pocket tigers: The sad unseen reality behind the wildlife film. *Times Literary Supplement*, 6.

Milton, K. (2002). *Loving nature: Towards an ecology of emotion*. London: Routledge.

Ministry of Land, Infrastructure, Transport and Tourism. (2006). *Survey of the current state of the kaso regions*. http://www.mlit.go.jp/singikai/kokudosin/keikaku/jiritsu/9/03.pdf (accessed 20 September 2011).

Mitman, G. (1999). *Reel nature: America's romance with wildlife on film*. Cambridge, MA: Harvard University Press.

Molenberg, D.P. (2007). Unpublished interview with Enrique Peruzotti.

Monbiot, G. (2002, December 17). Planet of the fakes. *The Guardian*. http://www.guardian.co.uk/media/2002/dec/17/broadcasting.comment (accessed 29 October 2012).

Monbiot, G. (2006). *Heat: How to stop the planet from burning*. London: Random House.

Morcombe, J. (2010). Fairlight's Abbott ex-challenger now off to contest Blair, Qld. *Manly Daily*. http://manly-daily.whereilive.com.au/news/story/it-wont-be-dull-fairlights-abbott-challenger-now-off-to-contest-blair-Qld (accessed 20 January 2012). This webpage is no longer available. Printed hard copy of webpage available from author.

Morgan, M. G., Kandlikar, M., Risbey, J., & Dowlatabadi, H. (1999). Why conventional tools for policy analysis are often inadequate for problems of global change. *Climatic Change*, 41, 271–281.

Morton, K. (2005). The emergence of NGOs in China and their transnational linkages: Implications for domestic reform. *Australian Journal of International Affairs*, 59(4), 519–532.

Morton, T. (2007). *Ecology without nature: Rethinking environmental aesthetics*. Cambridge, MA: Harvard University Press.

Moser, S. C. (2010). Communicating climate change: History, challenges, process, and future directions. *Wiley Interdisciplinary Reviews—Climate Change*, 1(1), 31–53.

Moser, S. C., & Dilling, L. (2004). Making climate hot. *Environment: Science and Policy for Sustainable Development*, 46, 32–46.

Mould, R. F. (2000). *Chernobyl record: The definitive history of the Chernobyl catastrophe*. Bristol: Institute of Physics Publishing.

Muramatsu, N., & Akiyama, H. (2011). Japan: Super-aging society preparing for the future. *The Gerontologist*, 51(4), 425–432. http://gerontologist.oxfordjournals.org/content/51/4/425.full (accessed 29 October 2012).

Murata, K. (2007). Pro- and anti-whaling discourses in British and Japanese newspaper reports in comparison: A cross-cultural perspective. *Discourse and Society*, 18(6), 741–764.

Murthy, D. (2011). Twitter: Microphone for the masses? *Media, Culture & Society*, 33(5), 779–789.

Nakamoto, M. (2011, 26 December). Report slams response to nuclear crisis. *FT.com*. http://www.ft.com/cms/s/0/a5711ed0-2fc4-11e1-8ad0-00144feabdc0.html#axzz2Act7O7FT (accessed 29 October 2012).

Nash, C. (2010). Fields of conflict: Journalism and the construction of Sydney as a global city, 1983–2008. Doctoral dissertation, University of New South Wales.

National Academies, The. (2010, 19 May). Strong evidence on climate change underscores need for actions to reduce emissions and begin adapting to impacts. http://www8.nationalacademies.org/onpinews/newsitem.aspx?RecordID=05192010 (accessed 13 February 2012).

Natural England. (2010). *Monitor of engagement with the natural environment: The national survey on people and the natural environment*. Peterborough: Natural England.

Nelson, J. (2008). Economists, value judgements, and climate change: A view from feminist economics. *Ecological Economics*, 65, 441–447.

Neumayer, E. (2007). A missed opportunity: *The Stern Review* on climate change fails to tackle the issue of non-substitutable loss of natural capital. *Global Environmental Change*, 17, 297–301.

Newport, F. (2010, 11 March). Americans' global warming concerns continue to drop. *Gallup Politics*. http://www.gallup.com/poll/126560/americans-global-warming-concerns-continue-drop.aspx (accessed 4 June 2011).

New York Times & *CBS News*. (2011). *New York Times/CBS News Poll: September10–15*. http://s3.documentcloud.org/documents/250094/new-york-times-cbs-poll-results.pdf (accessed 28 December 2011).

Nightingale, N. (2012, 17 January). Unpublished interview with Morgan Richards.

Nip, J. Y. (2004). The queer sisters and its electronic billboard: A study of the Internet for social movement mobilization. In W. van de Donk, B. Loader, P. Nixon, & D. Dieter (Eds.), *Cyberprotest: New media, citizens, and social movements* (pp. 239–260). London: Routledge.

Nippon Hoso Kyokai (NHK—Japan Broadcasting Cooperation). (2011, 15 March). [Video file]. http://www.youtube.com/watch?v=yQHmF-btcCI (accessed 19 November 2012).

Nisbet, M. C. (2009). Communicating climate change: Why frames matter for public engagement. *Environment: Science and Policy for Sustainable Development*, 51, 12–23.

Nisbet, M. C., & Mooney, C. (2007). Science and society: Framing science. *Science*, 316(5821), 56. doi: 10.1126/science.1142030.

Nishi, Toshio. (2012, 6 April). On the cesium road. *Hoover Digest*, 2. http://www.hoover.org/publications/hoover-digest/article/113111 (accessed 30 October 2012).

Nolan, D. (2008). Tabloidisation revisited: The regeneration of journalism in conditions of "advanced liberalism". *Communication, Politics and Culture*, 41(2), 100–118.

Northwest old growth forest legacy campaign [Video file]. (2009, 3 April). http://www.youtube.com/watch?v=Ck3OlvI-Soc (accessed 2 January 2012).

Norton, D. W. (2010, September). *Constructing "climategate" and tracking chatter in an age of Web n.o* [sic]. Washington, DC: Center for Social Media, American University School of Communication. http://www.centerforsocialmedia.org/sites/default/files/documents/pages/david_norton_climategate.pdf (accessed 3 January 2012).

O'Brien, K., St Clair, A., & Kristoffersen, B. (2010a). *Climate change, ethics and human security*. Cambridge, UK: Cambridge University Press.

O'Brien, K., St Clair, A., & Kristoffersen, B. (2010b). The framing of climate change: Why it matters. In K. O'Brien, A. St Clair, & B. Kristoffersen (Eds.), *Climate change, ethics and human security* (pp. 3–22). Cambridge, UK: Cambridge University Press.

Odagiri, M. (2012). *NHK—pluses and misuses of domination*. Tokyo: Best Sellers.

Old growth forest defenders—Sisters of the Siskiyous [Video file]. http://www.youtube/com/watch?v=odhA6C8mlpg (accessed 2 January 2012). This video is no longer available on YouTube. Original video available from author.

Old growth timber of Washington state [Video file]. (2009, 29 August). http://www.youtube.com/watch?v=8YMcAuZhNqM (accessed 2 January 2012).

O'Neill, S., & Nicholson-Cole, S. (2009). Fear won't do it: Promoting positive engagement with climate change through visual and iconic representations. *Science Communication*, 30(3), 355–379.

Ono, A. (2005). *Environmental sociology of mountain villages*. Tokyo: Nosan Gyoson Bunka Kyokai.

Opal Creek—Oregon Cascades [Video file]. (2007, 4 September). http://www.youtube.com/watch?v=IC9SiqPlj80 (accessed 2 January 2012).

Opel, A., & Pompper, D. (Eds.). (2003). *Representing resistance: Media, civil disobedience and the global justice movement*. Westport: Praeger.

O'Reilly and Associates Inc. (1993). *Global Network Navigator* (GNN). http://oreilly.com/gnn/ (accessed 8 January 2012).

Oreskes, N. (2004). Beyond the ivory tower: The scientific consensus on climate change. *Science*, 306(5702), 1686.

Organisation for Economic Co-operation and Development Nuclear Energy Agency. (2002). *Chernobyl: Assessment of radiological and health impacts: 2002 update of Chernobyl: Ten years on*. http://www.oecd-nea.org/rp/reports/2003/nea3508-chernobyl.pdf (accessed 29 October 2012).

Oughton, H. D. (2011). Social and ethical issues in environmental risk management. *Integrated Environmental Assessment and Management*, 7(3), 404–405.

Paine, M. (2008, 22 October). Forest violence filmed [Includes video file]. *themercury.com.au*. http://www.themercury.com.au/article/2008/10/22/34071_tasmania-news.html (accessed 22 January 2012).

Pantti, M., Wahl-Jorgensen, K., & Cottle, S. (2012). *Disasters and the media*. London: Peter Lang.

Parsons, C. (1982). *True to nature: Christopher Parsons looks back on 25 years of wildlife filming with the BBC Natural History Unit.* Cambridge, UK: Patrick Stephens.

Paton, J. (2011). Climate change: The economics and politics of justice. *Australian Journal of Political Science, 46*, 353–358.

Peach, S. (2011, December 9). Did Muller's "BEST" study cool the heated global warming rhetoric? Yale Forum on Climate Change and the Media. http://www.yaleclimatemediaforum.org/2011/12/did-mullers-best-study-cool-the-heated-global-warming-rhetoric/ (accessed 6 January 2012).

Pearse, G. (2009). Quarry vision: Coal, climate change and the end of the resources boom. *Quarterly Essay, 33*.

Perfil. (2006, December 6). Para Duhalde, los ambientalistas "obstaculizan el diálogo". http://www.perfil.com/contenidos/2006/12/06/noticia_0041.html (accessed 14 January 2012).

Perko, T. (2011). Importance of risk communication during and after a nuclear accident. *Integrated Environmental Assessment and Management, 7*(3), 388–392.

Perlman, D. L., Adelson, G., & Wilson, E. O. (1997). *Biodiversity: Exploring values and priorities in conservation.* Boston: Blackwell Science.

Pernetta, J., & Hughes, P. (1990). *Implications of expected climatic changes in the South Pacific region: An overview.* UNEP Regional Seas Report no. 128. Nairobi: United Nations Environment Programme.

Peruzzotti, E. (2005). Demanding accountable government: Citizens, politicians, and the perils of representative democracy in Argentina. In S. Levistky & M. V. Murillo (Eds.), *The politics of institutional weakness: Argentine democracy.* University Park: Pennsylvania State University Press.

Peruzzotti, E., & Smulovitz, C. (2006). *Enforcing the rule of law: Social accountability in the new Latin American democracies.* Pittsburgh: University of Pittsburgh Press.

Peters, H. P. (1995). The interaction of journalists and scientific experts: Co-operation and conflict between two professional cultures. *Media, Culture & Society, 17*(1), 31–48.

Pew Research Center. (2010). State of the news media: Major trends. http://stateofthemedia.org/2010/overview-3/major-trends (accessed 12 January 2012).

Pew Research Center. (2011a, December 1). Modest rise in number saying there is "solid evidence" of global warming. http://www.people-press.org/2011/12/01/modest-rise-in-number-saying-there-is-solid-evidence-of-global-warming/ (accessed 8 December 2012).

Pew Research Center (2011b). State of the news media 2011: Overview. http://pewresearch.org/pubs/1924/state-of-the-news-media-2011 (accessed 16 December 2011).

Pew Research Center. (2011c). State of the news media 2011: Key findings. http://stateofthemedia.org/2011/overview-2/key-findings (accessed 13 January 2012).

Pickerill, J., Gillian, K., & Webster, F. (2011). Scales of activism: New media and transnational connections in anti-war movements. In S. Cottle & L. Lester (Eds.), *Transnational protests and the media* (pp. 41–48). New York: Peter Lang.

Pidgeon, N., Kasperson, R. E., & Slovic, P. (Eds.). (2003). *The social amplification of risk.* Cambridge, UK: Cambridge University Press.

Pliskin, N., & Romm, C. T. (1997). The impact of e-mail on the evolution of a virtual community during a strike. *Information and Management, 32*(5), 245–254.

Porritt, J. (1984). *Seeing green*. Oxford: Basil Blackwell.

Porritt, J., & Lawson, N. (2011, 7 December). Climate change: Is David Attenborough right? *RadioTimes*. http://www.radiotimes.com/news/2011-12-07/climate-change--is-david-attenborough-right (accessed 20 October 2012).

Potter, E. (2009). Climate change and the problem of representation. *Australian Humanities Review, 46*, 69–79.

Powell, R., & Chalmers, L. (2005). *The estimated economic impact of the proposed Port Stephens–Great Lakes Marine Park on commercial activities*. Armidale: Centre for Agricultural and Regional Economics Pty Ltd.

Powell, R., & Chalmers, L. (2006). *The estimated economic impact of Batemans Marine Park on commercial activities*. Armidale: Centre for Agricultural and Regional Economics Pty Ltd.

Prendergast, D. K., & Adams, W. M. (2003). Colonial wildlife conservation and the origins of the Society for the Preservation of the Wild Fauna of the Empire (1903–1914). *Oryx, 37*(2), 251–260.

Press Association. (2012, 17 February). David Cameron in France to sign nuclear power deal. *The Guardian*. http://www.guardian.co.uk/politics/2012/feb/17/david-cameron-france-nuclear-power (accessed 19 November 2012).

Preston, P. (2011, 28 August). The future may be online, but many will slip through the net. *The Observer*. http://www.guardian.co.uk/media/2011/aug/28/future-may-be-online-but-many-will-slip-through-the-net (accessed 19 January 2012).

Price, M., & Thompson, M. (Eds.). (2002). *Forging peace: Intervention, human rights and the management of media space*. Edinburgh: Edinburgh University Press.

Proctor, J. D., & Pincetl, S. (1996). Nature and the reproduction of endangered space: The spotted owl in the Pacific Northwest and southern California. *Environment and Planning D: Society and Space, 14*, 683–708.

Quantcast. (2011). Drudgereport Network. http://www.quantcast.com/drudgereport.com (accessed 2 January 2012).

Queensland Times, The. (1909, August 21). Among the mountains: On the main range. *The Queensland Times*, p. 3.

Queensland Times, The. (1910, 15 January). National parks. *The Queensland Times*, p. 4.

Queensland Times, The. (2010a, 6 August). Facebook hack attack unnerves pollie. *The Queensland Times*, p. 6.

Queensland Times, The. (2010b, 20 July). Green goes it alone on preferences. *The Queensland Times*, p. 5.

Queensland Times, The. (2010c, 23 August). Greens buoyed by strong showing in region. *The Queensland Times*, p. 5.

Queensland Times, The. (2010d, 22 May). Greens pick Petersen. *The Queensland Times*, p. 3.

Queensland Times, The. (2010e, 30 July). I'm not a criminal. *The Queensland Times*, p. 1.

Queensland Times, The. (2010f, 19 July). State knocks back toxic dump plan. *The Queensland Times*, p. 3.

Queenslander. (1919, 8 February). The national park. *Queenslander*, p. 41.

Rahu, M. (2003). Health effects of the Chernobyl accident: Fears, rumors and the truth. *European Journal of Cancer, 39*, 295–299.

Ramana, V. M. (2011). Nuclear power and the public. *Bulletin of the Atomic Scientists, 67*(4), 43–51.

Rankin, M. B. (1993). Some observations on a Chinese public sphere. *Modern China*, 19(2), 158–182.

Rayner, G. (2011, 27 November). Who's to blame for climategate? *The Telegraph*. http://www.telegraph.co.uk/earth/copenhagen-climate-change-confe/6672875/Whos-to-blame-for-Climategate.html (accessed 3 January 2012).

Raz, G. (2009, 22 November). Scientist explains earth's warming plateau [Audio recording and transcript]. National Public Radio (NPR). http://www.npr.org/templates/story/story.php?storyId=120668812&ft=1&f=1007 (accessed 15 March 2011).

Redhead, S. (2006). The art of the accident: Paul Virilio and accelerated modernity. *Fast Capitalism*, 2(1). http://www.uta.edu/huma/agger/fastcapitalism/2_1/redhead.html

Reese, S. D. (2003). Framing public life: A bridging model for media research. In S. D. Reese, O. H. Gandy, & A. Grant (Eds.), *Framing public life: Perspectives on media and our understanding of the social world* (pp. 7–31). Mahwah: Lawrence Erlbaum Associates.

Rettie, R. (2003). Connectedness, awareness and social presence. Paper presented at PRESENCE 2003, Aalborg, Denmark.

Reuters. (2012, 10 April). UK in nuclear decommissioning deal with Japan. http://www.reuters.com/article/2012/04/10/japan-britain-nuclear-idUSL6E8FA3JP20120410 (accessed 19 November 2012).

Reynolds, I. (2011, 15 March). Toshiba shares hit by fears over future of nuclear. Reuters. http://www.reuters.com/article/2011/03/15/us-japan-quake-toshiba-idUSTRE72E1H220110315 (accessed 15 July 2011).

Ribbon, A. (2002, 3 November). Airport ad gets the axe. *The Mercury*, p. 3.

Rigg, K. (2011, 26 January). Skepticgate: Revealing climate denialists for what they are. *The Huffington Post*. http://www.huffingtonpost.com/kelly-rigg/skepticgate-revealing-cli_b_814013.html (accessed 5 October 2011).

Rignot, E., Velicogna, I., Van den Broeke, M. R., Monaghan, A., & Lenaerts, J. (2011). Acceleration of the contribution of the Greenland and Antarctic ice sheets to sea level rise. *Geophysical Research Letters*, 38, L05503. doi:10.1029/2011GL046583.

Rintoul, S. (2009, 12 December). Town of Beaufort changed Tony Abbott's view on climate change. *The Australian*. http://www.theaustralian.com.au/politics/the-town-that-turned-up-the-temperature/story-e6frgczf-1225809567009 (accessed 29 October 2012).

Rival, L. (1998). Trees, from symbols of life and regeneration to political artefacts. In L. Rival (Ed.), *The social life of trees: Anthropological perspectives on tree symbolism* (pp. 1–38). Oxford: Berg.

Roberts, J., & Nash, C. (2009). Reporting controversy in health policy: A content and field analysis. *Pacific Journalism Review*, 15(2), 35–53.

Rojecki, A. (2011). Leaderless crowds, self-organizing publics, and virtual masses. In S. Cottle & L. Lester (Eds.), *Transnational protests and the media* (pp. 87–98). New York: Peter Lang.

Rojek, C. (2001). *Celebrity*. London: Reaktion Books.

Romm, J. (2010, 11 January). FoxNews, WattsUpWithThat push falsehood-filled *Daily Mail* article on global cooling that utterly misquotes, misrepresents work of Mojib Latif and NSIDC. *ThinkProgress*. http://thinkprogress.org/climate/2010/01/11/205330/foxnews-wattsupwith-that-climatedepot-daily-mail-article-on-global-cooling-mojib-latif/ (accessed 17 January 2011).

Rose, D. (2010, 9 January). The mini ice age starts here. *Mail Online*. http://www.dailymail.co.uk/sciencetech/article-1242011/DAVID-ROSE-The-mini-ice-age-starts-www.here.html (accessed 4 January 2012).

Rose, N., & Miller, P. (1992). Political power beyond the state: Problematics of government. *The British Journal of Sociology*, 43(2), 173–205.

Rosenthal, E. (2011, 15 October). Where did global warming go? *The New York Times*. http://www.nytimes.com/2011/10/16/sunday-review/whatever-happened-to-global-warming.html?pagewanted=all (accessed 16 October 2011).

Rosenthal, J. (2011). Level 7 major nuclear accidents: Chernobyl death toll and Fukushima. *The Huffington Post*. http://www.huffingtonpost.com/john-rosenthal/level-7-major-nuclear-acc_b_852666.html (accessed 16 July 2011).

Rossiter, D. (2004).The nature of protest: Constructing the spaces of British Columbia's rainforests. *Cultural Geographies*, 11(2), 139–164.

Rousey, J. (2010, 12 July). Media not excited anymore about debunked climategate scandal. http://www.mediaite.com/online/media-not-excited-anymore-about-debunked-climategate-scandal (accessed 12 March 2011).

Rowe, W. T. (1993). The problem of "civil society" in late imperial China. *Modern China*, 19(2), 143–148.

Roy, P., & Connell, J. (1991). Climatic change and the future of atoll states. *Journal of Coastal Research*, 7, 1057–1075

Rudd, K. (2006, October). Faith in politics. *The Monthly*. http://www.themonthly.com.au/faith-politics-kevin-rudd-300 (accessed 11 November 2011).

Rudd, K. (2007). *Climate change: Forging a new consensus*. National Climate Change Summit, Parliament House, Canberra.

Russell, C. (2008). Climate change: Now what? *Columbia Journalism Review*, 47, 45–49.

Ryan, Y. (2010). COP 15 and Pacific Island states: A collective voice on climate change. *Pacific Journalism Review*, 16(1), 193–203.

Saad, L. (2011, 28 March). Water issues worry Americans most, global warming least. *Gallup Politics*. http://www.gallup.com/poll/146810/water-issues-worry-americans-global-warming-least.aspx (accessed 28 December 2011).

Said, E. (1993). *Culture and imperialism*. New York: Vintage.

Sandman, P., & Lanard, J. (2004). *Crisis communication: Guidelines for action*. Fairfax: American Industrial Hygiene Association.

Saving Agarikon (Fomitopsis officinalis) from an Oregon old growth forest about to be clearcut [Video file]. (2009, 30 September). http://www.youtube.com/watch?v=cT4yBdLKljo (accessed 2 January 2012).

Scalmer, S., & Goot, M. (2004). Elites constructing elites: News Limited's newspapers, 1996–2002. In B. Hindess & M. Sawer (Eds.), *Us and them: Anti-elitism in Australia*. Perth: API Network.

Schedler, E. P., Glastra, F. F., & Kats, E. (1998). Public information and field theory. *Political Communication*, 15, 445–461.

Scheufele, A. D. (2002). Examining differential gains from mass media and their implications for participatory behavior. *Communication Research*, 29(1), 46–65.

Schiermeier, Q. (2011, 27 May). Wildlife threatened by Fukushima radiation. *Nature*. http://www.nature.com/news/2011/270511/full/news.2011.326.html (accessed 12 November 2012).

Schlosberg, D. (2007). *Defining environmental justice: Theories, movements, and nature*. New York: Oxford University Press.

Schmid, D. S. (2011). When safe enough is not good enough: Organizing safety at Chernobyl. *Bulletin of the Atomic Scientists*, 67(2), 19–29.

Schmidt, G., & Wolfe, J. (Eds.). (2009). *Climate change: Picturing the science*. New York: W. W. Norton and Company.

Schmitter, P. C. (1974). Still a century of corporatism? In F. B. Pike & T. Stritch (Eds.), *Social-political structures in the Iberian world* (pp. 85–130). Notre Dame: University of Notre Dame Press.

Schwarze, S. (2011, 19 November). An incapacitated truth: Voice, strategy, and networks in climate change communication. Paper delivered at National Communication Association meeting, New Orleans.

Seaton, A. V. (n.d.). *The occupational influences and ideologies of travel writers: Freebies? Puffs? Vade mecums? Or belles lettres?* Newcastle upon Tyne: Centre for Travel and Tourism in Association with Business Education Publishers.

Secord, J. A. (1996). Epilogue: The crisis of nature. In N. Jardine, J. A. Secord, & E. C. Spary (Eds.), *Cultures of natural history* (pp. 447–459). Cambridge, UK: Cambridge University Press.

Seddon, G. (1997). *Landprints: Reflections on place and landscape*. Cambridge, UK: Cambridge University Press.

Segerberg, A., & Bennett, W. L. (2011). Social media and the organization of collective action: Using Twitter to explore the ecologies of two climate change protests. *The Communication Review*, 14(3), 197–215.

Seon-Kyoung, A., & Gower, K. K. (2009). How do the news media frame crises? A content analysis of crisis news coverage. *Public Relations Review*, 35, 107–112.

Shabecoff, P. (2000). *Earth rising: American environmentalism in the 21st century*. Washington, DC: Island Press.

Shapiro, J. (2001). *Mao's war against nature*. Cambridge, UK: Cambridge University Press.

Shapiro, M. (2004). *A sense of place: Great travel writers talk about their craft, lives and inspiration*. San Francisco: Travelers' Tales Books.

Shepard, A. (2011). Online comments: Dialogue or diatribe? *Nieman Reports*. http://www.nieman.harvard.edu/reportsitem.aspx?id=102647 (accessed 20 September 2011).

Shilton, K. (2010). Participatory sensing: Building empowering surveillance. *Surveillance & Society*, 8(2), 131–150.

Sich, S. A. (1996, May/June). Truth was an early casualty. *The Bulletin of the Atomic Scientists*, 32–42.

Silverstone, R. (1984). Narrative strategies in television science—a case study. *Media, Culture & Society*, 6, 337–410.

Sklair, L. (2001). *The transnational capitalist class*. Oxford: Blackwell.

Slovic, P. (2000). *The perception of risk*. London: Earthscan.

Smith, D., & McCloskey, J. (1998, October/December). Risk communication and the social amplification of the public sector risk. *Public Money and Management*, 41–50.

Smith, J. (2011, 7 April). A long shadow over Fukushima. *Nature*, 472(7). http://www.nature.com/news/2011/110405/full/472007a.html (accessed 29 October 2012).

Snickars, P., & Vonderau, P. (Eds.) (2009). *The YouTube reader*. Stockholm: National Library of Sweden.

Soble, J. (2011, 28 October). Japan's timid media in spotlight. *FT.com*. http://www.ft.com/intl/cms/s/0/256c607c-016c-11e1-ae24-00144feabdc0.html#axzz2Act7O7FT (accessed 29 October 2012).

Society of American Travel Writers Foundation. (n.d.) Lowell Thomas Travel Journalism Competition awards for work published in 2009–2010. http://www.satwf.com/2010_Lowell_Thomas_Travel_Journalism_Winners/2010_List_of_Winners (accessed 29 October 2012).

SockPuppet. (2010). Why is this porn advocate complaining about porn?!!!? [Web log comment]. *Iain Hall's Sandpit*. http://iainhall.wordpress.com/2010/07/10/why-is-this-porn-advocate-complaining-about-porn/ (accessed 21 July 2010).

Source Watch. (2011). Matt Drudge. http://www.sourcewatch.org/index.php/Matt_Drudge (accessed 10 April 2011).

Spalding, M., Wood, L., Fitzgerald, C., & Gjerde, K. (2010). The 10% target: Where do we stand? In C. Toropova, I. Meliane, D. Laffole, E. Matthews, & M. Spalding (Eds.), *Global ocean protection: Present status and future possibilities* (pp. 25–40). Cambridge, UK: UNEP-WC-MC..

Spark, A. (2010, 8 June). A few simple questions must be answered. *The Queensland Times*, p. 8.

Spash, C. L. (2007). The economics of climate change impacts à la Stern: Novel and nuanced or rhetorically restricted? *Ecological Economics*, 63, 706–713.

Spencer, M. (2010, February). Environmental journalism in the greenhouse era. *Fairness and Accuracy in Reporting (FAIR)*. http://fair.org/extra-online-articles/environmental-journalism-in-the-greenhouse-era/ (accessed 11 December 2011).

Stein, L. (2011). Environmental website production: A structuration approach. *Media, Culture & Society*, 33(3), 363–384.

Stern, N. H. (2007). *The economics of climate change*: The Stern Review. Cambridge, UK: Cambridge University Press.

Stewart, K. (2007). *Ordinary affects*. Durham: Duke University Press.

Street, J. (2004). Celebrity politicians: Popular culture and popular representation. *The British Journal of Politics & International Relations*, 6(4), 435–452.

Stuart, N. (2010). *Rudd's way: November 2007–June 2010*. Carlton North: Scribe Publications.

Sullivan, A. (1985). *Greening the Tories: New policies on the environment*. Policy study 72. London: Centre for Policy Studies.

Sutton, S. G., & Tobin, R. C. (2009). Recreational fishers' attitudes towards the 2004 rezoning of the Great Barrier Reef Marine Park. *Environmental Conservation*, 36, 245–252.

Svampa, M., & Antonelli, M. A. (Eds.). (2010). *Mineria trasnacional: Narrativas del desarrollo y resistencias sociales*. Buenos Aires: Biblos.

Swyngedouw, E. (2007). Impossible "sustainability" and the post-political condition. In R. Krueger & D. Gibbs (Eds.), *The sustainable development paradox: Urban political economy in the United States and Europe* (pp. 13–40). New York: Guilford Press.

Swyngedouw, E. (2010). Apocalypse forever? Post-political populism and the spectre of climate change. *Theory Culture & Society*, 27(2–3), 213–232.

Tanaka, M. (2012). A year since the Earthquake. *Mitsutoyo Report*, No. 254 http://www.mitutoyo.co.jp/new/report/no254/kantogen/index.html (accessed 13 February 2013).

Tanner, L. (2011). *Sideshow: Dumbing down democracy*. Melbourne: Scribe Publications.

Tasmania's ancient forests—under threat [Video file]. (2007, 19 June). http://www.youtube.com/watch?v=O4tw7toP9js (accessed 2 January 2012).

Tasmania's forests: A global treasure, a national responsibility. [Video file]. (2006, 6 November 6). http://www.youtube.com/watch?v=5mNc6-jIJkY (accessed 2 January 2012).

Tasmanian Department of Tourism, Sport and Recreation. (1995). *Strategies for growth*. Hobart: Tasmanian Department of Tourism, Sport and Recreation.

Tasmanian forestry contractors attack protestors [Video file]. (2008, 21 October). This video is no longer available on YouTube. Original video available from author.

Tertrais, B. (2011). Black swan over Fukushima. *Survival*, 53(3), 91–100.

The PM & the frog—Australia still logging old growth forests [Video file]. (2009, 19 July). http://www.youtube.com/watch?v=anetSXEHuu8 (accessed 2 January 2012).

The Tally Room. (2010). Blair—election 2010. http://www.tallyroom.com.au/election-2010/blair (accessed 11 June 2010).

The Upper Florentine trilogy: Part one [Video file]. (2007, 27 August). http://www.youtube.com/watch?v=LlrdvYr6wvQ (accessed 2 January 2012).

Thomassin, A., White, C. S., Stead, S. S., & Gilbert, D. (2010). Social acceptability of a marine protected area: The case of Reunion Island. *Ocean & Coastal Management*, 53(4), 169–179.

Thompson, C. (2008, 7 September). I'm so totally, digitally close to you. *The New York Times Magazine*, p. 42.

Threats to spotted owls [Video file]. (2007, 14 November). http://www.youtube.com/watch?v=LJL7XD5DU7A (accessed 2 January 2012).

Toller, V. (2005, 2 May). Marcha masiva contra dos papeleras sobre el Uruguay. *Clarín*.

Toller, V. (2007, 29 August). Unpublished interview with Silvio Waisbord.

Tomlinson, J. (1999). *Globalization and culture*. Cambridge, UK: Polity Press.

Tomlinson, J. (2007). *The culture of speed: The coming of immediacy*. London: Sage.

Torgerson, D. (1994). Strategy and ideology in environmentalism: A decentered approach to sustainability. *Industrial and Environmental Crisis Quarterly*, 8(4), 295–321.

Torgerson, D. (2000). Farewell to the green movement? Political action and the green public sphere. *Environmental Politics*, 9(4), 1–19.

Toman, M. A. (2006). Values in the economics of climate change. *Environmental Values*, 15, 365–379.

Tosi, M.C. (2006, 15 July). Gualeguaychú mostró que su espíritu combativo sigue intacto. *La Nación*.

Tourism Tasmania. (2008). Tasmania: Discover Australia's natural state. http://www.discovertasmania.co.uk (accessed 8 February 2009). This webpage is no longer available. Printed hard copy of webpage available from author.

Tourism Tasmania. (2011). Motivations research. http://www.tourismtasmania.com.au/__data/assets/pdf_file/0009/47178/motivations.pdf (accessed 8 February 2013).

Tourism Tasmania. (2012). Visiting journalist program. http://www.tourismtasmania.com.au/media/vjp (accessed 8 May 2012).

Tourism Tasmania North America. (2008). Discover Tasmania: Australia's natural state. http://www.discovertasmania.com/__data/assets/pdf_file/0008/57590/MotvBrochure_NAM_web.pdf (accessed 13 March 2012).

Trettin, L., & Musham, C. (2000). Is trust a realistic goal of environmental risk communication? *Environment & Behavior,* 32(3), 410–426.

Tsouvalis, J. (2000). *A critical geography of Britain's state forests.* Oxford: Oxford University Press.

Tuchman, G. (1978). *Making news: A study in the construction of reality.* New York: Free Press.

Turner, G. (2004). *Understanding celebrity.* London: Sage Publications.

Turner, G., Bonner, F., & Marshall, D. P. (2000). *Fame games: The production of celebrity in Australia.* Cambridge, UK: Cambridge University Press.

Ungar, S. (2000). Knowledge, ignorance and the popular culture: Climate change versus the ozone hole. *Public Understanding of Science,* 9(3), 297–312.

Unger, J., & Chan, A. (1995). China, corporatism, and the East Asian model. *The Australian Journal of Chinese Affairs,* 33, 29–53.

United Nations. (1998). The United Nations Scientific Committee on the Effects of Atomic Radiation (UNSCEAR) report: Sources, effects and risks of ionizing radiation. New York: United Nations.

United Nations Environment Programme. (2009). *From conflict to peacebuilding: The role of natural resources and the environment.* Nairobi: United Nations Environment Programme.

United Nations Conference on Sustainable Development Rio+20 Communiqué. (2012, 17 June). *The future we choose: Declaration from the high-level dialogue on global sustainability.* http://www.un.org/gsp/sites/default/files/event_attachments/20120618%20RioDeclaration.pdf (accessed 30 January 2013).

United Nations International Strategy for Disaster Reduction. (2012). *Disasters through a different lens: Behind every effect there is a cause.* New York: United Nations International Strategy for Disaster Reduction.

Urkidi, L. (2010). A global environmental movement against gold mining: Pascua-Lama, Chile. *Ecological Economics,* 70(2), 219–227.

U.S.A. Department of Commerce. (2010). *Mining industry overview and exporting opportunities in Latin America.* Special report compiled by the Trade Winds Forum 2010–The Americas Team. http://www.globalvirginia.com/Information%20Documents/Latin_America_Mining_Guide.pdf (accessed 18 February 2013).

Van Aelst, P., & Walgrave, S. (2004). New media, new movements? The role of the Internet in shaping the "anti-globalization" movement. In W. Van de Donk, B. Loader, P. Nixon, & D. Dieter (Eds.), *Cyberprotest: New media citizens and social movements* (pp. 97–122). London: Routledge.

Van de Donk, W., Loader, B., Nixon, P., & Dieter, D. (Eds.). (2004). *Cyberprotest: New media, citizens and social movements.* London: Routledge.

Vanderford, M. L. (1989). Vilification and social movements: A case study of pro-life and pro-choice rhetoric. *Quarterly Journal of Speech, 75*(2), 166–182.

Van Dijk, T. (2002). Political discourse and political cognition. In P. A. Chilton & C. Schäffner (Eds.), *Politics as text and talk: Analytic approaches to political discourse* (pp. 203–237). Amsterdam: John Benjamins Publishing Company.

Vexnews. (2010). Hajnal Ban: With friends like these she doesn't need enemies. http://www.vexnews.com/news/9676/hajnal-ban-with-friends-like-these-she-doesnt-need-enemies/ (accessed 12 December 2011).

Virilio, P. (2007). *The original accident*. Cambridge, UK: Polity Press.

Vivanco, L. (2002). Seeing green: Knowing and saving the environment on film. *Visual Anthropology, 104*(4), 1195–1204.

Vivanco, L. (2004). The work of environmentalism in the age of televisual adventures. *Cultural Dynamics, 16*(1), 5–27.

Voyer, M., Gladstone, W., & Goodall, H. (2012). Methods of social assessment in marine protected area planning: Is public participation enough? *Marine Policy, 36*, 432–439.

Vromen, A., & Coleman, W. (2011). Online movement mobilisation and electoral politics: The case of Getup! *Communication, Politics & Culture, 44*(2), 76–94.

Vromen, A., Gelber, K., & Gauja, A. (2009). *Powerscape: Contemporary Australian politics*. Crows Nest: Allen & Unwin.

Waisbord, S. (2000). *Watchdog journalism in South America*. New York: Columbia University Press.

Waisbord, S. (2011). Can NGOs change news? *International Journal of Communication, 5*, 142–165.

Wakeman, F., Jr. (1993). The civil society and public sphere debate: Western reflections on Chinese political culture. *Modern China, 19*(2), 108–38.

Walker, J. (2009, 19 December). Scientists "crying wolf" over coral: Hopes fade in Copenhagen, rise on the reef. *The Australian*, p. 1.

Walker, J. S. (2004). *Three Mile Island: A nuclear crisis in historical perspective*. Berkeley: University of California Press.

Walsh, J. E. (1983, March). Three Mile Island: Meltdown of democracy? *The Bulletin of the Atomic Scientists*, 57–60.

Walter, M., & Martinez-Alier, J. (2010). How to be heard when nobody wants to listen: Community action against mining in Argentina. *Canadian Journal of Development Studies, 30*(1–2), 281–301.

Wang, L. (2005). *Lü meiti: Zhongguo huanbao chuanmei yanjiu (Green media: Environmental communication in China)*. Beijing: Tsinghua University Press.

Wang, S., & He, J. (2004). Associational revolution in China: Mapping the landscapes. *Korea Observer, 35*(3), 485–534.

Warner, M. (2002). Publics and counterpublics. *Public Culture, 14*(1), 49–90.

Warner, R. (2008). A comparison of ideas in the development and governance of national parks and protected areas in the US and Canada. *International Journal of Canadian Studies, 37*, 13–39.

Washington Post–ABC News. (2009, 10–13 December). Washington Post–ABC News poll. http://www.washingtonpost.com/wp-srv/politics/polls/postpoll_121509.html (accessed 10 March 2011).

Watts, A. (2009, 28 November). "Climategate" surpasses "global warming" on Google. *Watts Up with That?* http://wattsupwiththat.com/2009/11/28/climategate-surpasses-global-warming-on-google-autosuggest-still-blocked/ (accessed 12 January 2012).

Weber, M. (1968). *Economy and society: An outline of interpretive sociology*. New York: Bedminster Press. (Original work published 1914).

Wei, Z. (2000). Zoujin xianshi, zaisi huanbao (Up close to reality, rethinking environmental protection). In *Ba qian li lu yun he yue: 2000 Zhongguo daxuesheng lüseying wenji (Eight thousand li of road, clouds, and moonlight: Collected works of Chinese university students' green camp in 2000)*. Beijing: Green Camp Publication. Document on file with author.

Weible, C. (2008). Caught in a maelstrom: Implementing California marine protected areas. *Coastal Management, 36*, 350–373.

Wescott, G. (2006). The long and winding road: The development of a comprehensive, adequate and representative system of highly protected marine protected areas in Victoria, Australia. *Ocean & Coastal Management, 49*, 905–922.

Westcott, C. W. (1991). Australia's distinctive national parks system. *Environmental Conservation, 18*(4), 331–340.

Wet Tropics Aboriginal Plan Project Team. (2005). *Caring for country and culture—The Wet Tropics Aboriginal Cultural and Natural Resource Management Plan*. Cairns: Rainforest CRC and FNQ NRM Ltd.

White, G. (1999). *Sacred summits: John Muir's greatest climbs*. Edinburgh: Canongate.

White, J. (1997). What is a veteran tree and where are they all? *Quarterly Journal of Forestry, 91*(3), 222–226.

Whitlock, G., & Carter, D. (1992). *Images of Australia: An introductory reader in Australian studies*. St Lucia: University of Queensland Press.

Wikipedia. (2012). 2009 United Nations Climate Change Conference. http://en.wikipedia.org/wiki/COP15 (accessed 14 March 2012).

"Wild Tasmania" trailer [Video file]. (2007, 29 September). http://www.youtube.com/watch?v=V4FPwDMxA1E (accessed 2 January 2012).

Wilson, E. O. (1984). *Biophilia: The human bond with other species*. Cambridge, MA: Harvard University Press.

Wilson, K. M. (2000). Drought, debate and uncertainty: Measuring reporters' knowledge and ignorance about climate change. *Public Understanding of Science, 9*, 1–13.

Winter wren bird song in old growth forest of North Cascades [Video file]. http://www.youtube.com/watch?v=jooQVThca3U (accessed 2 January 2012).

Wolfenden, J., Cram, F., & Kirkwood, B. (1994). Marine reserves in New Zealand: A survey of community reactions. *Ocean & Coastal Management, 25*, 31–51.

Wolfsfeld, G. (1997). *Media and political conflict: News from the Middle East*. Cambridge, UK: Cambridge University Press.

Wolfsfeld, G. (2004). *Media and the path to peace*. Cambridge, UK: Cambridge University Press.

World Intellectual Property Organization. (2003). World Intellectual Property Organization WIPO Arbitration and Mediation Center Administrative Panel decision: The Crown in right of the

State of Tasmania trading as "Tourism Tasmania" v. Gordon James Craven case no. DAU2003-0001. Australia Domain Name Decisions, Australian Legal Information Institute, University of Technology Sydney and University of New South Wales Faculties of Law. http://www.austlii.edu.au/cgi-bin/sinodisp/au/cases/cth/AUDND/2003/4 (accessed 22 January 2012).

World Nuclear Association. (2012). Reactor database. http://www.world-nuclear.org/ (accessed 14 January 2013).

Worsley, S. (2010, 16 June). SMS. *The Queensland Times*, p. 9.

Wynne, B. (1996). May the sheep safely graze? A reflexive view of the expert–lay knowledge divide. In S. Lash, B. Szerszynski, & B. Wynne (Eds.), *Risk, environment and modernity* (pp. 27–43). London: Sage Publications.

Yamamura, E. (2012). Effect of free media on views regarding nuclear energy after the Fukushima accident. *Kyklos*, 65(1), 132–141.

Yang, G. (2002). Civil society in China: A dynamic field of study. *China Review International*, 9(1), 1–16.

Yang, G. (2005). Environmental NGOs and institutional dynamics in China. *China Quarterly*, 181, 46–66.

Yang, G. (2006a). Activists beyond virtual borders: Internet-mediated networks and informational politics in China. *First Monday*, 11(9). http://firstmonday.org/issues/special11_9/yang/index.html (accessed 14 January 2013).

Yang, G. (2006b). Between control and contention: China's new Internet politics. *Washington Journal of Modern China*, 8(1), 30–47.

Yang, M. (2004). Spatial struggles: Postcolonial complex, state disenchantment, and popular reappropriation of space in rural southeast China. *The Journal of Asian Studies*, 63(3), 719–755.

Yardley, J. (2004, 10 March). Dam building threatens China. *The New York Times*.

Yardley, W., & Perez-Pena, R. (2009, March 16). Seattle paper shifts entirely to the Web. *New York Times*. http://www.nytimes.com/2009/03/17/business/media/17paper.html?_r=0 (accessed 28 March 2009).

Yokouchi, N., Abe, S., Shibata, I., Minamide, M., & Kato, H. (2012). Newspaper reports on East Japan Great Earthquake in four countries: Comparative analysis with articles during one month after the disaster. *Sociotechnica*, 9, 1-29. http://shakai-gijutsu.org/vol9/9_1.pdf (accessed 13 February 2013).

Young, S. (2011). *How Australia decides: Election reporting and the media*. Cambridge, UK: Cambridge University Press.

Yusoff, K., & Gabrys, J. (2011). Climate change and the imagination. *Wiley Interdisciplinary Reviews: Climate Change*, 2(4), 516–534.

Zehr, S. C. (2000). Public representations of scientific uncertainty about global climate change. *Public Understanding of Science*, 9(2), 85–103.

Zelter, A. (1998). Grassroots campaigning for the world's forests. In L. Rival (Ed.), *The social life of trees: Anthropological perspectives on tree symbolism* (pp. 221–232). Oxford: Berg.

Zhao, Y.. (2000). From commercialization to conglomeration: The transformation of the Chinese press within the orbit of the party state. *Journal of Communication*, 50(2), 3–26.

Zhaxiduojie, H. (2002). Sanjiangyuan de huhuan (The call of the three river source). *Friends of Nature Newsletter,* 4. http://www.fon.org.cn/index.php?id=3009 (accessed 2 April 2003). This webpage is no longer available. Printed hard copy of webpage available from author.

Zhou, Y. (2005). Living on the cyber border. *Current Anthropology,* 46(1), 779–803.

Contributors

WENDY BACON is Professor at the Australian Centre for Independent Journalism at the University of Technology, Sydney. She is an investigative journalist and a contributing editor to *New Matilda*. She is a board member of the Pacific Media Centre at the Auckland University of Technology. She was Australian Director of the Global Environmental Journalism Initiative 2009–2011.

DANIEL BROCKINGTON is Professor of Conservation and Development at the IDPM, University of Manchester. His research covers diverse aspects of conservation, development and celebrity. His recent books are *Celebrity and the Environment* (2009, ZED books) and *Nature Unbound* (2008, Earthscan, with Rosaleen Duffy and Jim Igoe).

CRAIG CALHOUN's world-recognized work connects sociology to culture, communication, politics, philosophy and economics. Before becoming LSE Director in 2012, he was University Professor at New York University. He is the author of several books including *Nations Matter, Critical Social Theory, Neither Gods Nor Emperors* and *The Roots of Radicalism*.

CATHERINE ANN COLLINS (PhD, University of Minnesota) is Professor of Rhetoric and Media Studies at Willamette University, and writes and teaches in the areas of environmental and visual rhetoric and narrative and trauma theory. Recent publications are in *Continuum, Interdisciplinary Aspects of Climate Change*, and *Challenges in International Communication*.

SIMON COTTLE is Professor of Media and Communications and Head of the School of Journalism, Media and Cultural Studies at Cardiff University. His latest books are *Disasters and the Media* (2012) with Mervi Pantti and Karin Wahl-Jorgensen, *Transnational Protests and the Media* (2011) co-edited with Libby Lester, and *Global Crisis Reporting* (2009). He is Series Editor of the Global Crises and the Media Series for Peter Lang Publishing.

ROBERT COX is Professor Emeritus of Communication Studies and the Curriculum in the Environment at the University of North Carolina at Chapel Hill. He is author of *Environmental Communication and the Public Sphere* (2013) and is a former president of the Sierra Club, the oldest environmental organization in the United States.

TANJA DREHER is Lecturer in Media and Communications and Convenor of the Major in International Media and Communication at the University of Wollongong. Tanja's research focuses on the politics and practices of "listening" in contexts including feminist anti-racism work, media and climate change and the impacts of Indigenous media in the non-Indigenous public sphere.

WILLIAM GLADSTONE is a marine biologist who undertakes research that supports decision making for environmental management and marine conservation, and has a particular interest in cross-disciplinary research. He has worked on marine conservation programs in various parts of Australia, and in the Red Sea and Gulf of Aden, the Gulf, Oman, and Coral Triangle (Indonesia–Malaysia–Philippines). He is Professor and Head of the School of the Environment, University of Technology, Sydney.

HEATHER GOODALL is Professor of History at University of Technology, Sydney. She has published on Indigenous people's cultural and environmental relationships to land and water in colonial and contemporary Australia. She teaches in Communications and Social Inquiry programs in Aboriginal, Indian Ocean, colonial and environmental histories. Her recent work has been on conflicting perceptions of environmental change and rivers.

MYRA GURNEY teaches communication theory and professional writing in the School of Humanities and Communication Arts at the University of Western Sydney. She is currently completing a PhD examining the discursive characteristics and politics of the climate change debate in Australia.

BRETT HUTCHINS is Associate Professor and Co-Director of the Research Unit in Media Studies at Monash University in Melbourne. His most recent articles appear in *Media, Culture & Society*, *Journalism: Theory, Practice & Criticism* and *Information, Communication & Society*. He is co-author of *Sport Beyond Television: The Internet, Digital Media and the Rise of Networked Media Sport* (Routledge, 2012) with David Rowe.

KUMI KATO, PhD is currently Professor in Environmental Studies at Wakayama University, Japan, and Honorary Associate Professor at the University of

Queensland, Australia. She is a member of the Kangaloon Group for Creative Ecologies and Sustainability Frontiers. Her main interest, ethical expression for the natural world, was the subject of a radio documentary, "Waiting for the Tide".

CLIO KENTERELIDOU is tenured Senior Teaching Fellow and Instructor in the School of Journalism and Mass Communication, Aristotle University of Thessaloniki, Greece. She holds a PhD in Political Communication (A.UTh., Greece), a Master of Arts in Communications Studies (University of Leeds, UK) and a First Degree in Economics (A.U.Th., Greece).

LIBBY LESTER is Professor of Journalism, Media and Communications at the University of Tasmania. Her recent research has appeared in *Media, Culture & Society, Environmental Communication, Journalism: Theory, Practice and Criticism,* and in the books *Media and Environment: Conflict, Politics and the News* (2010) and *Transnational Protest and the Media* (2011, co-edited with Simon Cottle).

ALEX LOCKWOOD is Senior Lecturer in Journalism and member of the Centre for Research in Media and Cultural Studies at the University of Sunderland. His research draws together interests in environmental responsibility, media, creative practices and affect.

LYN MCGAURR completed her PhD in the journalism program at the University of Tasmania. Her research interests include environmental conflict and the media, source-media relations, travel journalism, cosmopolitanism, place branding, and representations of environmental risk in news journalism. She has professional experience in journalism, book editing and public relations.

MICHAEL MEADOWS worked as a print and broadcast journalist for ten years before moving into journalism education in the late 1980s. He is based at the Centre for Cultural Research at Griffith University's Nathan Campus in Brisbane. His research interests include media representation of minorities, community media audiences, policy and practice, and media and images of landscape.

CHRISTINE MILNE is a senator for Tasmania and Leader of the Australian Greens. She is one of Australia's most experienced and respected environmental and community activists, with a career spanning thirty years. After leading the successful campaign to protect farming land and fisheries from the Wesley Vale Pulp Mill, Christine was elected to the Tasmanian parliament in 1989, and became the first woman to lead a political party in Tasmania in 1993. She was elected to the senate in 2004 and to the leadership in 2012 following the retirement of Senator Bob Brown.

ALANNA MYERS is a PhD candidate in the School of Culture and Communication at the University of Melbourne. Her thesis explores literary and cultural framings of nature as leisure space v. economic resource with reference to contemporary environmental activism. Her honours research analyzed climate change skepticism in the Australian media.

CHRIS NASH teaches and researches in the journalism program at Monash University. He has a professional background in radio, television and film journalism, and research interests in journalism's relationship to the built and natural environment, journalism as a disciplinary research practice in a scholarly context, and journalism and art.

DANIEL PALMER is Senior Lecturer in Art History & Theory in the Faculty of Art, Design and Architecture at Monash University. His writings on photography have appeared in journals such as *Photographies*, *Philosophy of Photography* and *Angelaki*, and he is the co-author of *Twelve Australian Photo Artists* (2009).

MORGAN RICHARDS holds a PhD in sociology from the University of Cambridge and is a Postdoctoral Research Fellow in the Centre for Critical and Cultural Studies, University of Queensland. She is currently working on *Wild Visions: The Rise of Wildlife Documentary*, which examines the British tradition of wildlife documentary.

ROBERT THOMSON is a Brisbane-based researcher.

KITTY VAN VUUREN is Lecturer at the School of Journalism & Communication at the University of Queensland, Australia. She teaches environmental communication and communication for social change. Her research interests include environmental communication, public opinion, and community participation with a particular focus on rural, regional and remote communities.

MICHELLE VOYER is a PhD candidate working on the human dimension of environmental policy, including the role and influence of the media in policy debates over environmental protection, especially marine protected areas and climate change. With a background in policy implementation, her interests lie in the practical application of government policy and science in a community setting.

SILVIO WAISBORD is Professor in the School of Media and Public Affairs at George Washington University. He is the editor-in-chief of the *International Journal of Press/Politics*. His most recent books are *Reinventing Professionalism:*

Journalism and News in Global Perspective and the *Handbook of Global Health Communication* (co-edited with Rafael Obregon).

GUOBIN YANG is Associate Professor of Communication and Sociology in the Annenberg School for Communication and Department of Sociology at the University of Pennsylvania. He is the author of *The Power of the Internet in China: Citizen Activism Online* (2009).

Index

T

Simon Cottle, *General Editor*

From climate change to the war on terror, financial meltdowns to forced migrations, pandemics to world poverty, and humanitarian disasters to the denial of human rights, these and other crises represent the dark side of our globalized planet. They are endemic to the contemporary global world and so too are they highly dependent on the world's media.

Each of the specially commissioned books in the *Global Crises and the Media* series examines the media's role, representation, and responsibility in covering major global crises. They show how the media can enter into their constitution, enacting them on the public stage and thereby helping to shape their future trajectory around the world. Each book provides a sophisticated and empirically engaged understanding of the topic in order to invigorate the wider academic study and public debate about the most pressing and historically unprecedented global crises of our time.

For further information about the series and submitting manuscripts, please contact:

Dr. Simon Cottle
Cardiff School of Journalism
Cardiff University, Room 1.28
The Bute Building, King Edward VII Ave.
Cardiff CF10 3NB
United Kingdom
CottleS@cardiff.ac.uk

To order other books in this series, please contact our Customer Service Department at:

(800) 770-LANG (within the U.S.)
(212) 647-7706 (outside the U.S.)
(212) 647-7707 FAX

Or browse online by series at:

www.peterlang.com